The Evolution of Modern Land Warfare

The Evolution of Modern Land Warfare

Theory and practice

Christopher Bellamy

London and New York

First published 1990
by Routledge
11 New Fetter Lane, London EC4P 4EE

Simultaneously published in the USA and Canada
by Routledge
a division of Routledge, Chapman and Hall, Inc.
29 West 35th Street, New York, NY 10001

Printed in Great Britain

British Library Cataloguing in Publication Data

Bellamy, Chris, *1955–*
The evolution of modern land warfare : theory and practice.
1. Land warfare, history 2. Military art and science
I. Title
355.02′09

ISBN 0–415–02073–5

Library of Congress Cataloging in Publication Data

has been applied for
ISBN 0–415–02073–5

For Andrea

'. . . a straggling road in Spain . . .'
Chesterton, *Lepanto*

Before everyone who wishes to become a commander-in-chief, there lies a book entitled 'The History of War'. It is not always, I must admit, very amusing. It involves toiling through a mass of by no means exciting details. But by their means we arrive at facts . . . and at the root of it lies the perception of how everything has happened, how it was bound to happen, and how it will again happen.[1]

Field Marshal Count von Schlieffen

In order for a man to become a great soldier . . . it is necessary for him to be so thoroughly conversant with all sorts of military possibilities that whenever an occasion arises he has at hand without effort on his part a parallel. . . . To attain this end . . . it is necessary to read military history in its earliest and hence crudest form, and to follow it down in natural sequence.[2]

General George S. Patton

No Social Democrat at all familiar with history . . . has ever doubted the tremendous importance of military knowledge, of military technique, of military organization as an instrument which the masses of people, and classes of the people, use in resolving great, historical conflicts.[3]

Vladimir Ilych Lenin

Contents

Figures

Tables

Acknowledgements

My thanks in particular to Colonel Mohammed Y. Effendi, retd, Pakistan Army, for his interest and help, especially in relation to the military history of South Asia, and to those at the US Army Command and General Staff College, Fort Leavenworth, Kansas, who suggested I undertook a special study of deep operations in historical perspective, which became the first 'Case study' (Chapter 5). Dr Bruce Menning, Dr Jake Kipp, Dr Roger Spiller, Dr Bob Berlin and Dr Robert Baumann and Colonel David Glantz, all commented on the initial draft and suggested further sources and lines of inquiry. They have thus given to that portion of the book an additional, most rigorous but helpful and stimulating, layer of supervision.

Although this book was substantially complete by October, 1987, the process of revision and fine tuning to current requirements took place after my arrival in Edinburgh. I therefore benefited from conversations, in other contexts, with John Erickson, who alerted me to philosophical issues affecting scientific revolutions, and to the utility of such devastating secret weapons as the block footnote. Carl Van Dyke, who shares my interest in the overwhelmingly rich body of nineteenth- and early twentieth-century military literature, much of which has been ignored or forgotten, provided stimulating ideas and alerted me to certain sources, and Lieutenant Colonel Joseph Lahnstein, USA, our Research Associate, always provided new and original insights, lightened with humour and a devastating turn of phrase. I am particularly grateful for Joe's help on 'local wars' and 'special operations' (Chapter 6). I also benefited from discussions with David McWhinnie, Lamancha Productions, with whom I worked in the complementary medium of film and video on military history.

Finally, I must thank Mr Carl Giles, the well known *Express* cartoonist, whose work I have admired for many years, for tracking down the cartoon and providing me with a special copy, a picture which, in my estimation, is worth about 60,000 words.

University of Edinburgh

Introduction

This book aims to provide a summary – in a handy and easy to understand form – of the way modern land warfare, military thinking, and concepts have evolved. It will therefore be of particular use to students and teachers of history, 'War Studies', 'Peace Studies', 'International Relations' and so on, and to all ranks of the armed forces concerned with warfare on or over land. In understanding how military operations are conducted and how they may be in future, and the complex interrelated issues involved, from politics to technology, history is a help, indeed a powerful and flexible ally. History is your friend.

This book is not intended to replace the vast body of literature on the history of war, on particular wars and battles, or military-scientific writing of any particular period, but rather as a signpost to them in the light of current concerns. There is no better advice than that of Napoleon, that 'the science of strategy is only to be acquired by experience and by studying the campaigns of all the great captains'.[1] This book cannot begin to substitute for first hand reading of reputable editions of the great military thinkers, from Sun Tzu through Jomini and Clausewitz to those of recent times: the late Ferdinand Otto Miksche[2] and Richard Simpkin spring to mind.[3] Nor does this book intend to cover as much ground as, for example, Theodore Ropp's *War in the Modern World*, which deals with the period since the Middle Ages but with more of a western European and American orientation, or with Archer Jones's *The Art of War in the Western World*, covering a similar geographical area to Ropp. Nor can it substitute for the Dupuys' *Encyclopedia of Military History from 3,500 BC to the Present*.[4] This book is of necessity highly selective and the selection is determined by issues which appear important today: the operational level of war (Chapter 3), the 'air-land battle' (Chapter 4), especially mobile formations operating ahead of the main body (Chapter 5), the fact that large-scale warmaking of this kind has recently been and is, likely to be confined to Asia, the apparent demise of large-scale warmaking as a means whereby the most developed nations settle political questions and the far greater prevalence of 'local' or 'guerrilla' conflicts (Chapter 6). For that reason, it really begins with Napoleon, although Frederick the Great's campaigns may have renewed relevance today as paradigms of 'limited' but nevertheless very violent, war. When dealing with Asia, there are reasons for going back further. Military science, like any science or discipline, has its own vocabulary, evolved over many centuries, indeed, millennia, but despite the subject's obvious grim importance the terms are often thrown around carelessly. This book aims to explain and define them. The book does not address the likely conduct of future operations, a subject on which I published a book in 1987,[5] but provides the essential background to them and current military writing and thinking, including some recent examples. In that sense, it is in the classical tradition of military history.

Military history is the oldest form of historical scholarship, with the possible exception of genealogy, and for a long time was arguably the most highly developed. Princes and potentates were recording their conquests and captives, artists portraying the military technology employed in meticulous and informed detail, long before it occurred to anyone to chart the daily lives of the labouring multitudes whose toil provided the surplus wealth needed to wage war.[6] At a time when modern history was only beginning to be studied in universities as an academic discipline, in the nineteenth century, military history was already highly developed, with a scientific methodology, and studied with a view to its lessons for the present and future. That war has been an immensely important part of human history is undeniable. As Miksche, one of the most original and irreverent military thinkers of recent times, quietly observed, 'world history could not be understood if historians were to leave out the wars'.[7]

The pedigree of military history as an aid to other branches of scholarship and endeavour is also impeccable. The crisp language of Caesar has for generations been considered the most suitable Roman text with which to begin the study of Latin. The archives of the Roman commander of fort Vindolanda, on Hadrian's Wall, which survived because they were written on wood, are the oldest original written archive in Britain and possibly Europe (AD 80–125).[8] The phenomenon of war has continued to fascinate cultured men of genius, many of whom have been trained in military skills. Leonardo da Vinci, who offered his services as a military engineer and designed numerous war machines, including a tank shaped as a flattened cone, and Suleyman the Magnificent's architect, Sinan, suffice to make the point.[9] Those who consider themselves scholars should therefore treat military history with respect. Surprisingly, this is often not the case.

The reasons are easy enough to understand. War is hell, grotesque, nasty, disgusting. Attempts to show that it is a necessary, even beneficial part of human activity have either been discredited or been beneath intellectual contempt, and this has not helped the scientific examination of the problem.[10] The idea that war is some kind of Darwinist bloodletting to encourage survival of the fittest is easily dismissed as throughout much of history war has been a relatively trivial cause of death compared with tempest, famine and disease. In recent times, war has tended to kill off many of the fittest and most able in the societies involved, rather than assured their survival.

However, this is no reason to ignore what happened in organized conflict. Yet many scholars do not wish to examine war, partly because of personal aversion to the subject, ideological constraints and, most strangely, because they fear that this may lead to 'drum and trumpet' replays and a 'fixation on trivial mechanical aspects rather than larger patterns and meaning'.[11] A lawyer or doctor who ignored or eschewed 'trivial mechanical aspects' would not last very long in his or her profession, and there is actually something to be said for 'drum and trumpet' military history. During the 1960s and 1970s the serious examination of the mechanics of military operations was very unfashionable in the west. This extended not just to 'trivial mechanical aspects' such as, for example, comparing the qualities of German and Soviet tanks in World War Two, but even the study of how battles and vast campaigns involving millions of men and women were fought. One suspects that in this period many would-be scholars without knowledge of military detail were attracted to 'Strategic Studies' as a fashionable subject without the rigorous imposed criteria of medicine, law, mathematics or physics, and that they found it easier to dismiss any attempt at serious examination as 'fixation on trivial mechanical aspects' rather than admit that they themselves knew nothing about the subject. If they do mention mechanical aspects, which can be extremely tricky, there is a risk that unfamiliarity with them will lead to misunderstanding or straightforward factual errors, which are unacceptable and do nothing to give the reader confidence in the writer's other assertions.[12] Admittedly, however, there is a risk that the serious study of military operations, the influence of terrain, the uncertainty or 'fog of war', the interaction of commanders' personalities, complex logistic questions and so on will become bogged down in what are often derisively called 'nuts and bolts', or with the 'aesthetics' of war.

The fascination with the horrendous phenomenon of warfare certainly has much to do with its aesthetics. Even a pacifist, as Tolstoy was in later life, could not but admit, although he well knew its horrors, that warfare had a certain terrible, bizarre, 'beauty', as he described the unfolding battle at Borodino.[13] The aesthetic allure of things military is not limited to the collection of buttons and badges, to war gaming, or to dressing up to re-enact past battles. The aesthetics of war manifest themselves more scientifically in the undeniable geometric attraction of maps portraying military operations; in the Faustian 'lines, circles, letters and characters' of treatises on gunnery, and the fearful symmetry of fortifications. Men of genius like Leonardo would have imagined most acutely the eviscerating agonies of fatal wounds. Yet even their intellects, sensible of the extremes of pleasure and pain, have not been able to resist the 'compelling fascination' of war and its challenge to man's ingenuity: 'You may loathe, you may execrate, but you cannot deny her . . . even those that hate her most are prisoners to her spell.'[14]

This refusal to look the phenomenon of war in the face has affected the other side of the coin – futures studies – as well, and in the same way. The journal of forecasting and prediction, *Futures*, has featured articles highlighting the paradox that, whilst much attention in futures studies is devoted to possible catastrophes from ecological imbalances, there are very few serious analyses of war and violence. This is ascribed to familiar causes: 'controversial topics, including realistic studies of war and violence', are to be avoided in a discipline that is striving for harmony. Endeavour in futures studies presupposes that there will be a future, and thus 'doomsday' appraisals are excluded. That has a sort of logic, but have the futurologists noticed that there were, for example, 162 major organized conflicts beginning between 1951 and 1985, but that none of them were nuclear and that mankind is still alive and shows no sign of eschewing organized violence? Lastly, there is the question of what is primly called 'legitimacy'. In other words, war is *not nice*. Therefore, its central aspects are to be 'sanitized' (a horrible word) or ignored and people who study it dismissed as psychopaths or as having a vested interest in continuing conflict. In this, as in so many other ways, the Arab historian Ibn Khaldun (1332–1406) was right: 'the past resembles the future as water resembles water'.[15]

This strange western attitude to the study of war has no parallel in Russia. The rigorous Imperial Russian school of military history, military geography, and military statistics survived the Revolution and the purges, and since the last war there has been a continual outpouring of writing on military subjects ranging from hundreds of war memoirs to highly scientific analyses of military operations and their implicit and explicit lessons for today. The Russians are masters of a span of scholarship which is neither musing on the broad socio-political aspects of war on the one hand, nor 'drum and trumpet' ('meanwhile, Durnford's platoon was having a hard time overrunning the machine gun position on the left flank') on the other. The mechanics of war ('military art') are examined, but in a detached and scientific way. For example, we read that the position to be attacked comprised two defensive zones, each with so many belts of so many trench lines each: the balance of forces with respect to overall numbers and key weapons systems was such and such; the terrain was waterlogged, reconnaissance was carried out in a certain way, certain measures were taken to guarantee surprise; at a certain point mobile forces were inserted to convert tactical success to operational.[16] As a significant new Soviet work, *The Evolution of Military Art*[17] stressed, these forms, procedures and 'mind-sets' have not evolved in a random fashion, but for good reasons over a long period. The Russians have no difficulty whatever in using military history as a means of discussing present-day issues, a point illustrated in detail in Chapter 5, and they seem to recognize instinctively the continuum of past and future: 'the idea of travelling into the future is wonderful, but the future is lifeless without the past.'[18] However, we must beware of being carried away by the pro-Russian euphoria about Mr Gorbachev's reforms current at the time of writing. In conversations with Soviet historians, they estimated that they had only used 6 per cent of their available military archives in published historical research, and history is easily rewritten and resculpted to suit the needs of the moment. They know that history can be their friend – perhaps too well.

Raymond Garthoff put his finger on the difference between the Russians' attitude and that of many westerners (especially Americans) in his introduction to some of General Pokrovskiy's perspicacious work, in 1959: 'indeed, many American readers will be perplexed by the fact that a Soviet nuclear weapons and missiles specialist introduces his work by referring to the struggle of the Romans and Carthaginians'.[19] I hope that after reading this book, they will not be so perplexed. I also make no apology for drawing on the rigorous Russian tradition of scientific examination of the mechanics of war, because I believe that, despite occasional abuses of history, it is the best.

While thus reaffirming the importance of military history, a reappraisal is also necessary. The western European, particularly Anglo-Saxon, experience of war is not necessarily the most important in the global evolution of warfare or the most relevant to current and future military operations. The enormous Russian and Soviet contribution to military theory and practice is still a closed book to most westerners, and inaccurate and unscientific attempts to sensationalize modern Soviet military power, lacking any sensible historical perspective, do not help. We really need to look at Russian writing on military history and strategy, not just because the Soviet Union is arguably now still, albeit precariously, the pre-eminent continental power, but as very intelligent appraisals of war and strategy generally, which happen to have been written by Russians.[20]

A reappraisal is even more necessary with regard to other great Eurasian empires: Persia (now Iran) furnished one of the greatest warlords of relatively recent times, Nadir Shah, but who has heard of him? The same is also true of the nation which probably still possesses the largest army on earth, the People's Republic of China.[21] Many would acknowledge Mao's expertise as a guerrilla leader, but who can speak with confidence of the Chinese experience of conducting war at the operational level, or name even a handful of generals from that nation's thousands of years of military history. Finally, there are the 'Prussians of Asia', the Vietnamese with the world's fourth largest army, whose victory in 1975 owes much more to their ancient military traditions than most have acknowledged. The Asiatic military traditions have sufficient in common to justify a special examination (Chapter 6).

This book therefore aims to fill three fundamental gaps. The first is to correct a straightforward cultural, even linguistic bias. Most history is not written in English and an additional effort is required to bring the military experience and accumulated wisdom of Russia, India, Persia, China, Vietnam, Japan to name a handful of fairly significant modern states, home to Americans and western Europeans.

The second gap is the surprising, almost wilful disregard or derogation of the study of military history and operations. I fully acknowledge that 'the parameters of the study of security problems have been too narrowly drawn'.[22] There is much more to security than military defence. Other threats to our security, to our very survival, may overshadow any purely military one: AIDS, the erosion of the ozone layer, the wastage of the planet's limited resources. Even the study of the phenomenon of war itself, which has been called 'polemology' (from the Greek, *polemos*: a war), is a much bigger subject than the study of military operations.[23] The Russians call the conduct of war 'military art' and the study of war in total, as a socio-political phenomenon (polemology), 'military science'.[24] Military art is the central and most important part of military science, of polemology. Many western War Studies courses place great stress on other aspects of war and its implications, but largely ignore this patently central element – the conduct of war. Thus these courses are shaped like a torus: a doughnut, or a popular kind of mint with a hole in the middle. The outer ring, the torus is, of course, very important, but don't forget the hole. Because if it were not for the hole the torus would not exist. There would be no need for the torus at all.

The third gap or objective is to correct or at least challenge a widespread scepticism about the value of military history to those engaged in military policy making and planning for the future. To be fair, this is understandable. Those in the Pentagon and the British Ministry of Defence charged with analysis and intelligence are so swamped by the volume of current material to be processed, a problem exacerbated by excessive demands for interminable briefings and the short time that officers spend in post, that there is no time to devote to long-term research or to acquiring a historical perspective. The result is like trying to do a jigsaw puzzle without most of the pieces. As Chapters 5 and 6 demonstrate, a historical perspective can make a direct contribution to seeing the wood rather than a riot of individual details of sensitive origin. In addition to its inherent interest, and its importance as a prominent part of the sum of human experience, military history (and not only recent military history) is of direct practical value today.

Chapter one

Ground rules

In the fertile coastal strip of Palestine there lies a beautiful plain. Under the peaches and blue ice-cream sky of a December late afternoon merging into evening, it is stunningly peaceful. Here over twenty cities were built, one on top of the other, for good reasons. Cities need food, and so they were built in the middle of a fertile, abundant plain. Cities needed to be defended, and so they were built on a piece of commanding ground. Cities need a reason for their existence, to trade and to thrive, and so they were built at the crossing of two great trade routes, the one running east–west, from the Mediterranean to the river Jordan, and the one running north–south, along the Mediterranean coast. It is therefore called a tel, an artificial mound, comprising city, upon city, upon city. It also became a place for battles, because geography focused armies upon it. The surrounding marshes concentrated them in the narrow plain by the city: so many battles, that this place became a byword for battles, indeed the ultimate battle. It is called Tel Megiddo – Armageddon.

It was here in the year 1469 BC that the Egyptian Pharaoh Thotmes (Thutmosis) III led perhaps 10,000 men in a rapid and unexpected march against rebel Palestinian chieftains. The chieftains had sent outposts to hold the Megiddo pass, a covering force which was easily scattered, leaving the King of Kadesh to face the Pharaoh on the Megiddo plain. Apparently, Thotmes' army advanced in a concave formation, its southern wing enveloping the rebels, while the northern was driven between the rebel flank and the city of Megiddo itself. The Egyptians therefore enveloped the rebel army, at the same time severing it from its base – the fortress. It is the first battle in history, as far as we know, to have been conducted in accordance with recognized strategic and tactical principles, indeed the first great recorded battle of history. The Egyptians made good use of the chariot, a revolutionary device embodying mobility, shock action, 'firepower' from its archers, and perhaps a small amount of protection. Nor was this the only battle of significance for the art of war to take place here. In 609 BC a Jewish army under Josiah was easily defeated in the biblical battle of Armageddon, less interesting than that of 1469 BC from a military viewpoint, if more so from a theological one. Over 2,500 years later, between 19 and 21 May 1918, the British under General Allenby defeated Turkish forces in the so-called 'second battle of Armageddon', the umpteenth of Megiddo, in which local Jewish settlers showed him the way through the marshes, a reminder of the continuing importance of terrain. Allenby's breakthrough was followed by a vigorous pursuit by cavalry, armoured cars and aircraft, which led many to place too much stress on light mobile forces, a misinterpretation of its lessons, in detail, if not in general. Ground, intelligence, surprise, breakthrough, pursuit; firepower, mobility, shock action; envelopment, cutting lines of communication. Since the beginning of the history of war, these factors have not changed.[1]

THE DARK SIDE OF CIVILIZATION

A city based on agriculture and trade and located in the Middle East is the right place to begin the study of war as a socio-political phenomenon, as well as military technique. War, as Clausewitz said, 'is an extension of politics by other, namely violent, means'. It is 'an act of violence intended to force the enemy to do our will'.[2] It presupposes the existence of communities organized on some scale, capable of formulating and articulating objectives, and of calculating that the possible risks are worthwhile when measured against the benefits of remaining on friendly terms with the neighbouring community of homo sapiens. War, is the dark side of civilization.

Zoologists and anthropologists have demonstrated convincingly that war has little if anything in common with any primeval violent instinct in man. It has still less in common with the response of rats or other communal species.[3] It has been argued that naturally carnivorous and predatory animals, like lions and wolves, have an inbuilt self-control mechanism to prevent them slaughtering members of their own species, whereas man as a primate does not, and that when his intellect furnishes him with a club or a battle fleet he is not inhibited from using it on other men. This may have some validity in the case of a bar-room brawl or a *crime passionel*, but hardly applies to the reasoned, cold, calculated application of organized military force, raised and equipped at great expense from the surplus wealth generated by agriculture and industry. It breaks down completely when one considers the difficulties of maintaining an organized military force in the field for any length of time (even 24 hours) and directing it towards even the simplest military objective. The author agrees wholeheartedly with anthropologists, that comparisons between an ape baring its teeth as a sign of aggression and organized violence in war are superficial and absurd. Anthropological studies provide no evidence that man is naturally violent towards man. In the modern context, the strange bond which develops between opponents in war, and the sympathy which often develops between individual combatants who come into contact when one or both are wounded or taken prisoner, supports this. Organized bodies of men in battle have more in common with sheep than wolves, and although individual aggression is obviously aroused in an atmosphere fraught with fear and danger, the impulse that leads societies to war is intellectual and social, and has nothing to do with the individual's genes.[4] Most men's instincts induce them not to attack the enemy in battle, but to go to ground or run away. Whether the youth of Sparta, the medieval knight, the conscript of 1914–18 or the modern professional soldier, men are required to undergo a long and deliberate training process to make them effective soldiers. With very few exceptions, they have to be thoroughly conditioned to kill their own kind.

It is generally acknowledged that warfare began after men had stopped being hunter-gatherers, and settled to cultivate the land. Cultivating land required an enormous investment of effort, and men were not to be deprived of the fruits of their labours easily. Close on the heels of agriculture came relatively large, organized communities, the first cities. These represented a surplus of wealth, and more often than not this process was accompanied by the growth of organized religion and a distinct ruling class. One does not have to be a Marxist to see that organized conflict has its origins in economic considerations.[5]

There are two sides to every armed conflict. It is not just a question of one side's inclination to coerce the other, but also the extent to which the latter can be coerced. A community which has nothing worth taking is not a very worthwhile target for aggression. If it is defeated and splinters into fragments living off the land, melting away into the landscape, it cannot be relied upon to 'do the aggressor's will', to paraphrase Clausewitz. Organized, centralized societies were at once better able to wage war, had more reason to do so, and at the same time were more vulnerable targets. A peasantry and urban society, used to discipline and committed to the land or property were more likely to do the will of the new ruler, as they had that of the old. The problem which modern organized armed forces face in dealing with terrorists and guerrilla fighters, presenting no specific large targets, with no identifiable or at any rate accessible power base, underscores this point. Four and a half millennia after the depiction of organized armed forces inflicting calculated humiliation on a defeated enemy, and then retiring to a celebration booze-up, on the Standard of Ur made in about 2500 BC, a bemused Engels wrote to Marx about the American Civil War:

> it is amazing how slight or, much rather, how wholly lacking is the participation of the population in it [the war]. In 1813, indeed, the columns of the French were continually interrupted and cut up by Colomb, Lutzow, Chernyshev and twenty other insurgents and Cossack leaders; in 1812 the population of Russia disappeared completely from the French line of march; in 1814 the French peasants armed themselves

and slew the patrols and stragglers of the allies. But here nothing happens at all. Men resign themselves to the fate of the big battles . . . the tall talk of war to the knife dissolves into mere muck.[6]

LEVELS OF CONFLICT: 'CONVENTIONAL' AND 'GUERRILLA' WARFARE

This book focuses primarily on the big battles, but if the populace as a whole in the area where those battles take place chooses to become involved, they have always been able to exert some influence on them, and sometimes the populace do not resign themselves to the fate of the big battles. The phrase 'war to the knife' comes from the Peninsula, which also gave us the Spanish word *guerrilla*. The Spanish guerrillas were a great nuisance to the French, but the big battles alone were decisive. The same is true of Soviet partisan operations in World War Two. Very large partisan forces operated in the German rear, conducting thousands of demolitions synchronized with Soviet main forces operations, but the Germans never lost control of the railways. In parts of Asia, where terrain and economy have suited, the 'big battles' have for hundreds of years represented vortices in a sea of less intense 'guerrilla war' (Chapter 6). There is a distinction between guerrilla warfare in conjunction with major conventional operations (the sense in which it was originally used), and terrorism in circumstances where a state of war cannot be said to exist. However, the moment armed clashes occur on any scale, principles of warfare – destruction of the enemy's will to resist, intelligence, fire and movement, encirclement, communications, logistics – apply. For this reason, the author is against the use of the term 'low-intensity' warfare: a bullet in the stomach is a bullet in the stomach, whether it comes from a member of a political group with a long name operating in an internal or 'unconventional interstate' conflict, or whether it comes from a member of 3rd Shock Army. Broadly speaking, there are two types of actions: police and anti-terrorist actions, at the lower end of the spectrum, and 'local wars'. The latter, while not total wars like World Wars One and Two, *are*, very obviously, wars, and conform to the same principles of military art as conflicts on a broader canvas or conducted for more uncompromising motives. The ground rules explored in this chapter apply equally to 'guerrilla' and 'low-intensity' conflicts, although the form in which they are expressed may appear different.

THE NATURE OF WAR

A thousand years, more or less, elapsed between the conflicts shown on the Standard of Ur, using technology specially designed or adapted for military purposes and probably some kind of deliberate tactics, and the recognizably professional Egyptian conduct of the Battle of Megiddo in 1469 BC.[7] Since then, it has been possible to identify certain key elements in the conduct of war. From one viewpoint, there are four principal elements. The first that springs to the modern mind is technology. The second is the complex question of how armies are organized to function most effectively, and, indeed, the ability of a state, regime, or movement to maintain and supply them at all. This has been extremely critical throughout military history, and will be called, for the moment, logistics/organization. The next element is the way armed forces are handled, which is best called military art. Military art has traditionally had two key levels, strategy and tactics. The terms come from the Greek *strategos*, meaning the office or command of a general, and tactics from Greek *taktika*, 'matters pertaining to arrangement'.[8] Clausewitz defined tactics as the use of armed forces in an engagement, and strategy as the use of engagements for the object of the war.[9] Wavell used the analogy of a game of cards, strategy is the bidding, and tactics the way the hand is played.[10] In the last century or so, this at first sight simple, though always elusive, division into two has become more complicated, and a third, intermediate level has emerged. Or, rather, it has been formally recognized, since it arguably always existed in some form. The latter is known as operational art (between tactics and strategy), and is explored in detail in Chapter 3. The last of the four elements is command, control, communications, and intelligence, an interlinked and mutually supportive cluster of capabilities described as C^3I ('C cubed I', the 'cubed' stressing the fact that the various parts are multiplied together, not merely added). Without communications, there can be no command, no control, no usable intelligence – and so on.

Ancient armies were relatively small and victory depended primarily on a combination of technology and appropriate technique (see Chapter 2), plus tactical skill on the part of the commander. Armies were rarely more than equivalent to a single corps in strength, could live off the land, and did not have to worry about lines of communication. Weapons, too, were simple: edged weapons could be repaired by artisans accompanying the army, and even arrows and javelins could be recovered. Ancient armies were therefore freed from logistic constraints, and everything, unless plague intervened (the destruction of Sennacherib's army) depended on the conduct of the battle itself. 'Strategy' in fact evolved from logistic considerations. It was only when armies became increasingly dependent on consumable ammunition, and larger amounts of forage and rations than the country could easily supply, that their movements and dispositions were constrained by the need to keep open supply lines and access to depots and magazines. Then, tactics – technique lost its overwhelming dominance and a new science, or aspect of military art, strategy – emerged.[11] The greatest 'strategists' either disposed of forces whose requirements were so simple that they could all but forget about lines of communications (Genghis Khan), or were adept at ensuring that their armies' essential requirements were met more effectively than the enemy's and, indeed, than the enemy believed possible (Marlborough, Suvorov, Napoleon).

In the Middle Ages, technology and technique combined, still held an extremely important place (for example, the exploitation of the longbow, adopted by the English from the Welsh) but a particular combination of logistic and organizational factors became dominant in Europe, while in Asia, other logistic and organizational factors applied. Ancient armies, resting on a social system sustained by slavery, had sometimes been able to acquire a highly professional character. The conditions of feudal and early modern society changed this, a reminder that military technique is dependent on social conditions in all sorts of ways. Medieval and early modern armies were only enlisted for short, finite periods and western leaders always found it difficult to pay them. This, together with technology/technique and physical toughening, was one of the cardinal reasons for the success of the eastern Empires (notably the Mongols) in the Middle Ages. The western levies were brave enough, but were hastily raised and were not accustomed to acting as a team for a prolonged period. The same was true as late as the sixteenth century, when western armies still comprised mercenaries who fought only as long as they were paid. This, according to Professor Elton, was true not only of the German *landsknechts*, but also of the Spanish *tercios* and French forces which came closer to national armies. In contrast, the forces of the Turkish Sultan, Suleiman the Magnificent (1494–1566) were supported by a superb organization and an efficient taxation system, and it was this which enabled them to become the envy, and terror, of Europe.[12] By the nineteenth century, and through to 1945, success in war was arguably dependent, more than any other single factor, on the logistic/organizational element: the ability to train, raise, and deploy mass armies and to feed and supply them in the field. The French in the Napoleonic Wars, the North in the American Civil War, the Germans from 1864 through to 1914, the British and French in 1914–18, with not a little help from the Russians and Americans, and the Soviet and Anglo-American forces in 1939–45; victory went, above all, to the side which mobilized most men and machines and kept them fit to fight.

Looking at the problem from a slightly different angle again produces four key elements. The perceptive British military commentator, Colonel Henderson (1852–1903) defined war as, first and foremost, a matter of movement; second, of supply (equating to the logistic/organizational aspect already identified); third, a matter of destruction (attrition), which broadly equates with technology although technology also affects movement; and fourth, the fact that war is:

> not merely a blind struggle between mobs of individuals without guidance or coherence, but a conflict of well-organized masses, moving with a view to intelligent co-operation, acting under the impulse of a single will, and directed against a definite objective.[13]

In other words, C³I, again, involving means of 'intercommunication and observation.[14] Taken in a broader context, we are reminded of the words of the great American military and naval thinker, Alfred Thayer Mahan, 'not by rambling operations or naval duels are wars decided, but by force, massed and handled in skilful combinations'.[15]

Many commentators on military affairs today seem to be almost exclusively preoccupied with technology. Technology, and the ability to use it, is indeed very important, and the next chapter is devoted to these questions. However, it is essential to put technology into perspective. Throughout most of military history, technology *per se* has been the least important determinant of victory. It helps to look at every military situation in terms of the two quartets:

- technology, logistics/organization, military art (strategy, operational art, tactics), C³I.
- movement, supply, destruction, C³I.

PRINCIPLES OF WAR

There is a long-standing argument about whether war is an art or a science. Clausewitz considered it to be one of the social sciences and that, more than anything, it resembled a game of cards. Jomini considered it a 'violent and impassioned drama', and certainly not an exact science.[16] This is underlined by the fact that there are no universally agreed principles of war, although the various lists that have been drawn up frequently have elements in common.

Jomini enunciated but 2 principles of war: to use freedom of manoeuvre to bring masses of one's own troops against fractions of the enemy's, and to strike in the most decisive direction. Clausewitz enunciated 4 'rules': to employ all available forces with utmost energy; concentration at the point where the decisive blow is to be struck; to lose no time and surprise the enemy; and to follow up success with utmost energy. He also enunciated 3 general principles for defence, 14 for offence, 8 for troops and 17 for use of terrain. Marshal Foch the great French commander in World War One also enunciated 4: economy of force, freedom of action, free disposal of forces and security. The great British military theoretician of the first half of this century, Major General J. F. C. Fuller, enunciated 3 groups of 3 – 9 principles in all – with typical cartesian precision. These and other sets of principles are shown in Table 1.1.

The principles most often cited are: the offensive, concentration of forces (mass), economy of force, manoeuvre, unity of command (co-operation), security, surprise, simplicity, planning and command. Although the US Navy had 12 principles to the Army's 9, one author, writing in the US Naval Institute Proceedings in 1967 proposed 5. The French strategist André

Table 1.1 Principles of war according to various authorities

Jomini	*Clausewitz*	*Foch*	*Fuller*	*British (Kingston-McLoughry)*	*British FSR*	*US Army*	*US Navy (Keenan)*	*Warsaw Pact 1956–9 (Savkin)*
1. Use freedom of manoeuvre to bring masses against fractions	Employ all available forces with utmost energy	Economy of force	1. *Mental* (a) Direction	Selection and maintenance of the aim	Maintenance of the aim	Mass	Objective	Mobility and high tempos of combat operations
2. Decisive direction	Concentrate where decisive blow to be struck	Freedom of action	(b) Concentration	Maintenance of morale	Offensive action	Mobility	Weighted distribution of force	Concentration of main efforts and creation of the necessary superiority at the decisive place and time
3. —	Lose no time and surprise enemy	Free disposal of forces	(c) Distribution	Offensive action	Concentration	Objective	Co-ordination of effort	Surprise
4. —	Pursue success with utmost energy	Security	2. *Moral* (a) Determination	Security	Economy of force	Economy of force	Initiative	Combat activeness (aggression: *aktivnost*)
5. —	—	—	(b) Surprise	Surprise	Co-operation	Simplicity	The unorthodox	Preservation of own troops' effectiveness
6. —	—	—	(c) Endurance	Concentration of force	Security	Security	—	Comformity of goal of operations to the actual situation
7. —	—	—	3. *Physical* (a) Mobility	Economy of effort	Surprise	Unity of command	—	Interworking (co-operation)
8.—	—	—	(b) Offensive action	Flexibility	Mobility	Organization for combat	—	—
9. —	—	—	(c) Security	Co-operation	—	Surprise	—	—
10. —	—	—	—	Administration	—	—	—	—

Beaufre suggested one combined principle based on Foch: 'to reach the decisive point thanks to freedom of action gained by sound economy of force'. Economy of force here means an optimal allocation of forces, divided between protecting oneself against the enemy's preparatory manoeuvre while carrying out one's own preparatory manoeuvre and decisive blow.[17]

ATTACK AND DEFENCE

It is already obvious that principles of war place much stress on the attack, or offensive, as the decisive form of war. We are immediately faced with a paradox, for, as Clausewitz said, the defensive is the stronger form of war; yet one cannot win purely by adopting the defensive. The defensive is the 'stronger' form, and is therefore often adopted by the weaker side, because the defender has the advantage of cover, and of choosing the ground, ground which he may know well, and can prepare. Nor is it just a question of ground; throughout history, all projectile weapons have been more accurate and lethal when shot or fired from a standstill than when on the move. Command and control of a defending force is also easier. 'Offensive' and 'defensive' are clearly unstable states: the defender is well placed to surprise the attacker by counter-attacks.[18] The term 'counter-attack' or, on a larger scale, 'counter-offensive', can be used as a general one for any kind of reciprocal action. It also has a more specific meaning, to retain lost ground vital to the defence. In the latter case, it is used alongside 'counter-penetration', which means to block an enemy incursion into a defended area, and 'counter-stroke', which means destroying enemy forces which are on the move. In many forms of war – notably naval war and guerrilla war, the side on the 'defensive' actually requires more strength and resources than the 'attacker', because the attacker can hit and run, whereas the 'defender' never knows where he will strike and therefore has to protect everything. Clearly, one can be on the offensive strategically and the offensive tactically, and vice-versa. A particularly strong combination was favoured by the German commander and theorist Helmuth von Moltke, the elder, (1800–91), mastermind of the Austro-Prussian (1866) and Franco-Prussian (1870) wars. Moltke perfected the combination of the strategic offensive with the tactical strength of the defensive: moving to threaten the enemy's lines of communication and thus forcing them to attack, forsaking the tactical advantages of the defensive (see 'Types of manoeuvre', below).

MANOEUVRE AND ATTRITION

Just as attack and defence are, at first sight, opposed concepts, which in fact exhibit constant interaction and harmony, so are manoeuvre and attrition. Some have gone so far as to assert that there are two distinct and mutually exclusive schools of thought, 'attrition theory' and 'manoeuvre theory'.[19]

This is wrong. War is an act of force, ultimately manifested in the ability to destroy or immobilize the enemy's forces and render them incapable of effective resistance. That is attrition. Manoeuvre means moving one's forces in such a way as to multiply their effectiveness and ability to inflict attrition. Mobility is the ability of a force to move: not only in terms of whether it is on feet, hooves, wheels, tracks, wings, or rotors, but in terms of how much fuel it has, the condition of roads and the soil, and so on. Movement is the act of moving. Manoeuvre is moving so as to gain an advantage over the enemy, whether getting into a better fire position, or getting astride the enemy's communications to force him to attack. But manoeuvre without the ability to strike is illusion. As Mahan said: 'Force does not exist for mobility but mobility for force. It is of no use to get there first unless, when the enemy arrives, you have also the most men – the greater force.'[20]

Let us use the example of a charity fight between a well-known ballet dancer, and a well-known heavyweight boxer. Both embody the capacity for movement and sheer physical strength in considerable degree. Although the element of movement is more pronounced, or more elegantly pronounced, in the ballet dancer, he would not be able to move like that were he not a very strong man. Similarly, the boxer, although the element of brute strength might appear most prominent, must also be enormously fit and agile. Those who oppose manoeuvre theory to attrition theory, and favour the former, will, of course, support the dancer. The dancer's elegant pirouettes around the outside of the ring will no doubt elicit applause, but very little else. However, if by moving swiftly in this way the dancer is able to get behind the boxer and deliver a decisive blow to the back of his neck, then he will have won. Manoeuvre will have enabled him to inflict attrition. If, by forcing the boxer to strike at air, he exhausts the boxer, then he may also be in a favourable position to win. If, on the other hand, the boxer strikes him repeatedly, his ability to move will be limited, and if the boxer hits him very hard indeed, no amount of elegant movement will stop him falling over. War obeys similar rules.

Whatever the peacetime predilection with movement, soldiers in real war place a premium on firepower. Whether it is British officers in the Indian Mutiny, photographed with privately purchased heavy-calibre revolvers festooning their persons; the Waffen SS pillaging machine guns from shot-down aircraft and loading them onto jeeps to enhance their firepower; the parachute regiment in the Falklands with far more than the statutory complement of general purpose machine guns per section, beefed up with anti-tank weapons for use against enemy trenches, firepower is what soldiers choose.

FIRE AND MOVEMENT

War on land is different from war at sea and in the air, in that one can remain still. Ships and aircraft are always moving, and therefore require sophisticated weapons systems and techniques to compensate for their movement as well as the possible movement of their targets. Although modern tanks can fire from the move, and the mounted archer able to shoot from horseback enjoyed certain advantages, fire delivered from a halt is much more accurate and effective. Those elements which can fire and move at the same time are terribly vulnerable when they stop (tanks, fixed-wing aircraft, helicopters); for some reason that seems to be the price they pay. Fire and movement are the basis of tactics. When you are moving, the amount of fire you can deliver is limited. Therefore, at every level land forces tend to break down into moving and firing elements. Within a section (American squad) there are riflemen and at least one machine gun: the riflemen fire when the machine gun moves, the machine gun when the riflemen move, and some riflemen give covering fire while others move, a procedure known as 'skirmishing'. At higher levels, sections, platoons and companies take turns at the two tasks. Within an artillery battery, one section may fire while the other moves, and in major attacks huge numbers of guns and rocket launchers fire as infantry and tanks assault. Although the tank can fire and move, in practice it has always needed heavier fire support from assault guns and conventional artillery. Fire and movement represent another of the cardinal dualities of war.

TYPES OF MANOEUVRE AND OPERATIONS

Various forms of manoeuvre give one side the advantage in strength at the decisive point, or place the other in an unfavourable or dangerous position. The simplest is to strengthen one portion of one's line and push the attack where local superiority exists, as the Theban general Epaminondas did against the Spartans in 317 BC. This requires something of a gamble, in not reducing or thinning the line too much. The same consideration applies in more sophisticated form in the case of NATO and Warsaw Pact forces in Europe today. The Russians, for example, have analysed the vast database provided by World War Two, with refined computer calculations, in order to determine just how far they can accept unfavourable 'force ratios' on 'passive sectors' in order to obtain the desired superiority at the key point, with a lesser (or non-existent) superiority overall.[21]

Before armies became so large and weapon ranges and dispersion so great that they filled the entire theatre of military operations (see Chapter 3), it was possible to overlap the enemy line. A force deployed to fight in one direction is always more vulnerable when attacked from the side. That is true of an individual, a tank, or an army. This overlapping or outflanking movement can be taken further, so as to reach the enemy rear. This is then called an envelopment. A variant of the flanking movement was the famous 'oblique attack' perfected by Frederick the Great of Prussia (lived 1712–

1786). Frederick's superbly drilled army was able to redeploy in this fashion faster that the opponent could respond.

Maximum advantage is obtained when part of one's own force can get round the enemy to attack his back. To execute such a manoeuvre, secrecy, speed of movement and some way of co-ordinating the forces on either side of the enemy are necessary. Not easy, but most effective if successful. An enemy force in linear formation could try to withdraw to the side, but would then be vulnerable as its side would be exposed and would not move very quickly.

At the tactical level, envelopment places the enemy under severe immediate threat from front and rear, and possibly flank as well. Envelopment has at least four cardinal advantages: the *surprise* factor of appearing behind the enemy, both psychological and practical (in that it will take time for the enemy to make arrangements to deal with the new threat); reinforcements and supplies are prevented from reaching the enemy; his retreat is cut off; and movement towards the enemy rear is accomplished more economically in terms of casualties and supplies as one does not have to fight through him but goes round him. At the higher (operational and strategic) levels of warfare envelopment may not have such immediately lethal effects, but instead it forces the enemy to attack the enveloping force or else find itself cut off (the second and third points, above) (Figure 1.1). Thus the crucial switch of forcing the well positioned defender into the open to attack, and placing the former attacker on the stronger tactical defensive, is achieved.

The enemy can also be enveloped from both sides, called a double envelopment. This is sometimes called the 'pincer movement', for obvious reasons. The archetype of this movement is Hannibal's victory at Cannae in 216 BC. Double envelopment (indeed, single envelopment also) can be achieved either by having one's centre fall back, while the wings remain static, (Cannae) or by having the centre stand firm and advancing the wings, as at Isandhlwana (Zulus against British, 1879) or Khalkhin-Gol (Russians against Japanese, 1939), and as intended in a single envelopment in the Schlieffen Plan (1914).

Figure 1.1 Single envelopment, double envelopment

If the enemy is completely prevented from withdrawing, which may also have the effect of preventing reinforcements or supplies reaching him, he is said to be encircled (Figure 2.1). An encircled enemy does not have to be completely surrounded by an unbroken cordon of troops: it is sufficient simply to cut off his lines of retreat and resupply, which can be done with relatively small groups of forces. However, when there are other groups of enemy forces about, that is not enough, for they may attack and try to break the encirclement. The encircling forces would then themselves be attacked from the rear. Therefore, when executing encirclements and inner front, facing the encircled enemy and an outer front, to prevent other forces reaching him, have to be established. These correspond to terms once used in the siege of fortresses and cities: circumvallation, to keep the fortress garrison in and reduce it, and contravallation, to prevent relief forces getting to him.

The two faces of encirclement should not be confused with the term double encirclement, which is when one encirclement is followed by or conducted in conjunction with a wider or deeper one (Figure 1.3). A classic example of this is the double encirclement at Stalingrad with operation *Uranus*, the smaller encirclement and then *Little Saturn*, the larger. In this case, there are not 2 but 4 faces: the inner and outer fronts of the inner encirclement, and the inner and outer fronts of the larger one.

In recent and modern warfare we encounter the term vertical envelopment, where the encircling force (or part of it) is delivered by air: parachute, glider, helicopter. To simply fly over the enemy would lose two of the cardinal advantages of envelopment: the enemy's appearance in the rear would be no psychological surprise, although it might be accomplished more quickly, limiting counter-measures; and the force would be subject to attrition from the enemy's air defences. Therefore, a vertical envelopment

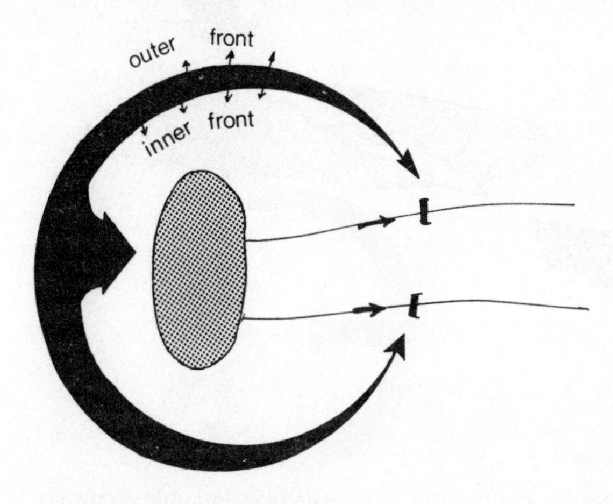

Figure 1.2 Encirclement

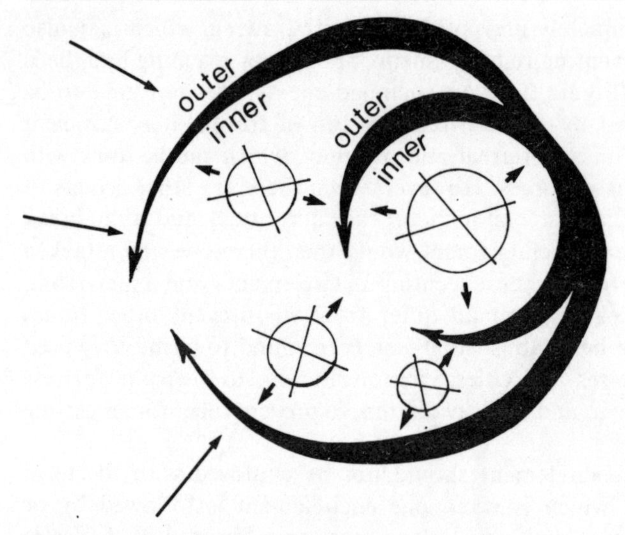

Figure 1.3 Double encirclement

ideally incorporates a measure of horizontal envelopment as well, as far as possible to avoid detection and, at least, the encircled force's own anti-aircraft striking power (Figure 1.4).

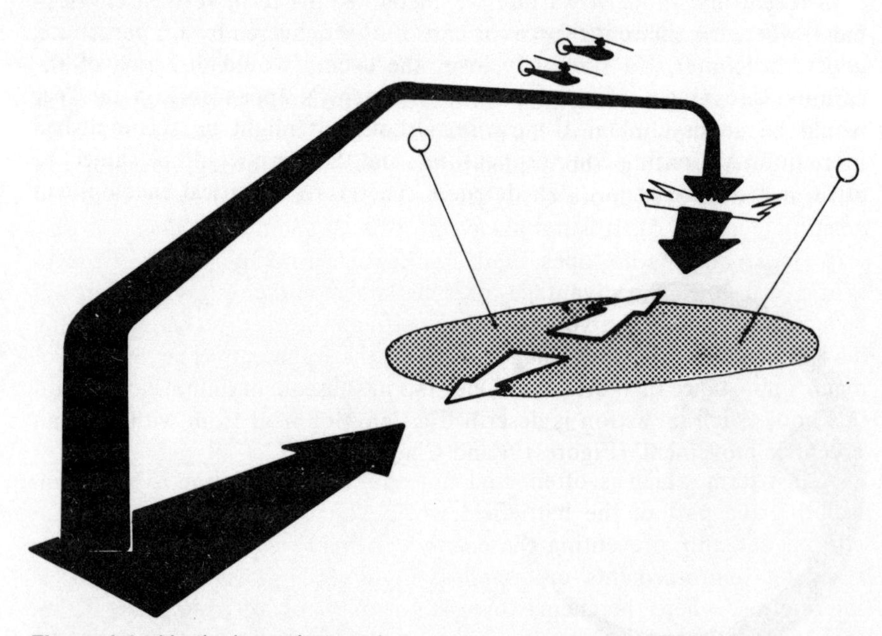

Figure 1.4 Vertical envelopment

Note: Although it is possible to overfly the objective, it is usually also desirable to move to the side, to avoid its own anti-aircraft fire, and achieve surprise

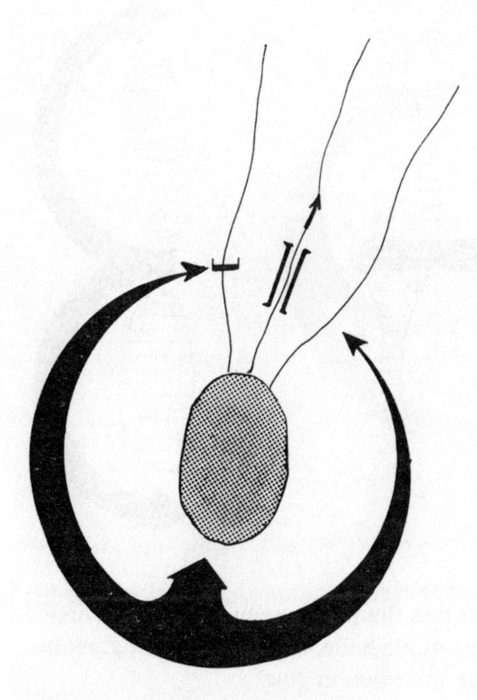

Figure 1.5 The 'golden bridge'

An encircled enemy may fight bitterly, and it may in some circumstances be adequate simply to force his withdrawal. In this case, he can be left an escape route, known as the 'golden bridge'.[22] In withdrawing, he can be cut down and destroyed more easily than when concentrated in an encircled position (Figure 1.5). It is not always possible to encircle the enemy in this way. There may be no 'open flank', as happened on the western front in World War One. The advantages of encirclement and envelopment can still be achieved, however, if you can break through the enemy's front, splitting his force in two, and then encircle each of the fragments. This became of major importance in World War One, and in subsequent military theory. In this book, such an action is described as 'envelopment from within' or an 'eccentric movement' (Figure 1.6 and Chapter 5).

A last term which is often used nowadays is interdiction. This means isolating the part of the battlefield where you wish to concentrate your efforts, not only preventing the enemy's withdrawal, but preventing him receiving reinforcements or supplies. Figure 1.7 shows an example of interdiction, where the enemy force is not only cut off from his rear, but from friendly formations on either side. Interdiction thus not only prevents withdrawal, reinforcement, and resupply from the rear, but also 'lateral reinforcement', as it is called. The term interdiction is more often used to

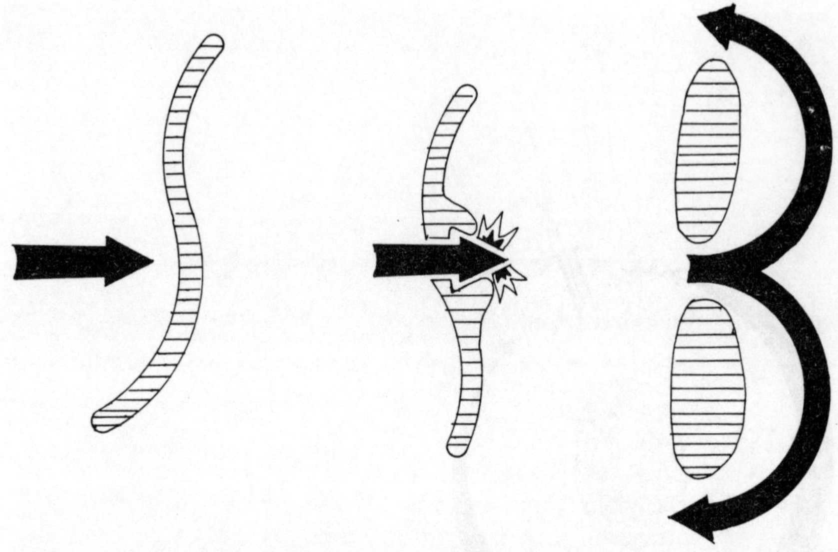

Figure 1.6 Eccentric movement

describe strikes by artillery and air forces than movements of ground forces, although some types of forces, for example helicopter-borne and parachute troops, are particularly useful in the interdiction role.

INTERIOR AND EXTERIOR LINES

A related question is that of interior and exterior lines of communications. Jomini explained this concept simply using the diagram in Figure 1.8. Within a strategic theatre, armies on lines A, B, C can concentrate against

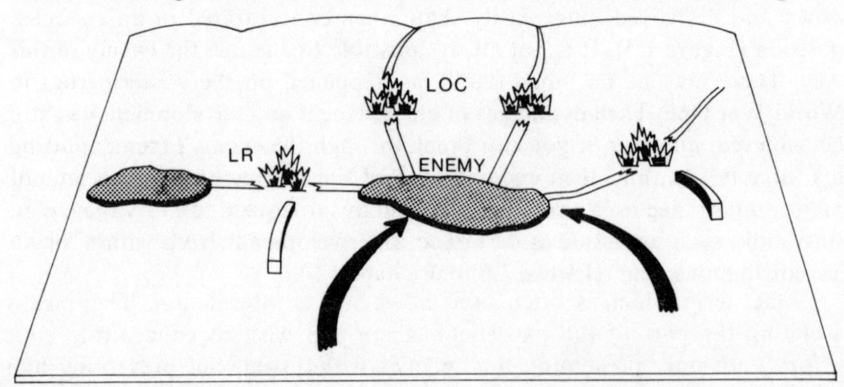

Figure 1.7 Interdiction

Note: Interdiction strikes isolate the enemy force which we wish to attack by cutting its lines of communication (LOC) and preventing lateral reinforcement (LR). It may also shield our own attacking forces against counter-attack, as shown here, by providing shields of fire, smoke, remotely delivered mines, and so on.

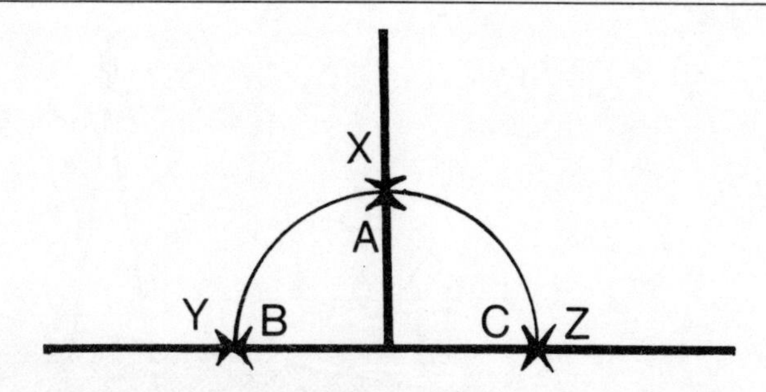

Figure 1.8 Interior and exterior lines

Note: Armies moving on lines A, B, and C can concentrate more easily to defeat armies on X, Y, and Z individually than those moving on X, Y, and Z can do the same to A, B, or C.

an army on any of lines X, Y, Z faster than any of those armies can come to another's assistance. It is not that armies X, Y, and Z are surprised by what happens, necessarily, just that they cannot react fast enough to do anything about it (the true meaning of 'surprise' in military operations). The classic example of the application of interior lines is Frederick the Great's defence of Prussia against numerous and powerful enemies, notably his campaigns of 1757 and 1758.[23]

There comes a point, of course, where a strategic 'central' position becomes an operationally or tactically enveloped position. As Moltke put it:

> The unquestionable advantages of the interior line of operations are valid only as long as you retain enough space to advance against one enemy . . . gaining time to beat and pursue him, and then to turn against the other. . . . If this space, however, is narrowed down to the extent that you cannot attack one enemy without running the risk of meeting the other who attacks you from the flank or rear, then the strategic advantage of interior lines turns into the tactical disadvantage of encirclement.[24]

Throughout history, the scale of military operations (see Chapter 5), and, consequently, the distance at which interior lines become encirclement, have increased. In both world wars, Germany was, arguably, operating on interior lines, and that undoubtedly helped it sustain the wars against colossally superior foes for many years. However, even there, at some point, interior lines became encirclement.

TERMS RELATING TO TACTICS AND GROUND

Two terms often used are enfilade and defilade. Enfilade means fire from a flank, which can have a far greater effect on an enemy force than fire from the front (Figure 1.9). Defilade means a position in a defile, depression, valley, or cutting in the ground. A weapon or weapons in such a position are well protected. The combination enfilade–defilade is particularly effective,

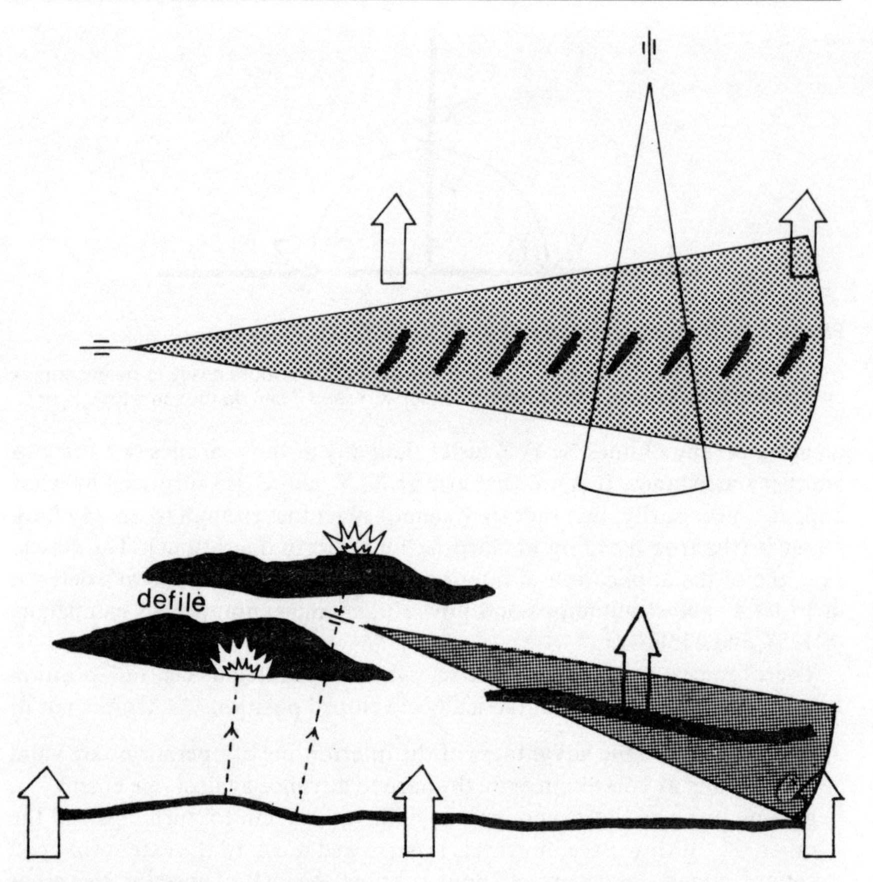

Figure 1.9 Enfilade and defilade

Note: The weapon in enfilade (top) can cover more enemy troops in its lethal arc of fire than can a weapon in front of them. A weapon in a defilade position (bottom) is well protected against the fire of attacking forces (and its flash and smoke may also be obscured), but may be able to engage advancing forces with devastating enfilade fire.

because the forces in defilade are protected against fire from the enemy's main position and can hit the enemy's vulnerable flank when he advances. These terms are only used at the tactical level.

A last important idea is that of the reverse slope. The British General Wellington was a great exponent of the reverse slope, (the side of a hill away from the enemy). The main force was deployed on the reverse slope, shielded from the enemy's artillery fire, at a distance from the crest corresponding to the effective range of its weapons. When the enemy eventually reached the crest they came over it to be met with highly effective fire from forces which had been waiting under cover (Figure 1.10). The reverse slope also became enormously important when artillery acquired the ability to fire at targets it could not actually see – indirect fire (Chapter 3).

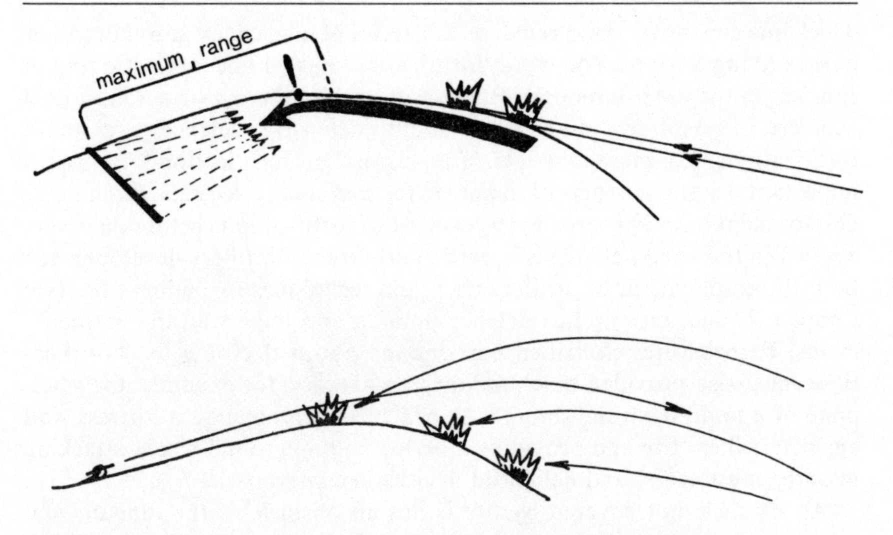

Figure 1.10 Reverse slope as utilized by infantry and artillery

Guns positioned on reverse slopes were invisible to the enemy, who would have to 'search' the entire reverse slope of the hill with fire, with little idea whether they had hit anything or not.

THE THREE KEY CHARACTERISTICS OF MILITARY SYSTEMS

If we examine the roles and characteristics of military systems in past wars, three key criteria present themselves: striking power (sometimes divisible into firepower and shock action), mobility, and protection. The medieval knight in armour on horseback enjoyed (for the time) considerable mobility and protection, but his striking power was limited when compared with the longbow, for example. A fortress is a system which sacrifices virtually all mobility for a high degree of protection. The ascendancy of the tank results from its combining all three in high degree, although none is as developed in a tank as in certain other systems. Thus, artillery has more striking power in terms of range and calibre, but sacrifices some mobility and protection, even if it is lightly armoured and tracked. The assault helicopter has more mobility and probably striking power, but less protection. Throughout history, different types of combatant and different military systems have had to balance these three elements, and have done so in different ways.

THE POWER OF POSITION: OBSTACLES AND FORTIFICATION

War is a matter of movement, but it is also a matter of position. Anything that impedes enemy movement is potentially advantageous. Fortifications may also preserve one's own ability to move, whether as a screen behind which one can move unmolested, or as a series of magazines, storehouses, or havens along a route. The word 'fortification' has an anachronistic ring in English, but is used in modern Russian to refer to the construction of field defences. Present-day discussions about 'defensive defence' are about fortifications, no more, no less. The history of fortification[25] is just as important as the history of manoeuvre, and many aspects of military development have appeared in the context of fortifications before they were applied in the open field. This is particularly true of artillery developments, both the employment of artillery itself and techniques for indirect fire (see Chapter 2) and anti-aircraft defence; mining and mine-warfare; entrenchments; barbed wire; electrified barriers and photo-electric cells. Fortifications have also provided much military vocabulary; for example, the *glacis* plate of a tank, from the smooth, sloped surface protecting a fortress wall against artillery fire and providing a perfect killing ground which attacking infantry must cross, and delightful devices like *chevaux de frise*.[26]

'An obstacle not covered by fire is not an obstacle' is the fundamental rule. Any barrier can be surmounted given time and no distractions. From ancient times, this rule has been applied in the design of fortifications, and provision has been made for crossfire from aggressively protruding towers and bastions, or fire down from protruding machicolations. The configuration of fortifications is tightly governed by the range of contemporary weapons. When effective artillery appeared, a new form of fortification was devised in renaissance Italy, based on the bastion, facilitating enfilade fire along all curtain walls. The complex geometry, with its associated recondite vocabulary, which reached its peak in the designs of Vauban (1633–1707), was necessary because one cannot fire through one's own walls. The invention of barbed wire overcame the problem: it kept enemy infantry out but you could also shoot through it. By then, however, permanent fortifications had all but disappeared underground, as can be seen from 'Palmerston's follies', which scar the chalk of southern England, and the French forts at Verdun, their armoured observation cupolas like giant medieval knights buried up to their helmets in concrete.

Field fortifications also have a long history: the Romans were adept at their construction. In themselves, small earth banks, ditches and sharpened stakes are not an obstacle, but in combination with troops behind them they multiply the latter's combat power many times. The development of accurate long-ranged firearms gave increased importance to cover from fire, as well as cover from view (Chapters 2 and 3). It is arguable that in both the world wars of this century the crucial theatres were theatres of fortress warfare. The prevailing or dominant view before World War One (with some disssenting voices) was that swift movement would dominate. The lessons of the siege of Port Arthur in the 1904–5 Russo-Japanese War were ambiguous: those who wanted to, cited it as evidence that permanent fortifications could not hold. On the other hand, taking Port Arthur had cost the Japanese several months and tens of thousands of lives. Fortifications were still a force multiplier. The opening phase of World War One was also held to support the anti-fortress lobby: the Belgian forts fell to massive siege mortars, but lasted longer than anticipated thus upsetting German timing critically. This led the French to pull guns out of their fortifications at Verdun, but as Verdun became the focus of determined German attack, this proved to have been an error. A hundred men in a Verdun fort could hold up thousands, and however disgusting the conditions inside the dank, subterranean forts, men preferred to stay there, under metres of concrete, than face the burning hell outside.[27] The entire western front in World War One, after the initial manoeuvre phase, acquired all the characteristics of a fortress action. The lessons of World War One suggested that concrete was an effective and humane substitute for flesh and blood, and this led to the construction of the Maginot line as a *couverture*, behind which France could mobilize, and protecting strategic vitals that were perilously close to the frontier (exactly the same situation as West Germany today). In the Soviet Union, a far-sighted and aggressive offensive doctrine was not seen as in any way inconsistent with the construction of huge fortified regions, which the Germans reckoned to be more of an obstacle than the Maginot line. These fortified regions, with concrete fire points, mines and *abbatis* of trees, would shield the mobilization and deployment of counter-attacking armies. The USSR's advance into partitioned Poland in 1939 meant that the frontier moved west, and the Russians were, unluckily, in the process of disarming the old fortified regions when the Germans attacked, an issue of profound relevance in the context of current debates on 'defensive defence' and 'defensive sufficiency'.[28] The experience of Spain (Chapter 4) and the horrendous cost of storming the Mannerheim line screening Finland seemed to suggest that permanent or semi-permanent fortifications had much life left in them, provided they were not taken completely by surprise.

Many stages of World War Two also exhibit the characteristics of a fortress war. The manoeuvre aspects of the war on the eastern front were recognized as a function, in part, of the lack of mines. Had mines been used in anything like the numbers they were used in North Africa, tank penetrations such as those in the Stalingrad counter-offensive would have been out of the question. In the south, the Germans came to a halt and held a series of bristling 'hedgehogs': at Sevastopol and Leningrad the Russians held out for months in classic and grisly sieges. The Kursk salient was a giant field fortification, with defences up to a hundred miles deep. The later stages of the war exhibit fortress characteristics. The Soviet advance into the Ukraine, Belorussia, Poland and Germany consisted of the breakthrough of successive prepared lines of defence. The breakthrough procedures developed during this advance – involving the deployment and targeting of artillery, use of infantry to crack the 'tactical zone' and tank formations to widen the penetrations – were explicitly designed to break through multi-layered fortifications. In the west, D-day was an assault on a fortress: even the journalistic terminology of the time – *Festung Europa* – recognized the fact. The channel coast was a fortress wall, and discussion about its defence reflected principles of fortress warfare: should every effort be made to throw the allies back into the sea immediately, or should the breaches be identified and a powerful mobile counter-attack mounted? Whereas medieval fortress engineers had built concentric castles, with many layers of defence, each higher than the one before, there was but one Atlantic wall.[29]

After the war, fortification continued to baulk large in military writing. Intelligence reports on Soviet fortified regions in the Far East opined that they had declined in importance somewhat, because of their vulnerability to nuclear weapons (by implication, nuclear weapons alone could blast a hole in them quickly enough to be of decisive significance).[30] The Russians sank old tank turrets in concrete along their border with China, and are now talking about fortified regions again; the Israelis built the Bar Lev line.[31]

Whilst large-scale permanent fortifications are still very much in evidence, field fortification remains fundamental to modern tactics. The lay-out of a modern defensive position, with overlapping arcs of fire and with regard to the ranges of different types of weapon, is an exercise in fortification. Wherever and whenever troops are employed in operations other than pure police and counter-terrorist actions, they dig. The author's experience has instilled in him an unconventional definition of the difference between 'internal security' operations and 'conventional war', based on military exercises. In the former, you do not dig. In the latter, you dig, move, and dig again. Even a shallow scrape in a patrol base increases the chances of survival under mortar attack many times. Trenches on a European battlefield, indeed in any theatre of major war, must have very substantial overhead cover and protection, and this poses construction problems. Prefabricated components may help, as may automatic sensors to detect an enemy approach.

Returning to more permanent fortifications, and their future role, they fit in well with the need for secure, pre-placed land-line communications, and the need to utilize less well-trained manpower in a time of manpower shortage. The new generation of strategic weapons, requiring bulky power sources (lasers) or explosive lenses to focus the power of a nuclear explosion, may only be deployable in static positions. Too many people dismiss the subject of fortification out of hand: it is a crucial part of the history of war, and a crucial element of modern land warfare.

ALL WARFARE IS BASED ON DECEPTION (Sun Tzu)[32]

Sun Tzu's timeless aphorisms concerning this pivotal aspect of war from the fifth century BC are inimitable: the exploitation of the electromagnetic spectrum has increased the importance of, and opportunities for, deception still further. As Bart Whaley has pointed out in his excellent work on the subject, the two areas of endeavour where deception has been most widely used and analysed are war and magic (conjuring).[33] The parallels between the use of deception in two professions are many. Deception arguably involves two symmetrical types of activity. The first is *dissimulation* (hiding the real). This comprises three elements, the first being *masking* (in the military context, camouflage or smokescreens, or radio silence). In Russian *maskirovka* means camouflage but its meaning has been extended to effectively cover all aspects of deception. The other two elements of dissimulation are *repackaging* (dressing a new unit in the distinguishing insignia of one it is replacing, dressing a new commander and his staff as low-ranking soldiers) and *dazzling* (making the same unit appear several times, to swamp and confuse the enemy's senses). Its opposite is *simulation*, (showing the false). This comprises *mimicking* (creating enough distinctive characteristics to give a passable replica, for example, a small group of soldiers drawn from a wide variety of units may make the enemy think they are facing a much larger formation comprising all those units in full);[34] *inventing* completely new but false artefacts (dummy aircraft; the machines used by Zhukov to fake the sounds of tank engines in the prelude to Khalkhin Gol in 1939, for example), and *decoying*. Throughout military history, a prominent example of the latter has been to fake left and move right. Thus, Sherman took Atlanta in 1864, MacArthur Inchon in 1950. The reaction of the target is crucial: if he is not paying attention, he may not be deceived. The enemy may believe the deception, but not act on it in the desired way. Therefore, in warfare deception should be based on what you want the enemy to do, not what you want him to think. Thus deception (arguably what you want the enemy to think) begins to become reflexive control (what you want the enemy to do). The study of deception, in the author's view, should play a much more promiment role in the training of senior commanders, and military history, once again, is the database.

CONCLUSION

This chapter has outlined in the barest detail the origins of war as a social and political phenomenon, key elements in the conduct of war and key military–technical terms. The next chapter addresses one of those key elements, and what exactly we mean by 'military–technical'.

Chapter two

Technics and warfare

The instruments of battle are valuable only if one knows how to use them.

Ardant du Picq, *Etudes de Combat*[1]

TECHNICS - TECHNOLOGY AND TECHNIQUE

The word 'technics', popular in English usage earlier this century,[2] has become unfashionable, which is a pity as it embraces all types of 'technical' question. Not merely technology: the construction of equipment, and the ability of that equipment to enhance human potential, but all aspects of the interrelationship between equipment and its operators. Anyone glancing at the range of glossy publications dealing with military and defence matters today should be forgiven for thinking that technology is the overwhelmingly dominant factor in war, and that the sophistication of that technology must, of itself, confer a decisive advantage. This has never been true in the past and is unlikely to be the case in ground operations in future conflicts. Excessive preoccupation with technology, and even its wider utility (technics), to the exclusion of tactics and practical soldiering, is not a modern problem: it was widespread in the more technical arms (artillery and engineers) before World War One.[3] Yes, the development of weaponry and other products of technology (means of transport, 'launch platforms' and so on) have obviously affected the nature and course of warfare profoundly and continuously. However, weapons development is only one corner of a triangle, of which the other two are a tactical 'doctrine' for using the weapon, and the training of the combatants, individually and collectively to use it. What at first sight may seem to be a technological superiority is often more a result of technique, and may even result from using a simpler or more robust technology.

Even a very simple weapon can have a profound effect if well used. Thus, the Romans enjoyed no 'technological' superiority over the Celts, whose metallurgy was in many ways more advanced. But militarily, the Roman short, stabbing sword (*gladius*) was more effective than the long, slashing blades of the Celts, because of the way it was used. Roman infantry acting in terrifying unison would push their opponents away with their shields, causing them to raise their arms and expose the boneless abdomen to a fatal thrust. The technique has marked similarities with that for use of the short stabbing assegai (*iKlwa*) introduced by the Zulu ruler Shaka in the early nineteenth-century. Previously, the Zulus had used a long throwing assegai but the broad bladed *iKlwa* (an onomatopoeic term from the sound as it was pulled from the victim's body) was used, like the Roman sword, in concert with the shield and thrust into the opponent's left armpit, giving a relatively quick and merciful kill.[4] Romans and Zulus both practised their soldiers exhaustively in these techniques. The combined effect of the technology-training-technique/tactics triangle was sufficient to ensure success against other armies at roughly the same level of technological development. When the Zulus came up against massed firepower from single-shot breech-loading rifles the technological disparity of course told, but it arguably took the machine gun to finally and overwhelmingly tip the balance.

These very basic examples illustrate the importance of weapons, training and tactics being in congruence. Moving from the tactical to the operational and strategic levels, it is necessary to combine this triangle with the best possible command, control, communications, and intelligence (C^3I: see Chapter 1), logistics and high mobility. The Romans were able to do this with their excellent system of roads and engineering works and their clockwork procedure for establishing fortified camps and ensuring that their forces were properly provisioned. The Zulus achieved great operational mobility through physical fitness and could cross country at incredible speed. They showed a sound knowledge of operational principles in the Zulu War of 1879, defeating one of the separated British columns in detail having acquired overwhelming local superiority, at Isandhlwana, and sending two *impis* (divisions) in an encircling move to Rorke's Drift to cut the main British line of communications at a river crossing. However, this operational manoeuvre violated the Zulu king's politico-strategic directive not to enter British territory. He wanted to fight a *defensive* war in which the British would be encircled and destroyed within his own borders. Zulus, having crossed into British territory, the latter were now justified in seeking total victory. Perhaps the most perfect confluence of weaponry, training, discipline, tactics, C^3I, mobility and operational and strategic ingenuity occurred in the Mongol military system (see Chapter 6).

REVOLUTION IN TECHNOLOGY EQUALS STRATEGIC WEAPON

The most fundamental leaps in weapons technology (not the only technological leaps which have revolutionized war) have been the harnessing of new forms of energy. Harnessing fire made metal weapons possible; the discovery of gunpowder was the harnessing of chemical energy as a propellant and explosive; nuclear power made possible nuclear weapons (indeed, it was first used in weapons) and may be harnessed into directed energy weapons. Rockets, which appeared in China around AD 1000 rather than guns, were the first devices to exploit the chemical energy of explosives, but did not have the revolutionary effect on warfare of guns. The first reliable references to guns are the spectacular picture in the de Millimet manuscript, dating from late 1326 or early 1327, and a comment from a Florentine manuscript from the same year.[5] Unsurprisingly, knowledge about guns spread extremely rapidly.[6]

The first two centuries of firearms' employment underscore a trend which is also pronounced in the case of nuclear weapons and newer technologies. When a fundamental scientific-technological revolution utilizing a previously unknown form of energy occurs (chemical energy in the case of guns, nuclear energy in the case of A- and H-bombs), it is first employed as a strategic weapon. It is generally so cumbrous that it is of little use for anything else. It is excessively vulnerable in the open field, and is so rare and expensive that it is the prerogative of only the most powerful, organized, and centralized rulers. Guns were principally of value in knocking down city walls, or gates (as in the de Millimet manuscript) or, safely emplaced behind fortifications, for defence of key installations. They were also so expensive that they were the preserve of kings – *ultima ratio regis* – and arguably contributed to the coalescence of modern nation-states. The fifteenth century includes two notable examples of the use of artillery as a strategic weapon. The Turks developed a powerful artillery for overcoming fortifications and cracking open centres of power and population. In 1453 they used it against a truly strategic target: Constantinople, the heart of the eastern Roman Empire. Charles VIII of France similarly used artillery as a strategic weapon in his campaign in northern Italy in 1494. Later, as guns became more efficient and more mobile, they became more useful as 'tactical', or battlefield weapons. For the first two centuries of their existence, guns were arguably not a major factor in the conduct of warfare in the field, only at the strategic level.[7] We see just the same process with the nuclear warhead, although the process accelerated. The first atomic bombs were very big, and had to be carried by large aircraft, were very scarce, and would not have been very effective against army targets anyway.[8] By the 1960s they had become small enough and plentiful enough to be used tactically. We may see a similar process with laser and charged-particle beam weapons, which require so much power that at the moment they must be based in large vessels with a powerplant which can provide it, like ships or earth satellites, or in big static installations, for strategic aerospace defence and so on. In time, they, like guns and nuclear bombs, will become handier.

PARALLEL DEVELOPMENT AND TECHNOLOGY TRANSFER

There have been certain advances in weapon technology which would make the defeat of an army of an earlier era, however proficient, a relatively easy matter. Even Genghis Khan would probably be beaten in a straight fight by a modern NATO division (although he would probably avoid one and adopt a guerrilla strategy); Napoleon would certainly have won Waterloo if his troops (and not Wellington's) had been equipped with the Minié rifle with which the allies were equipped in the Crimea forty years later. This was because the Minié had an effective range up to 1,500 metres, while the smoothbore musket with which Wellington's and Blücher's men were armed was not effective much beyond 100 metres. A few World War Two divisions with their complement of tanks and aircraft would probably have defeated the vast armies of a quarter of a century before, of 1914 (though not necessarily those of 1918). However, it is significant that all these examples are fictional.

Throughout recent history armies' weaponry has developed at roughly the same rate and at any one time none possesses an overwhelming technological superiority over the other. When Clausewitz (1780–1831) said that 'European armies are comparable in equipment, organization and training',[9] he was not dismissing the impact of new technology unthinkingly. He was well aware that weapons must 'suit the nature of the fighting, which in turn determines their design'.[10] However, as far as the situation of one force relative to the other is concerned, he was right.

Generally speaking, technological advances and ingenious improvisations are seldom confined to one side and never for very long. However, the fact that opposing sides' armaments progress at a similar rate does not prevent that progression itself altering the appearance, nature, pace and scope of warfare. Jomini who was born the year before Clausewitz (1779) but survived longer, into an age of rapidly accelerating technological development, understandably had more to say on technics and warfare. In his *Précis de l'Art de la Guerre*, first published in 1838, he forecast the increasing destructiveness of war and the reappearance of armour on the battlefield:

> The means of destruction are approaching perfection with frightful rapidity. The Congreve rockets, the effect and direction of which the Austrians can now regulate – the shrapnel howitzers, which throw a stream of canister as far as the range of a bullet – the Perkins steam guns, which vomit forth as many balls as a battalion – will multiply the chances of destruction as though the hecatombs of Eylau [1806], Borodino [1812], Leipsic [1813] and Waterloo [1815] were not sufficient to decimate the European races.
>
> If governments do not combine in a congress to proscribe these inventions of destruction, there will be no course left but to make one half of an army consist of cavalry with cuirasses in order to capture with great rapidity these machines; and the infantry, even, will be obliged to resume the armour of the middle ages, without which a battalion will be destroyed before engaging the enemy.
>
> We may see again the famous men-at-arms, all covered with armour, and horses also will require the same protection . . . artillery and

pyrotechny have made advances which should lead to think of modifying the deep formation so much abused by Napoleon.[11]

Although this seems quaint, and the Perkins Steam Gun did not lead to the downfall of western civilization as Jomini feared, Jomini clearly appreciated by this time that technological change could fundamentally alter the nature of warfare, and even that warfare could become so terrible that it would have to be outlawed by international agreement (by implication, that its inherent tendency to the extreme would curtail its utility as a political instrument). Interestingly, he saw technics – armour – as the answer to technics, and although he alluded to 'modifying' formations, he seems not to have anticipated the solutions which in fact precluded an exponential increase in battlefield casualties: vastly increased dispersion and quite different tactical forms. Like Clausewitz, Jomini realized that advantage would not be confined to one side. Years later, he wrote that 'If these arms [the Minié rifle] aided the Allies at the Alma [20 September 1854] and Inkerman [5 November], it was because the Russians were not provided with them . . . in a year or two all armies alike will be furnished with them.'[12] Jomini predicted that in spite of increased firepower, armies would not 'pass the day firing at each other', but that one side would always need to advance to attack the other. Generally, this proved correct, with horrific results. He was also right to note that the new weapons were more easily deployed in defence.[13]

Major technological inequalities only occur when there has been total 'cultural shock', as in the case of the European conquistadores in South and Central America in the sixteenth century, for example. Even then, if the less 'advanced' (at any rate, technologically advanced) party survives the initial encounter, history shows that they usually make up the difference very quickly. If a technological or organizational disparity remains, the weaker side tends to adopt guerrilla techniques.

Thus, the North American Indians rapidly acquired the same means of mobility – the horse – and small arms as the settlers and soldiers of European origin. Contrary to some popular impressions, the Indians used muskets or rifles comparable to those of their adversaries and sometimes, as the US Army found to its cost in the 1870s, even had better weapons (the commercially available Winchester '73 as opposed to the US Army's single-shot Springfield). As Carlo M. Cipolla, the eminent historian, noted, for centuries the Europeans only enjoyed a technological advantage at sea. This gave rise to the peculiar situation whereby 'the Europeans were vulnerable on land while they looked formidable and impregnable on their vessels'.[14] Overseas, the impact of European technological developments was not felt on land until the end of the eighteenth century, because of the difficulty of supplying a large army over long distances. Even then, European and white North American technological superiority only manifested itself significantly where technology was concentrated in high value systems under state control – principally artillery. The monolithic modern states did not triumph on land (as opposed to control over sea lines and coastal areas) because of a superiority in weapons technology itself, but because of organized, coherent planning for the future (foresight[15] being a great strategic advantage) and a systematic, inexorable advance. Non-European powers could get most of the new weapons they wanted (a historical fact which underscores the futility of current attempts to inhibit technology transfer), but this led to a more subtle form of European takeover: a monopoly over the manufacture and supply of such systems.

This phenomenon recurred in the 1980s with the Afghan Mujaheddin opposing the Soviet-backed Kabul government. A sinister-looking bunch of bearded and turbaned individuals, toting the latest anti-tank and anti-aircraft missiles, they had some (the US *Stinger*) which were technologically superior to anything issued to the withdrawing Russians. But the Mujaheddin proved disparate and disorganized, and this, combined with Soviet control of the air, prevented them capturing major objectives from the more cohesive and disciplined forces of the modern Soviet state. The Afghan guerrillas, imbued with an age-old genius for warfare did not, however, find it difficult to get accustomed to the new weapons and use them in a limited tactical context.

The most prominent instrument giving one side a decisive advantage in land warfare throughout most of history has not been a product of human ingenuity at all. Its relatively recent abandonment on a large scale, must, in terms of the history of war on land, be considered a colossal watershed. It was last phased out of service by the Soviet Army in the mid-1950s.[16] It is an animal: the horse, and its close relatives (onagers, asses, mules). There have been seven phases in the use of the horse family in war, and at the beginning of each the innovation proved utterly decisive. The first was the chariot, the four-wheeled version used, as we have seen, by the Sumerians at the beginning of organized warfare, the two-wheeled version by the Hyksos or Shepherd Kings in the conquest of Egypt. The next step was to ride the horse itself, the mounted javelin thrower, exemplified by the Assyrians. The heavier mounted lancer also proved decisive when first employed, for example, the Byzantine *klibanophoroi* or heavy cavalry. The Parthian mounted archer (the fourth phase) was well known to the Romans (see Chapter 6) and reappeared periodically with devastating effect, as in the Mongol conquests. A simple 'technological' advance was the appearance of the stirrup, coming from China in the fourth century AD, reaching eastern Europe by the seventh and by the eighth its western tip.[17] This permitted the evolution of the heavily armoured mounted knight, with a much more stable platform from which to wield heavier weapons. The sixth advance was the mounted pistolier or carabinier, firing from the saddle, who appeared during the sixteenth century and the seventh the mounted raider, doubling as a mounted infantryman, no longer a viable force on the battlefield itself but from the American Civil War to 1945 extremely effective for patrolling, operational manoeuvre, and pursuit (Chapter 5). The horse and its relations also remained vital to logistics for longer than its popularly imagined. A look at World War Two German newsreels soon dispels the impression that this was a 'mechanized war': horses are everywhere.

MAN THE MEASURE

The combination of the horse with such simple devices as the chariot and the stirrup, and the man operating in harmony with both, underscores another key point. Technology must match man: man is the measure. The story of the longbow is the classic example, the paradigm (see also below). The longbow, wielded with deadly effect by Welsh and, soon, English archers, could drive its arrow through plate armour, or a wooden door four inches thick, at a hundred metres or so, in the hands of a skilled archer. But to become an archer required a lifetime's athletic training. No one today would expect an olympic athlete to perform well or even adequately, after days or weeks of fitful sleep, poor food and rain-sodden nights in the open. The official records of sixteenth-century England are quite emphatic: archers went out of condition very quickly in the field. It was for this reason, and not because of any inherent superiority in range, accuracy, and certainly not rate of fire, that firearms supplanted the longbow in the sixteenth century.[18] The longbow in the right hands had also out-performed the much more technologically sophisticated crossbow. Different considerations led to the eclipse of the formidable Turkish composite bow in favour of firearms. Being of laminated construction, like the bows used in modern sport archery, they were very slow and expensive to manufacture.[19]

Technology in its own right has been a much more significant source of advantage at sea and in the air than on land. Even here, however, it has only been significant when harmonized with the physical, mental, and manipulative abilities of men. We can rapidly dismiss the Hollywood image of galley slaves kept in appalling conditions and flogged until they dropped. The Greek trireme, an astonishingly complex piece of machinery, recently expertly and lovingly reconstructed, required a team of well-motivated, fit and skilled young oarsmen and women to make it perform as ancient records indicate it should.[20] So crewed, its performance was indeed impressive. But the crew had to work as a team, 'hard work and much disciplined training', plus adequate nourishment being required to get their system's performance up to ancient Greek standards. Greek triremes, and the Venetian Galleys of 2,000 years later were crewed by volunteers, but even the Turks, who employed slaves and were not known for their humanitarian attitude to prisoners of war, realized that galley slaves needed to be kept fit. They must also have carried out some form of scientifically based training.[21]

Relative strength in the air and in space depends on technology itself to a greater extent than in the other two elements. Even in the air, however, the system is ultimately limited (so far) by the pilot. The speed at which air-to-air engagements take place has not increased greatly since the end of World War Two. Experience in Vietnam and the Middle East has shown that just below the speed of sound is the practical limit for air-to-air combat. Agility may be more important than speed, and agility is not necessarily the preserve of the more advanced aircraft. The introduction of the guided missile has added a critical new element: the missiles can compensate for vast disparities in the capabilities of the aircraft carrying them. In the Anglo-Argentinian conflict of 1982, the subsonic Harrier, seemingly outclassed by the supersonic Argentinian fighters, enjoyed unchallenged superiority because of the AIM-9 L missile with which it was equipped. The definitive comment rests with the Soviet Foxbat, once believed to be the world's highest and fastest flying combat aircraft. When a refugee pilot landed one in Japan, American experts were shocked when they found it to be made of steel, not titanium, and full of old fashioned valves, not solid-state electronics. They had to be reminded that it was still the world's highest and fastest flying combat aircraft, and less vulnerable to electronic counter measures to boot.

ARTIFICIAL INTELLIGENCE: THE END OF MAN THE MEASURE?

Throughout the history of warfare, then, technology and man have always striven for harmony, and man has imposed his limitations on technology. Does this mean that we are on the verge of the greatest revolution of all: the augmentation and in certain cases replacement of human intelligence by artificial intelligence (AI)? Only with AI can air combat advance beyond the transonic engagements permitted by human reflexes. Things will happen so fast in the cockpit that the pilot will have to let AI take over, retaining the ability to override if necessary. This is why the world's air forces, in particular, are taking an interest in AI. AI is essential to the operation of space systems. AI will act as a 'force multiplier', processing the detection, identification, targeting, and engagement of an enemy far faster than the human operator can. So will man cease to be the measure, and merely become the monitor?

Not entirely, to the extent that there must always be 'a human in the loop'. For reasons of politics, and global survival, there will have to be human checks, and wherever a human check is encountered, man will remain the measure. The extent to which weapons and related technology continue to be intertwined with human training will depend on the number of points at which a human remains in the loop and the level in the system at which they are positioned.

SIMPLICITY AND IMPROVISATION

Many of the temporary advantages enjoyed by one side because of weaponry and technics have either been very simple, or the result of accidents and improvisations. Frederick William's Prussian troops enjoyed a temporary advantage from 1718 because of the famous iron ramrod, which meant that they could load faster and more furiously than could the enemy without fear of the thing breaking off and putting their muskets out of action.[22] In the British-American War of 1812, the elite Rifle Brigade was deployed, armed with the technologically more advanced Baker rifle, and was out-shot by Americans using smooth-bore muskets which had a higher rate of fire.

Weapons have frequently proved most effective in roles which their designers did not intend. The widespread use of anti-aircraft guns in the anti-tank role, notably the German *Flak*-'88, is a prime example. The gun's anti-tank potential was first noticed in Spain, and some were used in this way in France in 1940 prior to their first widespread use against tanks in Libya in 1941. The lessons from Spain were quickly absorbed and the 1939 specification for the 1941 variant included a lower silhouette for the anti-tank role.[23] More remarkable was the unexpected value of anti-aircraft guns in the indirect fire ground artillery role. In the Battle of the Bulge (the Ardennes), in 1944–5, the Americans, surprised by the German attack but without a serious air threat to face, fired everything they had into the 'Bulge', including anti-aircraft guns. The recently introduced proximity fuzes on the anti-aircraft shells, designed to explode the shell as soon as it got within effective range of an aircraft, had an unexpected and unprecedented effect. German prisoners were stunned and shattered by the effect of an artillery bombardment where every round exploded at the optimum lethal height above the ground. The proximity fuzes were vastly more effective than traditional ground artillery airburst shells, timed to go off just before the end of the calculated time of flight – a somewhat hit and miss affair.

ASSIMILATING TECHNOLOGY

There has usually been a time lag between the idea for a new weapon or other revolutionary technological advance, its appearance and limited adoption, and its assimilation into the conduct of warfare through new tactical, operational, and strategic forms. The two key elements of modern air–land battle, the tank/armoured personnel carrier and the helicopter, were both envisioned by Leonardo da Vinci (the former to overcome the very problems with which we have seen Jomini wrestling 350 years later). If this is an extreme example, it was perhaps half a century between the general adoption of the arquebus in the sixteenth century and the development of the countermarch which permitted continuous rolling volley fire.[24] These processes are long drawn-out and overlap, with some surprising results. The English developed a coherent military system based on the longbow, deployed behind field defences with an armoured reserve for counterattack, based on tactics used earlier by the Welsh and the Scots. Its first major battlefield success was at Crécy in 1346, by which time guns had been around for at least twenty years. The development of an effective system around the longbow thus occurred *in parallel* with the first deployment of very *in*effective guns.[25]

The American Civil and First World Wars were conflicts where weapons technology and tactics were out of phase. The result: excessive casualties and inordinate battlefield effort for minimal gains. In the American Civil War, the weapons/tactics mismatch did not preclude large-scale operational manoeuvre, whereas in World War One, the existence of a continuous front did. In the American Civil War, both sides were armed with rifled muskets of unprecedented range and accuracy, but initially approached each other in Napoleonic war formations. In the early battles the sides stood in close ranks and fired, in volleys or at will, until one launched a charge to decide the issue with the bayonet. This proved so costly that the defensive generally adopted a dispersed formation, while by the end of the war both sides were practising crude infiltration tactics, making use of short 'rushes' and natural cover.

These conditions also placed more stress on entrenchments, which were widely used both in the defensive and as a base for offensive manoeuvre. Entrenchments had always been used in warfare, particularly in sieges. They featured in certain actions of the Napoleonic wars (for example, the lines of Torres Vedras), but not frequently in major field actions. Widespread entrenching seems to have been one of the few innovations to have emanated from the 'long peace' of 1815–53. It was probably connected in part with the advent of the rifled musket and the bitter and dirty colonial fighting undertaken by British, French, Americans and Russians during this period. If the use of elaborate entrenchments by both sides at the siege of Sevastopol (1854–5) was a continuation of the traditional prominence of entrenchments in siege warfare, the British and French were soon employing them in the open field. At Lucknow, on 4 July 1857, Sir Henry Lawrence's dying instructions were 'entrench, entrench, entrench'.[27] At Solferino, two years later, the French, many of whom had hard experience at Sevastopol and in North Africa, began to dig themselves in as soon as they had captured the Austrian centre, something which would never have happened on one of the first Napoleon's battlefields.[27] From then on, the spade was a prominent instrument of war, and in the Russo-Turkish war (1877–8) the Russians made it their practice to entrench on every new position as soon as it was taken.[28]

The superiority of the tactical defensive from the mid nineteenth century is borne out by quantitative research undertaken by Arnold and Trevor Dupuy. They computed the number of occasions when the attacker was successful against the number when he was numerically superior and the number in which his casualty rate was less than the defender's. The attacker's success rate and the number of occasions when he sustained fewer casualties than the defender both reached their nadir about 1860. The fact that the situation subsequently remedied itself was not the result of any further radial change in weaponry (although many later believed that quick firing artillery might restore an advantage to the attacker, since its concentrated fire could be applied at a time and place of the attacker's choosing). Rather, it resulted from commanders, staff officers and military theorists realizing what had happened and adjusting their procedures. The number of successful attacks after 1860 probably increased because commanders were now reluctant to commit forces to the attack at all unless they knew they had a good chance of success. Von Moltke (1800–1891) adapted his strategy to the greater tactical strength of the defensive by manoeuvring his armies onto the enemy's lines of communication and thus forcing the enemy to attack him – the strategic offensive combined with the tactical defensive (see Chapter 1).[29]

The same delay in assimilating new technological means also occurred with the tank and the heavier-than-air aircraft. Although both were employed in war almost as soon as they were invented, it can be argued that the tank represented a delay of nearly thirty years (1888–1916) in adapting the internal combustion engine to the needs of warfare. Furthermore, neither was fully assimilated into operational art and tactics until at least the mid 1930s, with the German revival of classic military principles which came to be known as Blitzkrieg, and more thoroughly in the mid-1940s, with fully motor-mechanized formations used by British, Americans, and Russians. As a general rule, the interval between appearance/adoption of a new weapon and its assimilation has been at least twenty to thirty years, maybe more. In elements other than land warfare, the weaponry itself may be more complex, but its assimilation a little simpler. Even then, it takes between 12 and 20 years from initial experiment to full operational status: for example, the aircraft carrier (early 1920s to early 1940s), the intercontinental bomber (late 1940s to early 1960s), although Tukhachevskiy was envisaging stratospheric bombers in the 1930s), the intercontinental ballistic missile (from 1946 to late 1950s), although initial research on which this was based, in Germany and the USSR, again went back to the 1930s).

ON MILITARY-SCIENTIFIC REVOLUTIONS

The influence of technology on warfare, the interaction of a number of technological changes, their assimilation and multi-faceted effect on tactics, operational art, strategy, and military thinking generally, has much in common with 'revolutions' in science. As Thomas Kuhn pointed out in his classic *The Structure of Scientific Revolutions*, the history of science is permeated by a number of apparent leaps or revolutions.[31] Kuhn argues that these are not usually the exclusive work of the person whose name is associated with them: the Copernican revolution, which placed the Sun rather than the Earth at the centre of the Solar System, Newtonian physics, Einsteinian physics, and so on. Nor are they changes which took place overnight. Rather, they are what Kunn calls *paradigms*. For Kuhn, a paradigm is a scientific achievement that embodies a whole bundle of theories, laws, procedures, practices. It thus becomes the dominant theory, and institutionalized practice within the scientific community.

Kuhn's idea of the paradigm is particularly relevant to revolutions in the area of 'military science' (embracing not just technology but all levels of military art, logistics, C^3I and so on). This is because military science is not merely recondite theory floating in a vacuum; it is about employing large numbers of sweating organisms with stomachs and brains in a very down-to-earth situation.

Many commentators on military history have, perhaps inadvertently, described changes in military theory and practice in precisely Kuhnian terms. Correlli Barnett, for example, noted that by the outbreak of World War One, three new factors had 'totally altered the terms of reference of warfare':

> The first of these factors was the technological and military revolution represented by the magazine rifle, the water cooled and belt-fed machine gun, smokeless propellants and quick firing artillery [the latter itself invoked a range of interrelated factors, which will be explained below]. The second factor was the problem of supplying and deploying and directing unprecedented numbers of troops, and this was linked with the related problem of making the right military use of the latest inventions, the telephone, wireless, the internal combustion engine, the flying machine.[32]

If we take the usual dictionary definition of a paradigm as a classic example, this is a paradigm of a Kuhnian paradigm. This mid to late nineteenth-century 'revolution in warfare', whose results were not fully apparent until well into World War One, is one of the most complex and fascinating paradigm changes in military science. The Russians created another Kuhnian paradigm, or, rather, confirmed its existence, when they called the possible changes wrought by the advent of the ballistic missile and the nuclear warhead the 'Revolution in Military Affairs'. This is perhaps even more of a paradigm because, unlike the 'nineteenth century 'revolution in warfare' mentioned by various western historians, the *revolyutsiya v voyennom dele* of the 1950s had a precise and unambiguous significance among an important scientific élite – the Soviet military.[33]

Another key paradigm is represented by the employment and effect of artillery in the early part of this century. Here we have an example of a major paradigm change which, after a long period of experiment and debate, followed by assimilation into standard practice, returned artillery to its logical place as the most lethal ground-based weapons system. Correlli Barnett mentions 'quick firing artillery' as part of the 'technological and military revolution': the paradigm is more fully expressed perhaps, as 'quick and indirect firing artillery of greatly increased lethality'.

During the Napoleonic Wars, artillery had been the greatest battlefield killer, which was logical enough as an artillery battery firing grapeshot could discharge almost as many balls as an infantry battalion's muskets, and at much greater range.[34] The introduction of the rifle and cylindro-conical bullet gave every infantryman a weapon which could, initially, strike at the same range as artillery. Because of its greater technological complexity and problems of scale, it took many years for the advantages of rifling, breech-loading and rapid firing to be applied to artillery. Even when they were, artillery and infantry alike were handicapped by the limitations imposed by the folds of the ground, which seldom permitted artillery to use its full range. The more numerous and less conspicuous infantry could pick off the gunners, as happened in the American Civil and Franco-Prussian wars. Methods of indirect fire, that is, directing the fire of artillery onto targets invisible to the guns themselves, had been used in siege warfare for centuries, and occasionally in the field. After the Franco-Prussian War of 1870–71, the Prussians began to explore the possibilities of indirect fire for field artillery, and from 1880 it became a subject for much debate among professional artillerymen. Indirect fire was not generally employed in the Russo-Turkish War of 1877–8, and the lethality of artillery relative to other arms reached its nadir (see Table 2.1). The advent of a system of indirect fire for use in fast-moving field operations is a classic example of the interaction of numerous aspects of technology and technique.

Indirect fire would not have been particularly effective at artillery's longer ranges, however, if it had only been able to fire solid cannon balls, or even the primitive common shell of Napoleonic times. More modern explosive shells, which had already begun to show their horrific potential in the Franco-Austrian (1859), Austro-Prussian (1866), and Franco-Prussian (1870–71) Wars, exploded lethally after plunging from above, rather than relying on their kinetic energy to carry them horizontally through carefully selected targets. They were themselves an 'area weapon', and complemented artillery's new role as an area weapon.

As with many other scientific revolutions, the final piece which brought the whole paradigm together was tiny and unspectacular. Earlier systems of indirect fire had relied on a line of stakes or other markers between the target and the concealed gun which was to fire at it. These techniques were obviously unsuitable for use in a fast-moving battle and for rapid lateral shifts of fire. The Germans developed the prototype of the modern 'dial sight', using a pointer on a graduated angle-measuring plate, by which the gun could be orientated in relation to a reference point which it could see, and thus swung in any direction through a given angle to fire at a target which it could not. The British used a primitive and improvised system of this type in the Boer War, but by this time the French, Germans, Austrians, and Russians all had specially designed indirect fire sights. In the Russo-Japanese War of 1904–5, the Japanese, well taught by the Germans, and the Russians, with their formidable technical artillery tradition, both made widespread use of indirect fire. Even now, however, the problem of the paradigm asserted itself: artillerymen who adopted covered positions for their guns were ordered forward onto ridges by generals who were still unaware of the possibilities of indirect fire.[35]

Without indirect fire, World War One could never have been fought as it was. Artillery which had to see its target over open sights would have been shot to pieces, and there was no way that the huge numbers of guns manufactured could have been crammed into such positions. Indirect fire techniques, however, meant that any target whose location could be plotted on the map could be hit. Indirect fire in turn led to greater emphasis on air reconnaissance and accurate maps. It made possible the dispersal of guns in depth as well as in breadth, and the concentration on a single target of all guns within range, over ranges vastly in excess of that of the human eye. Yet this crucial paradigm change is seldom mentioned by military historians.

The relevance of Kuhn's paradigm approach to changes in military theory and practice is evident from another analogy. In his discussion of the change from the Newtonian to the Einsteinian paradigm in physics, Kuhn points out that for practical purposes it is a question of scale or, in the case of physics, velocity: 'Newtonian dynamics is still used with great success by most engineers. If Newtonian theory is to provide a good approximate solution, the relative velocities of the bodies considered must be small compared with the velocity of light.'[36] As we shall demonstrate in Chapter 3, military science has seen quantum changes in the scale of operations which mirror the differences in velocity between those considered in Newtonian and Einsteinian physics. Kuhn also pointed out that paradigms change when scientists observe that the old paradigm does not work, when the old paradigm reaches a point of crisis.[37] The famous Schlieffen plan for the encirclement of the French armies, partially implemented in 1914, was derived from an old, traditional paradigm: encirclement of all the enemy's forces round an open flank. Schlieffen himself envisaged it as a 'giant Cannae', recalling the pincer movement executed by Hannibal against the Romans.[38] But, as with Newtonian and Einsteinian physics, the scale was quite different. For various reasons, the principles derived from ancient battles on a much smaller scale did not work. The following year, von Falkenhayn ruled out the possibility of encircling the vastly extended

Russian front in the east. Instead, he would have to break through the centre and encircle outwards (see also Chapters 1 and 3). Thus was created the concept of the Gorlic–Tarnow operation. There were precedents, but from now on the new paradigm would dominate.[39] Crisis in the old paradigm had led to the creation of a new one, the paradigm of operational manoeuvre which still seems most applicable today.

Modern academics are frequently unsympathetic to the problems faced by military men in assimilating new technology as part of a total paradigm change – 'new weapons: old mind-sets'. Frequently, however, they telescope history or pick on particular incidents rather than examining the detail of events. For example, we are told that 'cavalrymen dreamed of charges in the face of massed artillery, the machine gun and air power and kept their beloved arm in the field half a century beyond its utility'.[40] Cavalry used as a shock arm *on the battlefield* was arguably already passé by the time of von Bredow's 'death ride' in the Franco-Prussian war of 1870; most of the discussion of the role of cavalry for the next forty years revolved around its utility for operational manoeuvre and raiding (see Chapter 5) and using the horse primarily to transport riflemen rapidly to places where they were needed. At the end of World War One, and after it, there was actually a renaissance of cavalry in the latter role. The Russian Civil War saw successful employment of cavalry formations and the Red Army used cavalry, intelligently merged with armoured units and supported by aircraft, right through World War Two.

Military planners have undoubtedly been guilty of grafting new weapons technology onto the pre-existing politico-strategic framework, 'without realizing how that framework itself will be altered by the change in technique'.[41] If we mean the utility of major war as a political instrument, this is undoubtedly right. Most predictions of a future World War Three are incredible simply because there seems no way that any rational government could conceive such military action as likely to achieve any desirable political objective. If, however, by 'strategy', we mean the overall shape and conduct of operations, once major hostilities have started, the case is not proven. For example, one authority criticized those who 'unite the weapons' of the 1980s with the strategy of the 1940s – a synthesis which takes inadequate account of the impact of new military technology'.[42] I would agree that new military technology has probably made major war as a rational option obsolete. But in terms of the conduct of operations, the argument is less convincing. Its proponents, for example, assert that the French reliance on the Maginot line – 'the ultimate instrument of positional warfare' – occurred at a time when 'such static fighting was becoming increasingly obsolete'. 'Static' fighting (whatever that is supposed to mean, exactly) was far from obsolete, as Stalingrad and Dien Bien Phu would prove beyond doubt. What did these analysts make of the remark by General Gareyev of the Soviet General Staff in October 1988, when he said that a system of 'strongpoints' might form part of a 'new', more defensive Soviet strategy? And what did they make of the view of certain informed circles in the Pentagon, current at the time of writing, that the Russians were possibly moving towards 'Star Wars in concrete' ?[43] The problem with the Maginot line is of timeless relevance: political and financial considerations prevented the French and neighbouring Benelux finishing the scheme of frontier defences by taking it to the sea, and the Germans adopted classic military theory and went round the open flank.[44]

Criticisms that military experts and planners fail to assess the impact of new technology on war adequately would carry more weight if they were supported by detailed analysis of how, exactly, the 'new' weapon was likely to affect the conduct of war, in all its complex ramifications. To be fair, the military are sometimes guilty of the same omission. Will 'airmobile operations' in fact be possible on a central European battlefield or will the improved and highly sophisticated means of mine-warfare negate many of the vaunted means of battlefield mobility? Maybe generals are wrong to envisage a repetition of Patton's armoured cavalry actions of World War Two. But the academics, whose job it is to take a more synoptic approach, are equally wrong to dismiss the possibility that a future war might be more like World War One.

Historical analyses of such questions are frequently conditioned by hindsight. The parallel development of the longbow-based defensive system and the gun and the continued omnipresence of the horse have been mentioned. There is also, for example, a tendency to see the tank and, to a lesser extent, the aircraft, as breaking the trench deadlock on the western front in World War One. The most spectacular breakthroughs were made using infantry, intelligently and in concert with carefully prepared and orchestrated artillery fire (Brusilov's offensive, 1916, and the German offensive in the west of March 1918 – see Chapter 3). There are no technological panaceas – only the intelligent, studied and laborious adaptation of tactics and operational art to new means of warfare.

COUNTERING TECHNOLOGY: LETHALITY AND DISPERSION

It is possible to construct lethality indices for weapons based on such characteristics as rate of fire, number of potential targets per strike, incapacitating effect, effective range, accuracy, reliability, radius of action, and vulnerability (or, conversely, that dreadful modern word 'survivability', which this author has emphatically banned).[45] These indices can be plotted on a scale as in Figure 2.1. Because weapon lethality has increased at an accelerating rate, the scale is logarithmic. Against this, we have to consider the counter-measures which armies have taken, primarily spreading themselves out more and increased emphasis on using terrain cover. In ancient times, one man probably occupied 10 square metres of battlefield,

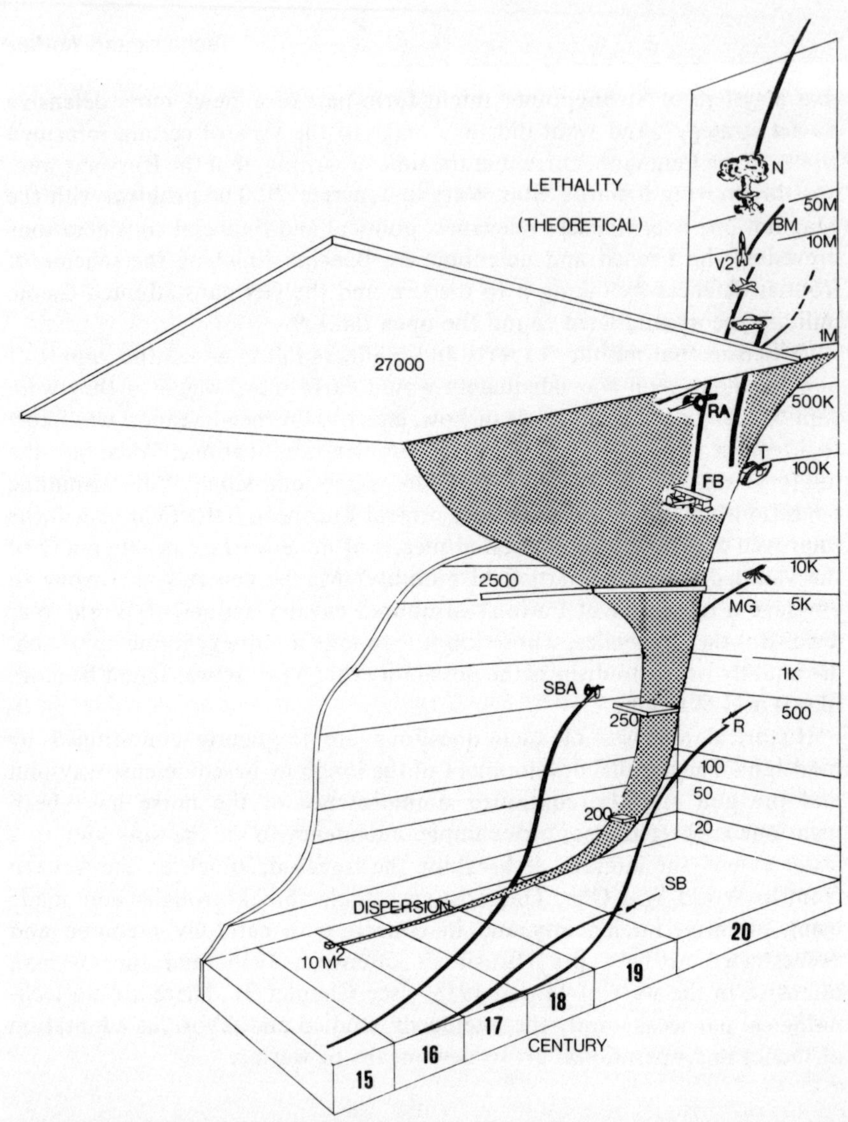

Figure 2.1 Lethality versus dispersion

Note: The lethality index (vertical scale on right) is the theoretical maximum killing capacity per hour: the dispersion of normal forces in combat is indicated by the shaded strip which represents the number of square metres per man in combat. Thus, for centuries, combat formations (the Greek phalanx, the Roman legion, the Spanish *tercio*) comprised men in close formation – 10 square metres each, no more. The black strips represent the lethality of weapons. Thus, from its appearance in the fourteenth century smooth-bore artillery increased in lethality fairly constantly, reaching its climax in the Napoleonic wars at the beginning of the nineteenth century. Smooth-bore small arms were first widely employed at the beginning of the sixteenth century, and continued in use until the mid nineteenth century. Although rifled small arms were manufactured as early as the seventeenth century, their lethality index was less than that of smooth-bores because they took much longer to load. With the appearance of the Minié rifle in the 1840s, they shot ahead, further increases in lethality being a function of breech loading,

and this increased little until the nineteenth century. In the American Civil War one man occupied rather over 200 square metres; in World War One, over 2,000 and in World War Two, over 20,000. Dispersion has in fact increased more than lethality, and this explains why, for example, casualty rates in combat were lower in World War Two than in World War One or the American Civil War.[46] However, there must be a limit to how far armies can go on being dispersed before they either lose the ability to act in a co-ordinated fashion or overflow the bounds of land theatres of operations.[47]

Although weapons have therefore become more lethal in absolute terms, this has not meant a constant increase in casualties in battle. Nor has the relative lethality of different types of weapon remained constant. The lethality of various weapons at various times is listed in Table 2.1. Even these statistics do not fully depict the impact of different types of armaments on the battlefield. If artillery has been the greatest killer in twentieth-century warfare, its effect in crushing morale, numbing thought and paralysing movement is incomparably greater. Air power is still more important from the viewpoint of morale and tactical and operational paralysis. As Table 2.1 suggests, the number of casualties inflicted on battlefield troops (as opposed to civilians) by aircraft has hitherto been surprisingly small. However, in all theatres, beginning in earnest with the Spanish Civil War, and continuing through the Germans' use of dive bombers in Poland, the effect on the enemy's capacity and will to resist of even a small number of aircraft has been disproportionately high.[48] Another important factor is the ratio of

---

magazines, and smokeless propellants. In reaction to these developments, dispersion, shown by the scaled square areas, increased rapidly, to 200, 250 and 2,500 square metres per man. The first of a range of vastly more lethal new weapons was the breech-loading, quick-firing rifled field gun, exemplified by the French 75 mm model 1897. Other new weapons, the tank and the fighter bomber, were potentially less lethal at first, but kept pace with increases in the lethality of artillery. The lethality curve for artillery continues upwards as a dotted line, as increases in range, precision guided and improved conventional munitions may have provided a revolutionary increase in lethality. The ballistic missile with a conventional warhead first appeared with the V2 in 1944, and the nuclear warhead the following year. Dispersion increased in proportion; by World War Two it was over 20,000 square metres per man. But can dispersion possibly keep pace with the increased lethality of weapons without either making co-ordinated military action impossible, or bursting the bounds of theatres of operations?

BM = Ballistic missile
FB = Fighter-bomber
MG = Machine guns
N = Nuclear weapons, starting with 20 kiloton Hiroshima bomb
R = Rifled small arms
RA = Rifled breech loading quick firing artillery with recoil mechanism and indirect fire capability
SB = Smooth-bore small arms
SBA = Smooth-bore artillery
T = Tank and tank gun

Source: based on information in Colonel T. N. Dupuy, *The Evolution of Weapons and Warfare* (Jane's, London, 1980), pp. 288–9 adapted and interpreted by the author.

Table 2.1 Relative effectiveness of different types of munitions, 1877–1967: (selected examples, per cent)

War or campaign	*Killed in Action, (KIA), Wounded in Action (WIA) or all*	*Inflicted on*	*By*	*Shells and Mortar Bombs*	*Aerial Bombs (H)*	*Small arms*	*Cold steel*	*Burns/ incediary*	*Pellet bombs and rockets*
1. Russo-Turkish War, 1877–8	WIA	Russians	Turks.	2.5	—	94.5	3.0	—	—
2. Russo-Japanese War, 1904–5	all	Japanese	Russians						
(a) Sieges		—		22.9	—	60.5	N/K[a]	—	—
(b) Open field		—		13.7	—	81.0	N/K	—	—
3. World War One, 1914–18	WIA	British	Germans	58.5	negligible	39.0	0.3	—	—
					—				—
4. Nomonham, 1939	WIA	Japanese	Soviets	53.0	—	35.9	N/K		
	KIA	Japanese	Soviets	51.2	—	37.3	N/K		—

5. World War Two									
(a) North Africa	all	US	Germans and Italians	75.0[b]	1.0	20.0	N/K	—	—
(b) Europe, 1st and 3rd armies 1944–5	WIA	US	Germans	63.4	4.9	24.6	N/K	1.2	
6. Korea	KIA	US	North Koreans	59.3	negligible	29.5	negligible	5.0	—
7. Vietnam War	all								
(a) 1965		North Vietnamese	US	60.0		28.0	—	2.0	22.0
(b) 1967		North Vietnamese	US	28.0		1.0	—	9.0	62.0

Notes:

(a) Casualties caused by agents other than artillery and small arms are given as 16.6 per cent for siege warfare and 5.3 per cent for the open field. One would expect a proportion of the former to be from mines, grenades, and weapons specially designed for siege and trench warfare, and most of those in the open field to be by bayonet.

(b) The high proportion of casualties attributable to artillery is almost certainly due to the rocky terrain, which enhances the lethal effect of shells striking it.

Source: V. A. Zolotarev, *Rossiya i Turtsiya: voyna 1877–78 gg (Russia and Turkey: The War of 1877–78)* (Nauka, Moscow, 1983), p. 56; Edward J. Drea, *Nomonhan: Japanese–Soviet Tactical Combat, 1939*. (Leavenworth Paper no. 2, Combact Studies Institute, US Army Command and General Staff College, Fort Leavenworth, 1981, p. 99); *Medical Statistics: Casualties and Medical Statistics of the Great War (History of the Great War)*, (HMSO, London, 1931), p. 40; Colonel James Boyd Coates, Jr, (ed.) *Wound Ballistics* (Office of the Surgeon General, US Government Printing Office, Washington DC, 1962), pp. 55, 755; Dr M. Lumsden, *Anti Personnel Weapons*, (Stockholm International Peace Research Institute, Taylor & Francis, London, 1978), p. 31.

killed-in-action (KIA) to wounded-in-action (WIA). Wounded men are more of a drain on their own side's resources than dead ones, because of the large infrastructure of medical personnel, both in the armed forces and possibly in the civilian sector, required to care for them. Area weapons like artillery tend to wound more than they kill: more precise weapons tend to be deadlier. One of the remarkable things about the casualty statistics published by the Russians after their announced intention to withdraw from Afghanistan in 1988 was the high proportion of killed in action. This is at least in part due to the Afghan guerrillas' continued use of the old Lee-Enfield rifle: a precision shooting weapon, unlike modern assault rifles.

THE SPECTRUM OF TECHNOLOGIES

If the lethality of weaponry, the dispersion of troops and the size of the battlefield (see Chapter 3) have all increased exponentially, there is a final factor which any politician contemplating the use of the military instrument must consider. Friedrich Engels (1820–1895) was one of the first and most eloquent commentators on the cardinal role of the industrial base in war. Modern armaments were the product of modern industry and the party with the stronger industrial base would win, especially since, as Engels believed, a future major war would be a protracted affair.[49] But even in Engels's time, the range of technologies involved was comparatively small. Metallurgy, particularly the working of iron and steel, governed the manufacture of gun barrels, armour, boilers, locomotives, and rails. The chemical industry governed the manufacture of propellants, involving technologies not dissimilar to those of the textile industry or manufacture of fertilizers. By World War One an automotive industry existed which applied its expertise to motor transport, aircraft and, later, tanks. The exploitation of the electronic spectrum was limited to a narrow band of radio frequencies and optical instruments. By and large, once battle was joined it was decided – or

Figure 2.2 The expanding spectrum of technologies

Note: In 1800 the spectrum of technologies directly utilized in land warfare was very narrow. By 1860 it had expanded considerably, though much of the innovation was centred around heavy industry (coal, stream, and iron/steel). Technology had advanced at a greater rate in the naval sphere, albeit of a broadly similar type. Armour plate is shown as a land system because it was used in fortifications. Further significant innovations by 1914 were the internal combustion engine, nitrate based explosives, a much more widespread use of optical instruments (indirect fire sights, range-finders) and (shortly) poison gas. World War Two saw a huge expansion and development of internal combustion engine systems, air systems, and the use of the electromagnetic spectrum. The latter has perhaps been the most pronounced expansion to date, and has also permeated every other type of technology (air systems, space systems, guided missiles and so on). The influence of the electromagnetic spectrum is felt in all systems today, so that the one-dimensional linear spectrum of 1800–1860 has become a two-dimensional one of infinitely greater extent. Where a technology used in an earlier era continues in use in a later one, it is referred by its initial letters only. The diagram does not illustrate major advances within any one technology (for example, the considerable advances in manufacture and stressing of gun barrels, breech loading as opposed to muzzle loading, and so on).

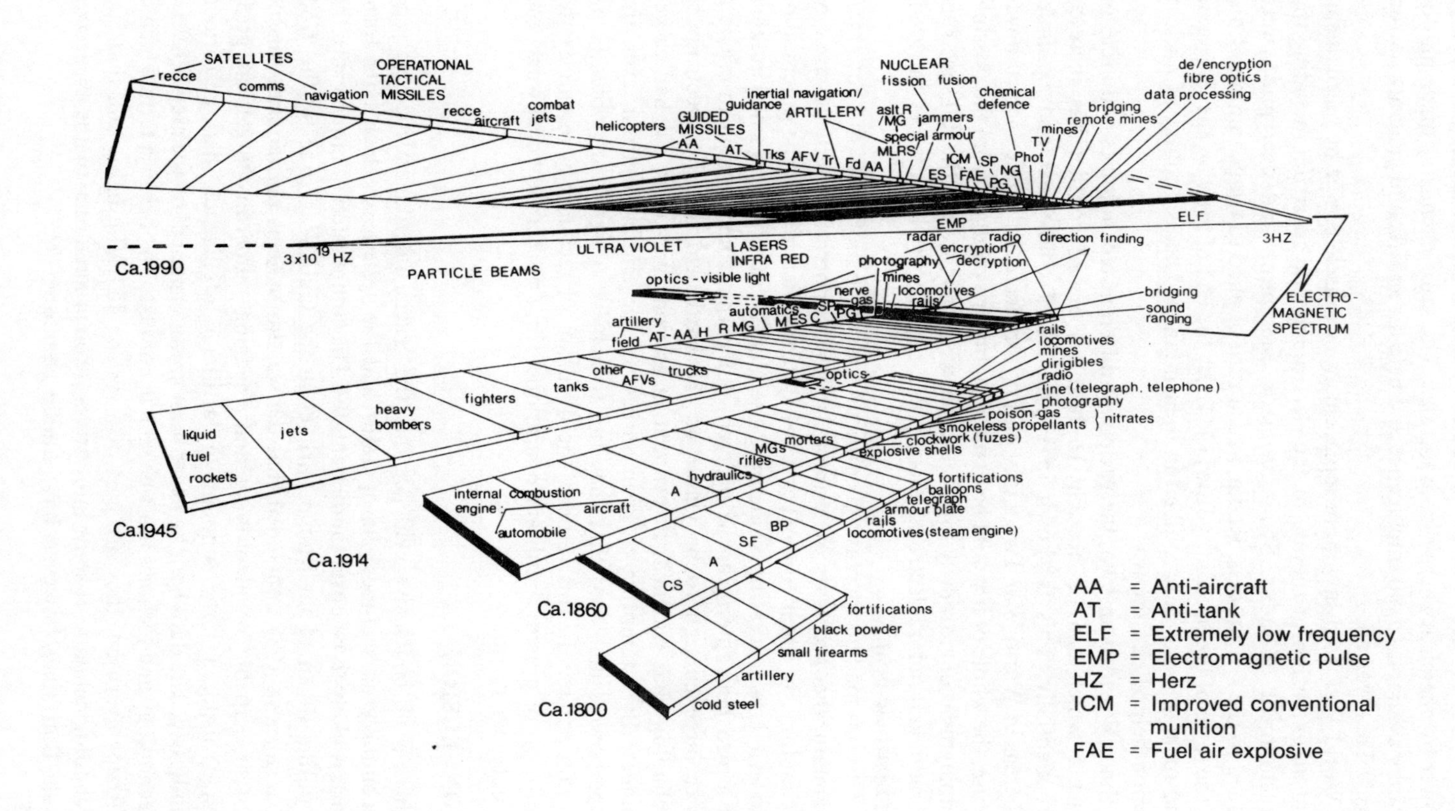

SATELLITES
recce
comms
navigation
OPERATIONAL TACTICAL MISSILES
recce aircraft
combat jets
helicopters
GUIDED MISSILES
AA
AT
inertial navigation/ guidance
ARTILLERY
Tks
AFV
Tr
Fd
AA
aslt R /MG
MLRS
NUCLEAR
fission
fusion
jammers
special armour
ICM
ES
FAE
SP
PG
NG
chemical defence
Phot
TV
mines
bridging
remote mines
data processing
fibre optics
de/encryption
ELF
3HZ
EMP
radar
radio
direction finding
Ca.1990
3x10^19 HZ
ULTRA VIOLET
LASERS
INFRA RED
PARTICLE BEAMS
photography
encryption/ decryption
optics - visible light
nerve gas
mines
locomotives
rails
automatics
SP
PG
bridging
sound ranging
ELECTRO-MAGNETIC SPECTRUM
artillery
field
AT-AA
H
R
MG
M
ES
C
rails
locomotives
mines
dirigibles
radio
line (telegraph. telephone)
photography
poison gas
smokeless propellants
nitrates
clockwork (fuzes)
explosive shells
other AFVs
trucks
optics
tanks
fighters
heavy bombers
jets
liquid fuel rockets
Ca.1945
MGs
mortars
rifles
hydraulics
A
internal combustion engine
aircraft
automobile
Ca.1914
fortifications
balloons
telegraph
armour plate
rails
locomotives (steam engine)
BP
SF
A
CS
Ca.1860
fortifications
black powder
small firearms
artillery
cold steel
Ca.1800
AA = Anti-aircraft
AT = Anti-tank
ELF = Extremely low frequency
EMP = Electromagnetic pulse
HZ = Herz
ICM = Improved conventional munition
FAE = Fuel air explosive

not – by the muscle of men and horses moving lumps of metal given deadly power by chemical energy from nitrates. The same chemistry made photography available to aid intelligence and help direct fire, and generate poison gas.[50] That was all.

World War Two saw more replacement of muscle power by mechanical, and an extension of the use of the electromagnetic spectrum including the development of radar. Semi-automatic encryption and decryption were added to the C³I process. Radar was in turn applied to trigger ammunition, in the form of the proximity fuse (1944). By the end of the war, the jet engine and rocketry were in use, albeit in their infancy. The technology of nuclear fission helped end the war.

Since World War Two, the variety of different technologies and areas of science which can be brought to bear on armed conflict has incresed exponentially. This is shown in Figure 2.2. To take just one example, whereas in World War Two a tiny part of the electromagnetic spectrum was in use, the whole of it would be in use in a modern conflict, from the possible employment of extremely low-frequency weapons through the full spectrum of radio and radar, electromagnetic pulse, millimetre waves, lasers, infrared, and particle beams.

This has made it increasingly difficult for even major industrial powers to maintain the necessary investment in every potentially crucial area of technology. In order to pay for a C³I system which utilizes the potential of modern data processing, something else must be sacrificed. A small number of armed helicopters can be ordered, but perhaps not enough to create a force large enough to be viable on the battlefield. As the spectrum of technology has widened, investment in any one part of it has had to get thinner. By attempting to keep abreast of the entire spectrum, even the superpowers may find that they fail to achieve a significant potential in certain areas.[51] In this respect, the diffusion of technology resembles the diffusion of potential military operations, from nuclear to 'local wars' and counter-terrorism.

CONCLUSION

There is no conclusive evidence that the crucial interdependence between technology or 'technics', its operator and the system of which both form part, is likely to be upset fundamentally. The commission of inquiry into the shooting down of an Iranian airliner in the Gulf War zone by the *USS Vincennes* in 1988 concluded that the warship's Aegis anti-aircraft missile system, capable of identifying and engaging numerous targets at great range, using a large amount of 'artificial intelligence' had performed faultlessly. The data had, however been 'misinterpreted' by the crew. Q.E.D. Technology and technics must be seen in context. To borrow a term which is perhaps overused, they are part of a *system.* If traffic lights fail, that is a technological fault. If some idiot drives through a red one, that is a *systemic* fault. Both may, however, have similar consequences.

Chapter three

The expansion of the battlefield and the operational level of war, 1800–1918

This chapter is an overview of the development of military theory and practice from Napoleon to the end of World War One, their interaction, in terms of the shape, scale, and scope of military operations, and the influence of technology. It is not a campaign or institutional history, but aims to portray the appearance and course of fundamental forces which have moulded warfare into its modern form.

With occasional exceptions, the development of military art in the pre-industrial era was driven by its most outstanding practitioners. Exceptions are Machiavelli (1469–1527) and the Englishman Henry Lloyd, a General, but not an outstanding one, who analysed the lessons of the Seven Years' War (1756–63) and exerted a great influence on Jomini and some of the later thinkers.[1] Outstanding practitioners might put their thoughts on paper, and that was the most significant form of military 'theory' in the Enlightenment. Examples are the work of the Italian Field Marshal Raimondo Montecuccoli (1609–1680); Maurice (or Moritz) de Saxe (1696–1750) and Frederick the Great,[2] and Suvorov.[3] Montecuccoli's *Aphorisms Relating to a Possible War with the Turks in Hungary* of the 1670s[4] is remarkable as an analysis of the lessons of recent wars as a basis for a prognosis for the future – quite different from Machiavelli's scholastic recapitulation of classical antiquity. The Enlightenment and the Scientific Revolution created thought patterns favourable to the development of a theoretical study of warfare.

Later, as military art, government, and society all became more complex, and as technology exerted a more pronounced but less predictable influence and the likely form of operations therefore became more uncertain, theorists became relatively more important. Napoleon was probably the last great captain to work out his scheme of battle as he went along. Henceforward, major land warfare was conducted, at least initially, in accordance with plans drawn up in a peace-time laboratory. For a long time, this development would not apply to guerrilla and counter-insurgency warfare, where natural warriors and the hardened professionals facing them in India, Africa, the Caucasus, Turkestan, the American west relied primarily on heuristic skills. But with regard to major inter-state and civil war, we see the appearance of famous strategists who, although they have frequently practised the art of war as well, are famous not because of the campaigns in which they fought, but because of their contribution to military theory.

In addition, the influence wielded by the great commander has become correspondingly less. Ludendorff, Foch, Pershing, Guderian, Rommel, Zhukov, Patton, Eisenhower, MacArthur, Montgomery – all may have been great commanders but no one could exert a decisive influence on the war as a whole. The strategists who shaped the concepts according to which these field commanders operated and the general staffs who consciously and sub-consciously interpreted them exercised an all-pervasive influence. Among these were the two classic interpreters of Napoleon: Jomini (1779–1869) and Clausewitz (1780–1831); the Prussian/German General Staff planners Von Moltke the Elder (1800–1891) and Schlieffen (1833–1913); those who laid the foundations of Soviet military thought – Mikhnevich (1849–1927), Neznamov (1872–1928), Svechin (1878–1938) and Frunze (1885–1925); the proponents of the air–land Battle – Fuller (1878–1966), de Gaulle (1890–1970), Guderian (1888–1954), Triandafillov (1894–1931) and Tukhachevskiy (1893–1937).

The nuclear era has produced its own theorists, but mercifully we have no means of gauging how correct their theories have been. The effect of nuclear weapons on warfare has, for obvious reasons, been debated in a theoretical environment and most of the key works on nuclear strategy seem to have little relation to the pre-nuclear tradition of military theory.[5] Rational discussion of the influence of nuclear weapons on the battlefield is difficult because of the mind-rending implications for escalation. Modern military theorists therefore tend to be sharply divided: those who deal with nuclear issues, frequently in an almost scholastic way, and those who deal with the conduct of 'conventional' war, believing that the conduct of warfare ends with the use of nuclear weapons,[6] a somewhat half-hearted way out. There is an argument that nuclear weapons simply represent an exponential increase in firepower, with implications consistent with all the other increases in firepower throughout the history of warfare. The problem, with nuclear, as with conventional weapons, is that we seem to have run out of battlefield.

Clausewitz's and Jomini's analyses of the campaigns of Napoleon and Frederick the Great mark the bifurcation of military theory and practice in the west (though both the analysts were themselves generals, which confers indisputable authority on their work). Subsequent theoretical and practical development has taken place in parallel and is interrelated, but has nevertheless been uneven and jerky. The Asian tradition conforms to the older pattern of practitioners setting down their experience, (Sun Tzu appears to have been a very effective General), and this has continued for longer than in the west, arguably up to the present day (see Chapter 6). In the west, the bifurcation, more or less coincides, tidily, with the Industrial

Revolution. It also coincides with demographic and organizational developments making possible the mobilization of more and more of the state's potential, leading to more and more total war. This is the fundamental difference between Frederick the Great and Napoleon.

The technique with which the wars of Frederick's time were conducted can in no sense be described as 'limited' or 'formalistic'. The ferocity and energy with which French and Prussians, Austrians, and Russians hurled themselves at each other, in energetic manoeuvres and blood-stained attacks pressed home to the end with a colossal proportion of the forces involved becoming casualties, represent the fullest exertion of force. Yet this was not inconsistent with limitation in the political sphere. The Austrians never intended to overthrow the regime in Prussia, let alone exterminate or extinguish it as a state, merely to restore it to a less prominent and threatening position. Frederick himself opposed popular resistance to invading armies.[7] The aims were limited; the instrument, vicious and sharp. The sharper the instrument, the more accurate, one hopes, the surgery. The era of 'total war', reaching its climax, arguably, in 1945 naturally stimulated interest in the origins of this phenomenon under Napoleon. But Frederick's campaigns may have more relevance to the modern era, where limited aims have to be accomplished, often at short notice and unexpectedly, by professional forces, within strict political limits. That does not mean that the military operations themselves will be any gentler: far from it.

NAPOLEON AND OPERATIONAL MANOEUVRE

Napoleon did not introduce a major new doctrine or weapons systems. He had an instinctive understanding of the realities of war, trained his troops and applied the results in battle. The use of skirmishers thrown out in front to engage the enemy with individual fire, and column formations had both evolved during the later eighteenth century and many of the characteristics associated with the Revolutionary (1789–1804) and Napoleonic wars had been envisaged or anticipated by Frederick and Suvorov.[8] Improvements in the power and mobility of artillery had been made before the French Revolution and appreciated if not fully utilized by Frederick the Great.[9] Napoleon brought them all together. Whereas the aim of eighteenth-century linear tactics had been to create the maximum possible small-arms firepower, Napoleon used massed artillery to blow a hole in the hostile battle formation, into which infantry and cavalry could be launched. Freed from the need to provide the main firepower, infantry's potential for shock action could be maximized. Another innovation was the introduction of autonomous divisions, each a fully combined-arms formation. This helped command and control and made the army as a whole more mobile. Conscription (from 1798) also offered the chance to mobilize the whole nation, translating its potential more directly into battlefield power. Autonomous divisions developed into 'army corps'. A corps could fight by itself, but the idea was not to use it in isolation but for one corps to be able to hold for long enough to fix the enemy while the others manoeuvred to concentrate on the battlefield and achieve a decisive local superiority.[10]

The genius of Napoleon therefore lay off the battlefield as much as on it, and particularly his ability to surround the enemy and concentrate superior forces at the decisive point. This brings us to a crucial development. For centuries, it had been sufficient to apply the term strategy to national defence priorities affecting the initial dispositions of forces and the way armies manoeuvred seeking the most favourable circumstances for engagement. Tactics meant the way a battle was fought when two hostile armies actually met. In the eighteenth century European armies manoeuvred strategically to a tactical fight on the battlefield. To this two-dimensional framework of thought Napoleon added a third intermediate dimension. This has been called grand tactics, but the term now used by Soviet, American, British, German and Israeli authorities is the operational level of war or operational art. Jomini clearly recognized the existence of a third, intermediate and in some ways qualitatively different level of war:

> Strategy is the art of making war upon the map and comprehends the whole theatre of operations. Grand Tactics is the art of posting troops upon the battlefield according to the accidents of the ground, of bringing them into action, and the art of fighting upon a map. Its operations may extend over a field of ten or twelve miles in extent. . . . Strategy decides where to act; logistics brings the troops to this point; grand tactics decides the manner of execution and the employment of the troops.[11]

'The art of posting troops upon the battlefield according to the accidents of the ground' across a battlefield of 'ten or twelve miles' would be considered 'tactical' today. However, in the conditions of the time, and in view of the concentration of entire armies within that space, it was 'grand tactics', leading to the destruction of an enemy army, and analogous to the operational level of war. Thus, Henderson described Napoleon's conduct of his finest battle, Austerlitz (2 December 1805) as a masterpiece of grand tactics.

The battle was fought on an undulating plain lying between two parallel chains of mountains sixteen to eighteen kilometres apart. Napoleon took up position with his left flank almost touching the northern range but a big gap between his right flank and the southern range. The allies fell into the trap, thinking it easy to drive between the southern mountains and Napoleon's army and envelop it, cutting his communications. Their camp fires the night before gave away the fact that they had taken the bait. Napoleon's plan was to attack the enemy right (his left) vigorously, to give their left outflanking force plenty of time to get on its way. Then, with 30,000 men in one powerful mass he would attack the hill in their centre, seize this commanding point and split their army in two. In order to lure the enemy round his right he placed very few troops in forward positions there, but kept the division with which he intended to hold the flanking attack well back from the battlefield the night before. The plan worked perfectly. By breaking through the enemy centre, isolating the two halves, he was employing a variant of the eccentric movement (see Chapter 1). This enabled him to defeat the Russo-Austrian army (an operational aim), which led directly to the break-up of the Third Coalition against France (a strategic aim).[12]

An example of the larger-scale operational manoeuvre by Napoleon is the Ulm campaign of 1805 (see Figure 3.1). The Marengo campaign of 1800 had shown the enormous effectiveness of operational level outflanking, but at the same time also the danger of fragmentation of forces and the need for

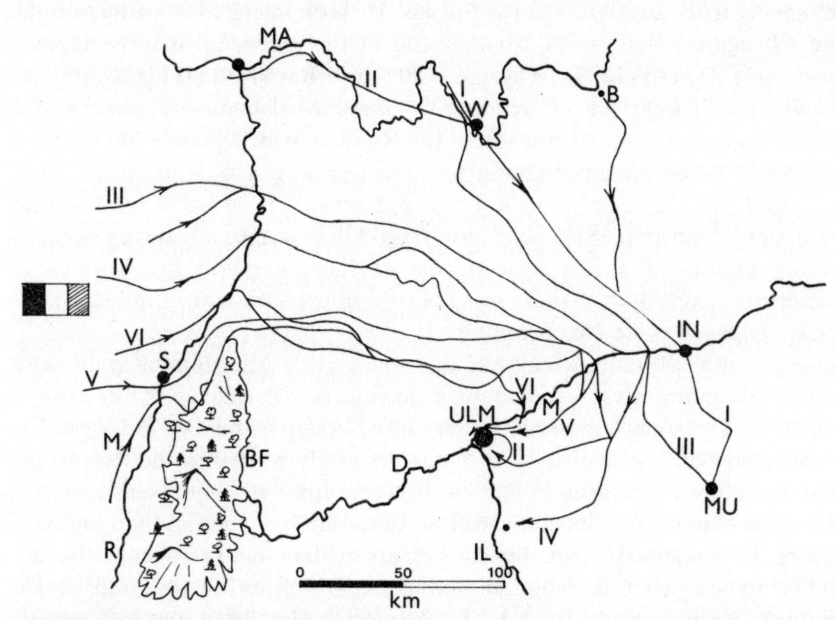

Figure 3.1 Ulm march-manoeuvre, 1805: operational encirclement of Mack's Austrian army

B = Bamberg
BF = Black Forest (obstacle)
D = Danube
IL = Iller
IN = Ingolstadt
M = Murat's Corps
MA = Mainz and River Main
MU = Munich
R = Rhine
S = Strasburg
W = Wurzburg

Roman numerals indicate corps numbers

operational art to remain subject to tactics, in the sense that the forces had to be able to concentrate if a battle was forced. In the large-scale approach march ('march-manoeuvre') in the Ulm operation, 200,000 men moved dispersed but able to be concentrated for a joint attack at any time. In the following year Napoleon's army advanced on the right of the Saabe in three columns forming a lozenge shape which Napoleon himself described as a 'battalion square of 200,000 men'.[13] The columns were about a day's march apart and connected only by a sparse network of couriers. This was in the main conditioned by the difficult terrain of the Franconian forest. However, the frontage was still rather small when compared either with Genghis Khan's advance on China in 1211 with four armies on a hundred-mile front, or with the separate armies manipulated by Moltke.[14]

In Spain in 1808, Napoleon began with an operational breakthrough, advancing from the Ebro against Burgos. He then intended to swing to right and left against the armies at either end of the extended Spanish deployment.[15] At Aspern (1809), Wagram (1809) and Borodino (1812), Napoleon found himself deprived of freedom of operational manoeuvre and could only assault frontally. At Borodino, the Russian army squarely blocked the way to Moscow and, possessing uncertain intelligence, Napoleon's army advanced in a single large mass along and beside the Moscow road. In the campaign of autumn 1813, Napoleon faced allied armies advancing concentrically and every thrust against one of them exposed his flank. The existence of continuous fronts in 1914–18 imposed the same limitations on armies engaged there (see below).[16]

Napoleon's ultimate defeat was due to over-reaching himself politically and in his grand strategy, and to a decline in the quality of the troops available. The superb material of the Grand Army formed at Boulogne for the invasion of England in 1803–5 was gradually wasted in the Peninsular War and all but destroyed in Russia. It is testimony to Napoleon's qualities as a great captain that he could still make unreliable material fight and win battles. He possessed the qualities which are evident in all great captains: the ability to see what is essential and disregard other things, and moral courage. He was ready to stake his last horse, last man and last gun to compel a decision. In war one must be ready to risk all to gain all. Napoleon had little time to train and drill his men as Frederick had done. He used no new weapons and devised no radically new tactical forms, although the Army Corps was crucial to the conduct of operational-level manoeuvre. He took the materials available and made them really formidable; the acme of the art of war, as of all art.

INTERPRETERS OF NAPOLEON: JOMINI AND CLAUSEWITZ

The campaigns of Napoleon have been exhaustively analysed, notably by Jomini and Clausewitz. Clausewitz's *Vom Krieg (On War)* is regarded by many as the greatest classic of military thought. Its identification of timeless verities of war such as the crushing influence of danger, the presence of friction making even the simplest thing extremely difficult and the uncertainty or fog of war, all make it a repository of eternal truth, although the truth is sometimes difficult to extract, because it was never finished. Its apocalyptic references to war 'tending to the extreme' make it grimly relevant in the nuclear age, and its very obscurity makes it a challenge to academics and a character builder for their students.[17] In addition to more specifically accurate predictions about the influence of technology on war, Jomini's *Précis de l'Art de la Guerre* is a much more manageable, elegant, and polished work. It is not surprising that Jomini was more popular among military men in the nineteenth century, but because it was more specific and less universal it has dated more quickly than Clausewitz. It was through the medium of Jomini that the essentials of the Napoleonic method were communicated to a nation which proved an oddly competent practitioner of them – the United States. It is no coincidence that probably the best translation of Jomini into English was made by two West Pointers in 1862.[18]

Although Jomini was a Swiss, he spent much of his career in the service of the Russians, and the dedication at the opening of *Précis de l'Art de la Guerre* is to the Russian Emperor. One sometimes hears academics ponderously asserting that Soviet military thinking today is 'Clausewitzian', for no other reason, apparently, than that Lenin agreed with Clausewitz about war being an extension of politics. Jomini was well aware of that, too: he just failed to encapsulate it in such an instantly memorable maxim. The first chapter of *Précis* . . ., 'De la politique de la guerre', is about precisely that, covering types of wars defined according to their political objectives. It would be truer to say, if anything, that Russian and Soviet military thought is Jominian, though even that would be a vast oversimplification. It was none other than Baron Jomini who wrote the Tsar at the end of March 1826, with advice 'on the question of teaching military science as it relates to a big war, in connection with the formation of a "school of strategy" ' (which finally opened in November 1832, as the Nicholas General Staff Academy).[19] Jomini anticipated modern Soviet concepts of Theatres of Military Operations (TVDs), and anybody familiar with the Soviet penchant for breaking through the 'tactical zone' to get into the 'operational depth' via corridors just wide enough to permit tanks to pass unscathed by anti-tank weapons on either flank cannot fail to sense a similarity with Jomini's approach; Soviet emphasis on the problems of switching one's 'main axis' recall Jomini's 'lines of operations', and so on. The links between Jomini and modern Soviet military thought are many, and unbroken.[20]

RECOGNITION OF THE OPERATIONAL LEVEL

Use of the term 'operations' to indicate a level between tactics and strategy has a long pedigree in the German army. Moltke the elder, who directed Prussia's armies to victory over the more battle-hardened and experienced armies of Austria and France, was a most cerebral soldier, the living expression of a new era of strategy and military art, in which meticulous planning and preparation ruled. He was known, with a sort of grudging affection, as 'the library rat' by his colleagues.[21] Moltke undoubtedly had a concept of an area of 'operations', between tactics and strategy, which for a long time had not been recognized, although he was against defining these terms too sharply. He saw the 'operation' as a relative term, determined not by the scale or forces involved but by the aim, using individual combats to achieve the aims set by strategy. This conforms to subsequent definitions of the operational level. In *On Strategy*, Moltke alluded to the operational level:

> The next task of strategy is the preparation of forces, the initial employment of the army . . . the orders for carrying this out are of decisive importance for the success of the entire enterprise. . . . The subsequent tasks of strategy, the application of the available means, are different, and so with operations. In this case, our will clashes with the will of the opponent.[22]

Thus in the period between 1857 and 1871, Moltke not only established the crucial importance of strategic deployment, the initial 'march-manoeuvre', which was to preoccupy military thinkers for the rest of the century and right through to World War Two; he also began to use the term 'operations' for a range of activity with aims between those of tactics and strategy. Since the terms 'operational' and 'operational level of war' have been adopted in western (Anglo-Saxon) armies some specialists have come to believe that their recognition of this level is comparatively new.[23] In fact, it has been recognized in British and American armies for a long time. Only the terminology was new.[24] Take the words of the eminent British military thinker John F. C. Fuller, in 1926, in describing grand tactics. Fuller defined these as the penetration or envelopment of an enemy, of which Napoleon was a past master.[25] According to Fuller, maximum physical destruction of the enemy at minimum cost on the battlefield was the preserve of minor tactics. However:

> The grand tactician does not think of physical destruction but of mental destruction, and when the mind of the enemy's command can only be attacked through the bodies of his men, then we descend to minor tactics, which though related, is a different expression of force.
>
> We see, therefore, that grand tactics is the battle between two plans energized by two wills, and not merely the struggle between two or more

> military forces. To be a *grand tactician*, it is essential to understand the purpose of each part of the military instrument.
>
> Grand tactics secures military action by converging all means of waging war towards gaining a decision i.e. the destruction of the enemy's plan.[26]

This may be what Sun Tzu was getting at when he said that it was supremely important 'to attack the enemy's strategy'.[27] Fuller's definition of 'grand tactics' makes two vital points. First, as noted, awareness of the third level between tactics and strategy is not new, even among the British. Second, the operational level of war is not only to be measured in terms of scale, operations by large formation but, exactly as Moltke had said, in terms of *aims*. However, the extension of the battlefield had much to do with bringing it into sharper focus. Richard Simpkin defined the operational level as the conduct of operations having as their aim objectives lying at one remove from an objective which could be couched in political or economic terms, that is, from a strategic aim.[28]

During the nineteenth century, and particularly at the beginning of the twentieth, the size of armies and of battles increased. This is shown graphically in Figure 3.2. The 'Battle of the Nations' at Leipzig in 1813 was probably the largest battle to date. Many nineteenth-century analysts used ancient and medieval statistics to suggest that 'mass armies' and 'armies of millions' were not new, but in the author's view the figures on which those assertions were based are quite unreliable. Solferino, with a total of half a million men involved, might have been bigger if the armies had had the opportunity to deploy instead of crashing into each other in a giant encounter battle. In addition to a dramatic increase in the size of armies, from the American Civil War onwards, dispersion increased sharply, causing battles to spread. The Russian attack on the Aladzha heights shows a dramatic extension, a result of tactical dispersion and the terrain. The telegraph was needed to co-ordinate the assault.[29] The Russo-Japanese war in 1904–5 witnessed battles greatly extended both in space (at Mukden, for example, a 120-kilometre front, and to a depth of 80 kilometres) and in time. Inevitably, 'battles' broke up into a series of smaller engagements all along an extended front and over a prolonged period. These engagements were nevertheless linked by a common purpose, a common design. If the individual engagements were tactical, what was the common scheme that linked them? It was hardly strategic, first, because it was aimed only at defeat of a single enemy army, and second, because the component actions were linked and mutually dependent, unlike the 'strategic' manoeuvres of separate forces in the past. Field Marshal Sir Douglas Haig, who commanded the British Army in France in World War One, compared the entire, gigantic Western Front Experience in 1914–18 to a single, very protracted battle, conforming to all the classic phases of a tactical engagement. Yet it would be ludicrous to call the conduct of this gigantic affair 'tactics'.[30]

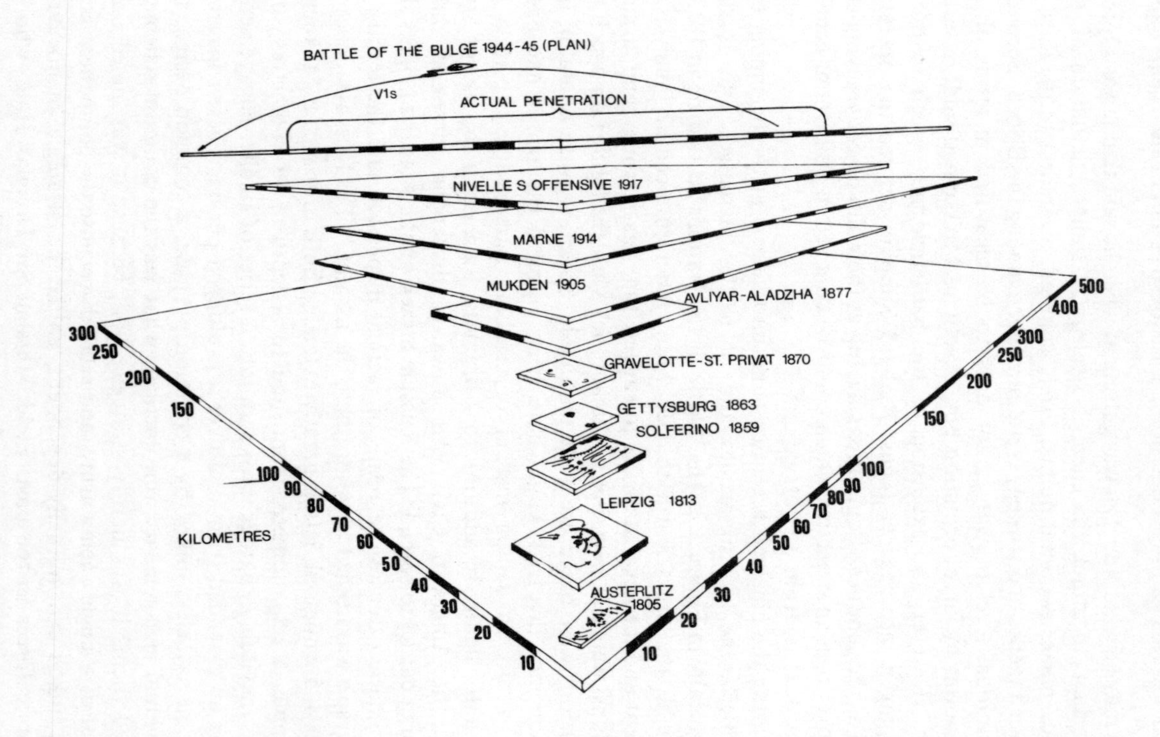

Figure 3.2 The expansion of the battlefield and the emergence of the operational level of war

By the end of the nineteenth century the implications of 'mass armies' and how they might be employed was a subject of widespread debate. The German General Sigismund von Schlichting (1829–1909) analysed the experience of the second half of the nineteenth century, notably in his *Principles of Modern Tactics and Strategy* (1897–99).[31] In common with many other thinkers of this period, of whom he was in the vanguard, Schlichting placed great emphasis on strategic deployment, which had come to depend to a large extent on railway networks and timetables, and the meeting engagement. His work was studied by a very forward-thinking and dynamic group of Russians at the General Staff Academy, who, unlike Schlichting, utilized the experience of the Russo-Japanese War, and began to envision a level of conflict above the tactical battle (combat). One of the instructors, the then Lieutenant Colonel Neznamov, used a public lecture in 1909 to identify changes in the art of war which were arising from the demands of mass industrial war. Much of what Neznamov said was taken from Schlichting, but was presented in a Russian context. Later, one of Neznamov's young students, Boris Shaposhnikov, recalled that Neznamov had been 'bringing us German views on *operational art*'.[32] It thus seems that the idea of the operational level of war, based on experience of the nineteenth century (including Napoleon) was gelling among a group of bright young staff officers in Russia and Germany some five years before the outbreak of World War One.

The concept of the operational level was first put on the map by another German, after World War One, General Baron Dr Hugo von Freytag-Loringhoven. Freytag-Loringhoven was described as 'the most distinguished soldier writer of Prussia' and 'the most distinguished living writer on militarism in theory and practice'.[33] So wrote a British commentator in 1917, when an English translation of his *Deductions from the World War* was, unusually, published while the war was still in progress. It is perhaps surprising therefore that Freytag-Loringhoven has been almost entirely neglected by English speaking military historians since. Freytag was a Baltic Russian by birth and served in the Russian army until the age of 21, which may partly explain the affinity between his ideas and those which the

Note: Napoleon's battle of Austerlitz took place in an area forming a rough parallelogram some 12 kilometres long and 4 to 6 wide. The 'Battle of nations' at Leipzig in 1813, involving a total of half a million men took place in an area some 14 kilometres square, although this includes considerable manoeuvre in the course of some 4 days. The battle of Solferino also involved half a million men. It mostly took place on a front about 8 kilometres wide, across a depth of 15 kilometres. Gettysburg raged in an area some 5 miles (8 kilometres) square. Gravelotte-St Privat, in the Franco-Prussian War, took place on a front of about 5 miles (8 kilometres) although the French left stretched another 3 miles. In the Russo-Turkish war, on the other hand, the Russians attacked the Aladzha heights in the Caucasus, conducting a wide encircling movement to attack the Turkish position in the rear. At Mukden in 1905 the battle extended over a 120-kilometre front to a depth of 80 kilometres. The 'miracle of the Marne' in 1914 involved co-ordinated actions across a 225-kilometre front and to a depth of 75 kilometres. Nivelle's offensive in 1917 took place in an area 130 kilometres square. As an example of a co-ordinated battle from World War Two, the 'Battle of the Bulge' of late 1944-early 1945 is selected. The Germans' thrust towards Antwerp was conceived as taking place in an area some 180 kilometres square: the actual penetration extended to 90 kilometres.

Russians evolved from their contacts with the Asiatic style of war, and its feel for scale. Subsequently in German service, he described the term operational thus:

> In the German Army, starting in the general staff, the employment of the term *strategisch* has fallen more and more into disuse. We replace it, as a rule, by the term *operativ*, 'pertaining to operations' and thereby define more simply and clearly the difference from everything tactical. All that pertains to operations as such takes place, on the whole, independently of actual combat, whereas in the term strategical things become easily confused . . . [speaking] of strategical conditions when it is merely a question of purely local matters . . . the term strategy ought to be confined to the most important measures of high command.[34]

Freytag-Loringhoven saw Napoleon as the initiator of operational-level manoeuvre warfare and also noted that the extension of the battlefield further affected the development of the third dimension.

> The incomparable military leader in him [Napoleon] will always have much to say to soldiers. He is the creator of modern large scale warfare. Moltke is based on him. The value which lies in studying his campaigns remains undiminished today, even if the numbers which came into action in the Great War [1914–18] were far in excess of those which he commanded; and even if the technological aids of modern times such as railways and radio telegraphy, telephones and the enormously more potent means of conducting war have given that conduct a form which is many times different from Napoleon's own.[35]

Freytag-Loringhoven also placed great emphasis on the role of Count von Schlieffen in training the German army and staff for the conduct of 'mass warfare', in which it had proved superior to other armies in 1914 and for a long time afterwards, and also in stressing historical experience in developing the concept of the 'battle of annihilation' through encirclement of the enemy, if possible on both flanks. This permeated Schlieffen's entire official and unofficial writings. In developing this point, Schlieffen did what many critics have done in literature and other fields in attributing intentions to the artist which were perhaps not there originally. This was particularly true of Schlieffen's famous study *Cannae*.

After World War One, another factor affected the crystallization of the idea of the operational level of war. Not only had battle itself changed, leading to a new level involving the co-ordination of several tactical engagements. Strategy had also taken on a new and higher meaning. The involvement of the total warmaking capacity of societies and the conduct of coalition war had arguably created a sort of 'super strategy' or 'grand strategy', aimed at the opponent's government, society, and economy and the fabric of alliances. This led Tukhachevskiy, for example, to argue that the three levels of war now comprised tactics, strategy, and 'war strategy' (grand strategy) – *polemostrategiya* (from the Greek polemos: a war).[36]

However, while this use of the term 'strategy' in its old sense might be adequate where military operations took the form of large-scale manoeuvre crowned by relatively brief and decisive pitched battles, (as Tukhachevskiy himself attempted to do in the 1920 drive to Warsaw), it was hardly suitable to describe the manipulation of armies and army groups in prolonged operations in contact with the enemy across a wide area. Neznamov's pupil, Shaposhnikov (later a Marshal of the Soviet Union and Chief of the General Staff of the Red Army) had, on the other hand, been using the term 'operational' to describe the level between tactics and 'strategy' (in all the latter's ramifications) as early as 1921.[37] It was the former Tsarist General, Svechin, who proposed the division strategy – operational art – tactics, during 1923–4, and published in his 'Strategy' which first appeared in 1926, a delineation which was formally adopted by the Red Army 'from the end of the 1920s'.[38]

THE AMERICAN CIVIL WAR, 1861–5

A generation of American officers raised on Jomini, including Lee, Sherman, and Grant, understood the importance of speed, manoeuvre, surprise, and decisive action and, in this century, Pershing, MacArthur, Bradley, and Patton carried on the tradition of American expertise in the attack. The American Civil War was fought over a vast area (620,000 square miles), three times the size of France. Strategy, operational art, and tactics were influenced by the topography to a greater extent than any other land conflict of modern times, except possibly the Russian Civil War. The theatre of war was divided into two major areas of military operations, eastern and western, by the vast Appalachian mountain range. Although the two capitals were in the east and that was where the war was decided, commanders neglected the western theatre at their peril.[39] It is not surprising that the Americans adjusted to European continental warfare speedily and successfully.

Grant's operation south of Vicksburg in 1864 has been called the most brilliant campaign ever fought on American soil. According to the modern American Field Manual 100–5 it 'exemplifies the qualities of a well-conceived, violently executed offensive plan'.[40] Instead of moving obviously against Vicksburg, Grant used the Big Black River to protect his flank as he moved towards the city of Jackson (see Figure 3.3). By threatening both Jackson and Vicksburg, Grant confused the enemy and prevented them uniting their forces against him. He thus defeated the enemy forces piecemeal in five separate engagements. Grant's forces covered 320 kilometres in 19 days, capturing Jackson and investing Vicksburg. They suffered half as many casualties as the opposition, reaffirming the general rule that if you go fast you reduce casualties.

During this war we also see the use of the 'strategic [operational] offensive–tactical defensive' advocated by Moltke. Lee and Jackson used it

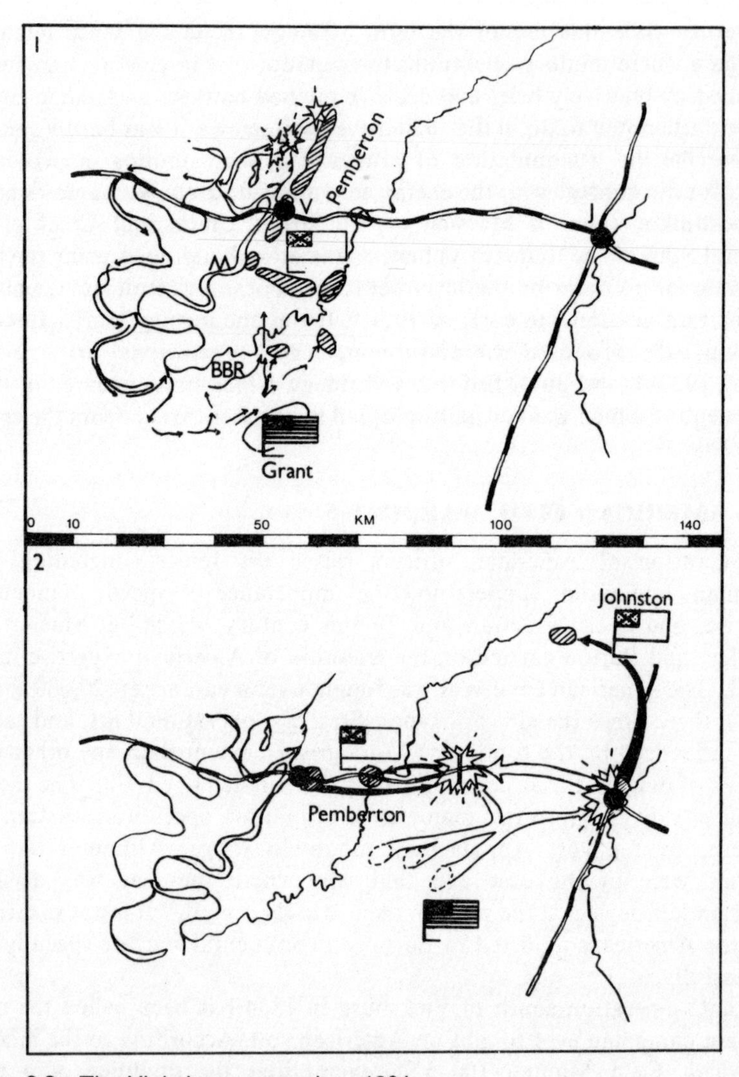

Figure 3.3 The Vicksburg campaign, 1864

BBR = Big Black River M = Mississippi River
J = Jackson V = Vicksburg

Source: FM100-5, adapted by the author

at Second Manassas and Chancellorsville in 1862 and 1863 respectively, and it was improved by Grant in the fighting around Richmond in 1864–5 and in the pursuit of Lee to Appomattox.[41] Moltke himself used it decisively in the operation leading to the two major battles of the Franco-Prussian War: Gravelotte-St Privat and Sedan. Behind an excellently executed cavalry screen, Moltke swung his forces in a wide envelopment that took them behind the main French army and onto the French line of communications.

Caught in the operational trap, the French had no alternative but to attack at once, and came to grief on the superior fire of the tactical defensive.[42]

The American Civil War saw the first prolonged and large-scale use of those critical strategic and logistic innovations: the railway and the telegraph. This in turn led to the development of deep raiding operations and saw increased emphasis on attacking communications and logistics. 'The original idea of raids received an impetus only when the old, solid highways heretofore used as lines of communications gave way to the railroad.'[43] Turning movements were constantly attempted in order to threaten something vital in the enemy's rear, usually communications, to force his withdrawal. Such a movement would give the turning force the significant advantage of then being able to fight on the defensive against an enemy forced to attack. If such a movement were conducted along a line of communications able to supply the turning army, it could be termed a penetration, in which the attacker could supply himself while the defender, with his communications cut, could not. This can be distinguished from a raid, which involves only transitory possession. However, both Union and Confederate generals realized that although lack of communications made such raids transitory, they still had spectacular political, psychological, and real operational impact. As Sherman observed 'railroads are the weakest things in war. A single man with a match can destroy and cut off communications'.[44]

The Confederates, who initially enjoyed superiority in cavalry, were quick to appreciate this and the raids conducted by Jeb Stuart, Forrest and Morgan during 1862 had a traumatic effect. The need to guard against these raids on communications caused a tremendous dispersion of Federal forces, one reason why the much more populous North was unable to field significantly greater numbers than the South on the battlefields themselves.[45] However, as an Austrian officer noted in 1908, Stuart, for example, never conducted a 'pursuing raid after a victorious battle' – the crucial interaction of main forces' breakthrough, exploitation, and pursuit which has characterized more recent armoured forces' actions, for example. This had also been advocated by Moltke.[46]

The North rapidly appreciated the value of such raids and Grierson's raid of April 1863 was one of the most successful of the war. Besides forming part of Grant's strategy of raids against the enemy's logistic base, the raid also distracted the Confederates' attention from the main Federal movement against Vicksburg.[47] However, Confederate raids continued to agitate Grant well into 1864, forcing the Federals to rely on a system of guarded convoys or more secure river communications.[48] Finally, however, it was the Union's raiding strategy which triumphed, both destroying the resources which would have enabled the South to continue the struggle and, by marching a 'well appointed army right through his territory', demonstrating 'to the world, both foreign and domestic' that the North could prevail.[49]

EUROPEAN DEVELOPMENTS: THE 'NINETEENTH-CENTURY REVOLUTION IN WARFARE'

The first use of railways to move troops operationally occurred in 1849, when a Russian corps was moved by rail from Warsaw to Vienna which was being threatened by Hungarian rebels. The idea seems to have come from the Austrians, and the Russians rather cautiously obliged. The decision had to be taken by the commander on the spot, as at that time there was no telegraph (invented in 1832) to the Emperor. The railway's potential for transporting and concentrating troops was explored in military writing during the 1850s, and railways were used – and attacked by both sides – in the Franco-Austrian War of 1859.[50] The railway and the telegraph revolutionized warfare, but did not necessarily take it to a higher level of creativity and ingenuity. Dependence on fixed, linear communications like these inhibited strategy and dictated the course of operations. Military intellectuals argued voluminously about the effect these innovations would have on war, and there could be no escaping the conclusion that armies constrained to railway lines and the number of trains that could be despatched down those lines would collide like 'rams', with less initial freedom than before. The side with the superior railway network and the superior mobilization plan could deploy a massive preponderance of force. Increasingly, the result of armed conflict would depend on the strategic deployment and the opening period of any war. Some argued that this would result in a decline of military art, and of the effect of military genius, and therefore could not occur. Armies would have to be limited in size. Others pointed out that no national government was particularly interested in preserving the purity of military art; all they cared about was winning, and if this meant putting millions of men into the field along immovable lines, so be it.[51]

The brilliant Prusso-German campaigns of 1866 and 1870, the Austro-Prussian and Franco-Prussian wars, did not evince the dreaded decline of military art. Meticulous planning there was, by the Germans, but the old strategic magic still worked. In the Austro-Prussian War Moltke used his maxim 'march separated – fight concentrated' ('Getrennt marschieren – vereint schlagen') with decisive effect in the Königgratz campaign.[52] These relatively swift campaigns mesmerized the military mind for nearly fifty years; although there was much detailed, intelligent and sometimes irreverent analysis, the historians' consensus that military staffs underrated the significance of the bloody impasses of the American Civil, Russo-Turkish (1877–8) and Russo-Japanese (1904–05) wars appears to be broadly correct.

The Russo-Japanese War is extremely important in the evolution of strategy and military thought, even today. Fortunately, it is exceptionally well documented, thanks to the presence of impartial but expert observers on both sides. Many of the instruments of warfare associated with World

War One – machine guns, barbed wire, entrenchments, continuous fronts, indirect fire artillery and prolonged battles comprising numerous subordinate linked engagements – were apparent. Contrary to popular belief, the Russian army did not perform at all badly, and at the end it was still intact and growing stronger having been pushed back up the Manchurian peninsula. The defeat of the Russian fleet at Tsushima and the tenuous link with European Russia over four thousand miles of incomplete single-track railway made resupply and reinforcement extremely difficult, while Japan was close by and her forces could move freely over the sea. It was this, and revolution at the heart of the state, which forced Russia to seek peace, not military defeat. If the Russians failed to perform as well in the field as the excellent army they sent deserved, it was because they, and in particular General Kuropatkin, had failed to grasp the full implications of the changed nature of war. Kuropatkin clearly endeavoured to maintain an 'operational line' in the best Jominian tradition. He spent the entire war endeavouring to bring the Japanese to a single, decisive 'battle of annihilation', not accepting that the great extension of armies and their dependence on logistics made achieving such a decision almost impossible. As in the American Civil War, armies could maul each other horribly, but they usually managed to retire, bloody but unbeaten, to continue the struggle. The problem of controlling separate armies in related actions along a wide front led to the further development of the operational level and the concept of the 'army group' or 'front' as the level of operational command.[53]

WORLD WAR ONE, 1914–18

The classic operational plan of the Great War is the famous Schlieffen plan for the invasion of France by Germany. The aim was to put great weight on the right flank while leaving merely covering forces on the left. Coincidentally, the French Plan 17 involved attacking the weak German left wing (Figure 3.4). The Germans lacked the nerve to strengthen the right sufficiently at the expense of the left, and this, combined with the unexpected delay imposed by the resistance of the Belgian forts probably contributed to the plan's failure. However, it is also likely that the lack of mobility of a non-mechanized army once it had left its railheads and the innate strength of the tactical defence as displayed at Mons and the Marne doomed the ambitious plan to failure even without these operational factors. Plan 17 met a similar fate. The unexpectedly rapid mobilization of the Russians also forced the Germans to draw off two corps at a critical moment.[54] However, the Germans did enjoy a spectacular success in the east, at Tannenberg in August 1914. The Russians launched a concentric attack on East Prussia with two armies, with the best Napoleonic intentions. However, superior German generalship and interior lines enabled them to engage the two Russian armies successively. Leaving only a cavalry screen

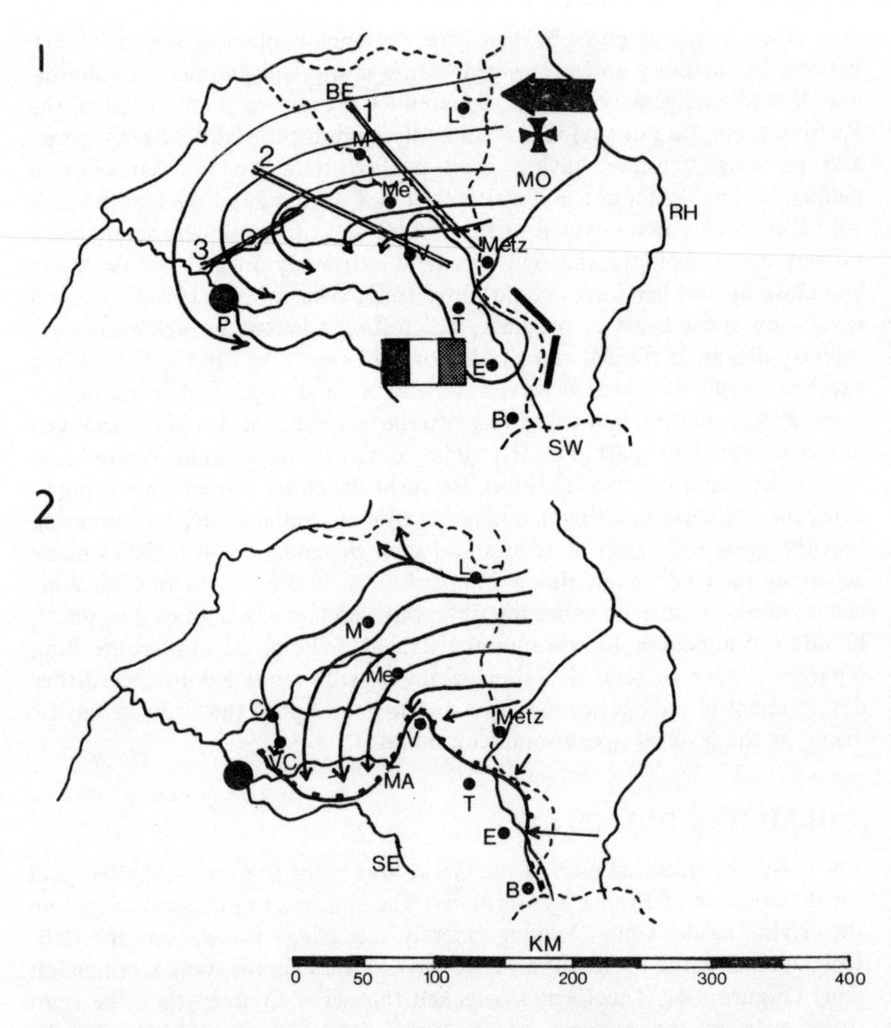

Figure 3.4 The Schlieffen plan: 1, as conceived; 2, as executed

Note: It can be seen clearly that the younger Moltke's weakening of the right wing and the detour inside Paris destroyed the plan's potentially devastating effect and permitted the establishment of a fortified line which might otherwise have been outflanked before it was established.

B	= Belfort	P	= Paris
C	= Compiegne	SE	= Seine River
E	= Epinal	SW	= Switzerland
L	= Liege	T	= Toul
M	= Maubeuge	V	= Verdun
MA	= Marne River	VC	= Villers-Cotterets
Me	= Mezieres	1	= Objective on day 22
MO	= Moselle River	2	= Objective on day 31
O	= Oise River	3	= Oise river holding line
RH	= Rhine River		

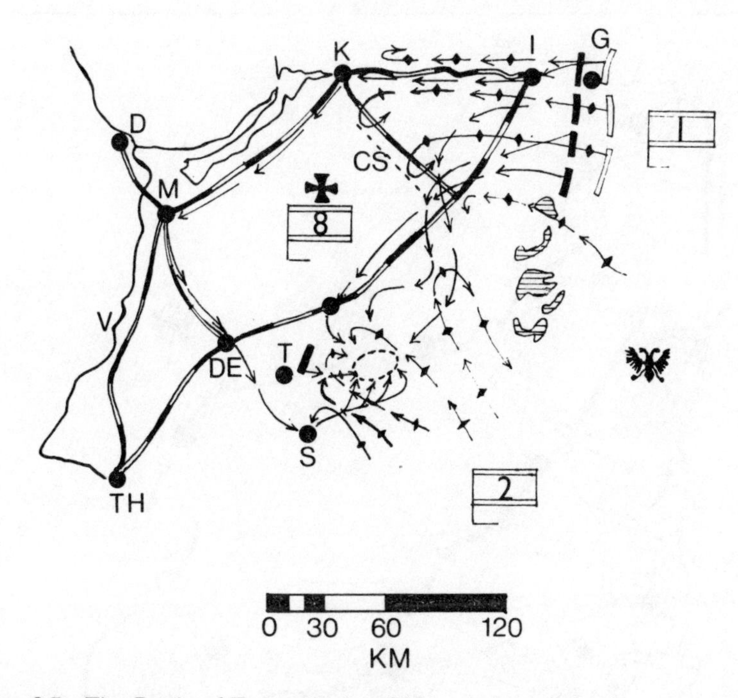

Figure 3.5 The Battle of Tannenberg, 1914

D	= Danzig (Gdansk)	K	= Konigsberg (Kaliningrad)
CS	= Cavalry Screen	S	= Soldau
DE	= Deutsch Eylau	T	= Tannenberg
G	= Gumbinnen	Th	= Thorn (Torun)
I	= Insterburg	V	= Vistula River

to confront Rennenkampf's first army in the north, the Germans under Hindenburg with Ludendorff as Chief of Staff destroyed the Russian second army under Samsonov and then turned again to defeat Rennenkampf (see Figure 3.5).[55]

After this, the consolidation of continuous fronts precluded large-scale operational manoeuvre in the west for some time.[56] In the east, the Germans attempted an operational breakthrough in accordance with a coherent concept. Realizing the impossibility of conducting an operational encirclement of the extended Russian front, whether by the Germans in the north or the Austrian KüK army in the south, in early 1915 von Falkenhayn developed the concept of the Gorlice-Tarnow breakthrough, as in Figure 3.6).[57] The plan was to surround and destroy the Russian third army and then turn on the rear of Brusilov's eighth army in the Carpathian passes to the south and threaten the rear of the Russian armies to the north.[58] It was a classic example of operational manoeuvre in what is often thought of as the positional period of the war. Falkenhayn even proposed that Brusilov's

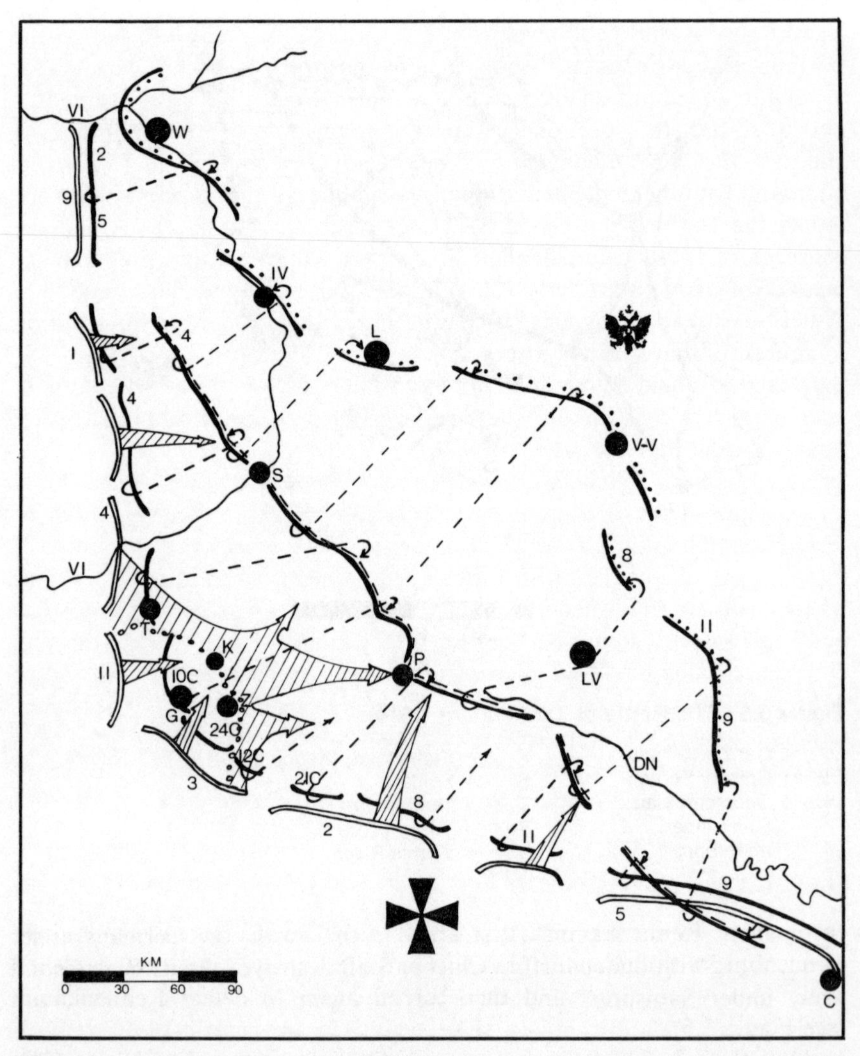

Figure 3.6 The Gorlice-Tarnow (Gorlitse Tarnuv) Operation, 2 May–23 June 1915

Note: Numerals refer to army numbers except where suffixed by C, in which case they refer to Russian Army Corps.

C = Cernovtsy
DN = Dnestr River
G = Gorlice
Iv = Ivangorod
K = Kalachitse
L = Lublin
Lv = Lvov
P = Przemsyl
T = Tarnow
VI = Vistula River
VV = Vladimir Volunsk
W = Warsaw
Z = Zmigrod

eighth army should be allowed to break through if it attacked, so that the German armies breaking through could smash it in the back in a classic 'revolving down' movement (see Figure 3.6). However, Austrian nerve was not equal to the prospect of Russians racing towards Vienna and Budapest and they settled for a conventional defence of the area.[59] Similar considerations apply to the implementation of the mobile warfare concept in Europe today; the conflict between the dictates of creative military art and the concerns of those whose land it is. Just as Napoleon had prepared his breakthroughs with artillery: 100 guns at Wagram, 200 at Borodino, 80 at Waterloo, so Falkenhayn also used crushing artillery superiority in addition to concentration of other forces and surprise. The Germans amassed 457 light and 159 heavy guns, not an enormous number by the standards of later battles of the war but the largest artillery concentration up to that time.[60] Furthermore, it was here that the Germans were the first to use the 'creeping barrage', advancing steadily just in front of the infantry.[61] This is an example of the need to evolve technique to exploit the potential of technology. The skilful co-ordination of artillery fire with the movement of infantry was one of the most difficult requirements of the Great War, but where it was accomplished it was usually crowned with success, as at Gorlice and in Brusilov's offensive a year later. The Gorlice-Tarnow breakthrough commenced along a 40-kilometre front on 2 May 1915, and by the end of June the Russians had been forced to withdraw up to 250 kilometres, although avoiding entrapment. The Russian and Soviet armies learned a great deal from this German operation and in it we can see the origins of World War Two and modern Soviet breakthrough technique.

GAS! GAS!

Meanwhile, events had occurred which underline the potentially decisive effect of new technology employed on a significant scale.

22 April 1915, was a glorious spring day. Air reconnaissance had disclosed considerable activity behind the German lines in the vicinity of Langemarck, near Ypres, but there was nothing abnormal in this. At 1700 hours the German heavy artillery began a furious bombardment towards Ypres, although the German field artillery was silent. Some French 75s began a lively fire, leading observers to think that the newly arrived Algerian division holding the line north of Langemarck was shooting itself in. Then those with a view of that sector saw two curious greenish-yellow clouds on the ground on either side of Langemarck in front of the German line. The clouds spread laterally, joined up and, propelled by a light wind, became a bluish-white mist 'such as is seen over water meadows on a frosty night'. Behind the mist, the Germans were advancing. Soon, even as far off as V Corps HQ 8 kilometres away a peculiar smell was noticed, accompanied by smarting eyes and tingling nose and throat. Meanwhile, French colonial

troops were noticed falling back into rear areas of V Corps, speech incomprehensible but pointing at their throats and coughing. It began to occur to officers in headquarters that this must be the gas of which some warning had been received. A week before the French had actually captured a German with a gas mask, who had told them of the dispositions of the 6,000 cylinders which the Germans had now opened to release chlorine gas. The French divisional commander had relayed this to the British and French High Commands, recommending that the installations be shelled or the threatened troops withdrawn, but nothing had been done. At about 1900 hours the French 75s, which had been keeping up a stiff fire, stopped. It was obvious that something 'very serious' had happened. The Germans, wearing crude respirators of moist gauze and cotton advanced through what had been the French line, threatening to envelop the Canadians to the east. By the early hours of the morning it was clear that the Germans had torn a gaping hole 6 kilometres wide in the Western Front. Gas had given them complete tactical and operational surprise, and if it had been used as part of a coherent operational plan might have brought about the rupture of that sector of the front and a breakthrough to the channel ports, winning the war in the west for the Germans. For various reasons, it did not. German senior commanders were distrustful of the effects of gas, and reserves able to rapidly exploit the breakthrough had not been provided. In addition, at the tactical level the crude respirators and training available to the Germans meant that the German troops were apprehensive of the effects of their own gas, and many British officers reported that they were stopped when they ran into it. Moreover, the gas was released in the evening leaving only a couple of hours of daylight to exploit its results. This was not the first time that gas had been used in war: on 27 October 1914 the British at Neuve Chappelle were bombarded by the Germans using 3,000 shrapnel shells containing the nose and eye irritant dianisdine chlorsulphate. On 31 January 1915, the Germans fired 18,000 shells containing bromide (tear gas) at the Russians at Bolimov in support of an attack. The extreme cold prevented it functioning effectively.[62]

World War One remains the only occasion where major industrial powers used chemical weapons against each other. Chlorine was supplemented by phosgene and in July 1917 the Germans deployed a new agent, the persistent blister agent mustard gas. The first nerve gas, tabun, was identified in 1936. Gas was used by the Italians in Ethiopia in the late 1930s, but in World War Two both sides refrained from its use, not because of moral scruples but because, as Churchill said, 'it does not pay them'.[63] Chemical weapons undoubtedly made the conduct of military operations more arduous and difficult, and this has been the main deterrent against their use since World War One. From 1915, onwards, however, chemical weapons were an essential part of the increasingly complex spectrum of combined-arms warfare.

COMBINED ARMS: INFANTRY AND ARTILLERY

The intelligent and precise use of artillery fire formed part of the scheme which the Germans evolved to restore infantry's mobility and success in the offensive. At the beginning of the war infantry units possessed rifles and a few heavy and static machine guns. Because of the clumsiness of early indirect-fire artillery technique, the first change was to enhance the infantry's own firepower with light automatic weapons, grenade launchers and, for the defensive, trench mortars. In September 1916 the French created the first modern combined-arms infantry battalion. Each section was organized around a light machine gun as a base of fire and was stiffened with grenade launchers. Each battalion was similarly beefed up with a company of eight heavy machine guns and its own 37 mm artillery piece.[64] The need for light artillery organic to the infantry unit, able to move with it and respond immediately to unexpected targets which could not be engaged by indirect fire artillery had first been recognized by the Russians in the organization of Brusilov's offensive in June 1916, where mountain guns were used for this purpose.[65]

The new offensive tactics reached their most fully developed form in the German offensive of March 1918. They were initially called 'Hutier tactics', based on the erroneous belief that they were invented by General Oskar von Hutier, but in fact they were the product of long evolution and co-operative effort. The usual term is 'infiltration tactics', suggesting bypassing points of resistance and pushing forward as far as possible. The word suggests a primary infantry effort, but in fact the Germans stressed the co-ordination (*zusammenwirken*) of all arms which was, as ever, the key to success.

In 1915 a French captain, André Laffargue, wrote a pamphlet *The Attack in Trench Warfare*, advocating a form of infiltration tactics. Although it did not become doctrine, it was disseminated and in early summer 1916 the Germans captured a copy. It was translated and had an immediate effect on German infantry tactics. The essence of Laffargue's scheme adopted by the Germans was a deep penetration; devouring the enemy rather than nibbling at him (metaphors referring to combat as if it were eating were popular throughout the Great War).[66] Such tactics were employed on occasions at Verdun, and in 1917 at Riga and Caporetto respectively.[67] On 1 January 1918 the German High Command published *The Attack in Position Warfare*, the doctrinal basis for the German offensives of 1918.[68] The German attacks succeeded tactically as a result of combining four main elements: Bruchmüller's artillery preparation, the combined arms assault, bypassing enemy centres of resistance instead of a linear advance, and attacks to disorganize the enemy rear area. The use of artillery was particularly important. The survey techniques developed in 1916–17 were used to pinpoint enemy artillery batteries, observation posts, headquarters, radio and telephone installations, reserve troop concentrations, bridges, and key approach routes. The development of the registration point (from 1916) enabled accurate fire to be delivered onto a target without preliminary adjustment on it, facilitating surprise.[69] A further technique for bringing down accurate fire without registration was developed by a German with a Russian name, captain Pulkowsky. This involved test firing each gun to determine its idiosyncrasies, called 'special influences', and recording the effects of daily factors such as meteorological data and ammunition condition (daily influences).[70] Similar meticulous calculations and zeroing of each gun were carried out by the Red Army before the Berlin Operation in 1945.[71] The junior German infantry and artillery officers were then carefully briefed together, thus greatly improving infantry artillery co-operation. This meticulous preparation was, paradoxically, forced on the Germans by lack of artillery resources compared with the allies and the need to be sparing with ammunition. Once again, the Russians had pioneered similar developments, also driven by lack of resources. Most notable was the work of Lieutenant Colonal Kirey, based on his experiences of artillery support for Brusilov's offensive.[72] However, Kirey's excellent work was not fully disseminated through the Imperial Russian army before its disintegration, although it was studied by the Red Army and forms the basis for their employment of artillery in the Great Patriotic War and even today.

In the 21 March 1918 German offensive, the bombardment began with a short gas attack, to force the British into their constricting gas masks, followed by the very effective mixture of gas and high explosive (the gas tended to settle in the trenches where troops sheltered from the HE). The Germans then moved the artillery bombardment backwards and forwards so as to confuse the British as to whether an attack was imminent. Meanwhile, the German automatic rifle teams moved as close as possble to the British positions. The confusion and panic caused by the bombardment was exacerbated by dense fog. Then German storm troopers with mortars and flame-throwers moved forward. The Germans attacked in two echelons: the first to pass between islands of resistance and to penetrate deeply to seize command and artillery positions while a second echelon eliminated the bypassed positions. Similar tactics had been used by the young German Erwin Rommel at Caporetto in 1917. Another crucial element was the pioneer units whose job was to make the battlefield passable for artillery and reinforcements.[73]

The artillery attack on the British rear caused disproportionate damage to the defence. The British defence lost coherence and fell back 38 kilometres in four days. Fuller observed that the British seemed to collapse and retreat from the rear forward. The British command learned of successful attacks of forward positions and ordered their remaining units to withdraw, even though they were successfully defending their bypassed positions, in order to re-establish a stable linear defence.[74] This has many lessons for the future European battlefield, where units will probably be isolated and bypassed, but the command *must not panic* and the units *must go on fighting*.

TANKS

Having reached March 1918, we must return to 1916 to consider the appearance of a new instrument of war, the tank. Armoured cars were in use at the start of the war and continued to be used, particularly on the eastern front and in the Middle East. The tank was initially designed as a solution to the problems of positional warfare, and its later adaptation as a means of exploitation, rather than cavalry and armoured cars, is an example of the tendency to use military technology for purposes other than that intended by its inventors. A British invention, 48 tanks were committed on the Somme on 15 September 1916, but only about half succeeded in getting into action. The decision to employ this potentially decisive new instrument in limited numbers has been widely criticized, and this underlines the value of new technology employed *en masse* and with surprise. However, it is arguable that, in the peculiar conditions of the western front, experience of their use could only be gained in the field against a real opponent, and that only by their actual employment with infantry in battle could proper tactics of combined infantry and tank fighting be evolved. It may be that the experience of the Somme was essential to the full-blooded employment of large tank forces in 1918, such was the 'learning curve'.[75] The first highly successful use of tanks on a large scale was at Cambrai in November 1917. Bearing in mind the relatively good condition of the ground, 324 tanks were launched with six infantry divisions, and achieved a strikingly successful breakthrough of the tactical zone. The plan was to pass cavalry swiftly through the gap, and exploit rapidly into what would later come to be called the operational depth. Yet this opportunity was not seized. It must be stressed that at this time cavalry was probably the only arm that could have carried out rapid operational exploitation: tanks were extremely slow, designed principally to ovecome the mud and barbed wire of the battlefield, no more. Their early employment more resembled the methodical progression of an artillery bombardment than tank warfare as it would later develop. Though not fulfilled, the Cambrai plan contained the essentials of the deep air-land battle: rupture of the defended front and exploitation by mobile forces.

Tanks did not play a significant role in the German breakthrough in 1918, but they were part of the combined arms scheme which enabled British, French, and Americans to launch a series of successful offensives from 15 July 1918. Infantry units now possessed greater mobility and firepower and acted in concert with tanks. Aircraft had a very limited ground-attack role but were more important for reconnaissance. This focused on anti-tank threats to the advancing forces. We can therefore see the interacting and mutually supportive roles of all types of forces and the embryo of modern combined arms warfare.

THE WAR IN PERSPECTIVE

Germany's ultimate collapse owed much to grand strategic considerations and the naval blockade, but every level of war is interlinked, and the increasingly competent operations of the British and the appearance of the Americans on the western front helped tip the balance. The Amiens operation (8–13 August 1918) was notable for its achievement of surprise and for the massed employment of armour and aviation to achieve a breakthrough. On a carefully selected weak sector of the German front 4th British Army and 31st Corps, 1st French Army, with 18 infantry and 3 cavalry divisions, nearly 2,700 guns, 511 tanks and about 1,000 aircraft attacked on a 32-kilometre front. They penetrated between 10 and 18 kilometres on the first day. This was arguably the beginning of a progressive collapse of German forces on the western front, leading to their capitulation on 11 November.

The conduct of the war by Britain and France, particularly the former, has been widely criticized. British generals and particularly Haig have been consistently accused of unimaginativeness and pig-headedness ever since. Yet it is very difficult to see what else they could have done. No First World War general (especially cavalry generals, like Haig) ever chose 'attrition' if he had any other option. They would have loved nothing more than a single, brilliant, imaginative stroke, with cavalry forces pouring round a flank or through a gap. But conditions precluded this absolutely. Throughout 1918, Haig was pressing for 'munitions of mobility' to exploit a breakthrough which he realized would occur, sooner or later.[76] It had become apparent that the war could not be won by a sudden decisive stroke, such as Nivelle had planned in 1917. Rather, the British pinned their faith 'in a continuous succession of attacks, each with limited objectives, pressed one after another as rapidly as was consistent with due preparation and economy of troops'.[77] This approach was the only feasible one, and it is noteworthy that after the war the Red Army adopted the same approach with the concept of 'successive operations' in the 1920s. Only later would weaponry and means of mobility catch up to the extent that it would be possible to think in terms of a single, 'deep operation', in place of the successive hammer blows.

The principal problem faced by attackers in World War One was not so much in accomplishing an initial tactical breakthrough, but in following it up and passing large numbers of troops, heavy equipment, munitions and supplies over broken and cratered terrain. Furthermore, the enemy, having withdrawn a short distance, could rush reinforcements to the threatened sector along communications (roads and railways) that were still intact far faster than could the advancing force, cut off from theirs. Thus, in September 1918, Haig stressed that if the allies were to advance swiftly after having broken the Hindenburg line, they had to prevent the Germans from destroying the railways. The Germans depended on these lines for supplying their front: after the breakthrough, the allies would need them as they advanced against the Germans in the long-awaited war of movement. Therefore, mounted troops needed to be pushed forward as fast and far as possible not only to exploit the breakthrough and keep the Germans off balance, but also to seize and preserve the future lines of advance.[78]

A concluding word is necessary on the eastern front. The battle lines there never acquired the solidity of those in the west, but neither did the means to penetrate them acquire the same density. Movement of large forces was constrained not, perhaps, so much by the existence of a solid front as by the difficulty of dragging formations across the vastness of the theatre. 'In the west, the armies were too big for the country; in the east, the country was too big for the armies.'[79] When the new Soviet government made peace, the Imperial Russian Army had fought well. They had lost to the enemy only Russian Poland, Lithuania, and part of Latvia, a tiny amount, compared with the Red Army's rapid 1941 exit from a quarter of all European Russia (the Ukraine, White Russia, the Crimea, the Don Valley, the North Caucasus, most territory around Leningrad, and all territory north-west, west and south of Moscow). As Freytag-Loringhoven observed, in spite of all the problems, the Imperial Russian Army had remained, to the end 'a redoubtable adversary'.[80]

The experience of the Great War is worthy of the most meticulous and painstaking study. Many of the clichés about it ('lions led by donkeys') are, the author believes, gross oversimplifications or completely wrong. It is hard to disagree with the man who commanded the British Army in its only major, prolonged, first-division continental contest this century, a contest which it won, with fewer casualties than France, Germany, Austria, or Russia:

> The huge numbers of men engaged on either side, whereby a continuous battle front was rapidly established from Switzerland to the sea, outflanking was made impossible and manoeuvre was difficult, necessitated the delivery of frontal attacks. This factor, combined with the strength of the defensive under modern conditions, rendered a protracted wearing-out battle unavoidable before the enemy's power of resistance could be overcome. So long as the opposing forces are at the outset approximately equal in numbers and moral and there are no flanks to turn, a long struggle for supremacy is inevitable.[81]

Chapter four

Air-land battle, 1918–77

This chapter traces the development of warfare in what is essentially its modern form, involving co-ordinated action by mechanical and armoured vehicles and aircraft, in three dimensions (laterally, to great depth, and vertically), and on three levels simultaneously (strategic, operational, and tactical).

RESTORING MANOEUVRE

World War One presented a paradox. After the initial mobile period, operational manoeuvre had been made impossible by tactical stasis. The great efforts to restore tactical mobility were not, however, accompanied by parallel efforts to restore operational manoeuvre, as shown by the failure to exploit the breakthroughs by the allies at Cambrai and by the Germans in March 1918. As the war ended, efforts were underway to remedy the situation. The war was expected to go on into 1919, and Fuller addressed the issue in his *Plan 1919*:

> The problem was no longer how to break the enemy's front . . . but, instead, how to maintain continuity of forward movement, in other words, how to effect an unbroken pursuit. As long as the enemy was only punched back, and after each blow was given time to recover, he could always find time to destroy communications – roads and railways – before he withdrew. Without roads and railways the pursuit of a determined enemy was not possible.[1]

Fuller's project for 1919 therefore embodied two elements; a force of fast-moving tanks to rush through the enemy's defence and go straight for division, corps, and army headquarters, and a more normal combined arms attack on the troops in front. In other words, one would attack the brain and stomach, the other the body.[2] This has been the basic principle behind all 'deep-attack' concepts, including German 'Blitzkrieg' (see below), the Soviet OMG (see Chapter 5), and the American 'Deep Attack' concept today.

If manoeuvre in the west had been restored by ingenuity and effort, its reappearance in the east during the Russian Civil and Russo-Polish wars was conditioned by the vastness of the theatre of war and the paucity of men and *matériel*. Penetration and encirclement were no longer so difficult and fluid manoeuvre was the rule.

It is easy to dismiss the Russian Civil and Russo-Polish wars as anachronisms in which the vast spaces permitted the reappearance of horsed formations which would not have looked unfamiliar to Genghis Khan. In fact, the combatants made great efforts to combine movement with firepower, and the Russo-Polish war produced outstanding examples of generalship. The 27-year-old Army Commander Tukhachevskiy's operation against Poland in July–August 1920, culminating in the battle of Warsaw on 12 August, seized many people's imagination, including that of Fuller. Tukhachevskiy has been criticized for a reckless advance without paying proper attention to logistics, but his opponent, Pilsudski, was remarkably complimentary: 'It was not a bad plan, not a bad plan at all, which I might have adopted myself.'[3] He also praised Tukhachevskiy's generalship:

> Tukhachevskiy understood the art of communicating to his subordinates his own energy and well directed activity. This fine display of generalship will always speak in favour of M. Tukhachevskiy's ability as a commander to conceive bold strokes and to carry them out energetically. . . . Tukhachevskiy made a most skilful disposition of his forces . . . only a general of the highest order could have undertaken so bold and coherent a deployment of his troops. . . . It is beyond doubt that Tukhachevskiy was no ordinary general. In his march beyond the Vistula he made a reality of his own designs and creative intelligence . . . he therefore passed in thought and later in reality through an extent of territory equal to half Europe. This was no mean achievement.[4]

What defeated Tukhachevskiy was not the speed of his advance, but nagging delays in it. He aimed to take Warsaw on 12 August, which implied the need to be in position by 9 or 10 August. But he was not in position then and the delay gave the Poles the chance to re-group and concentrate their forces with a speed which Tukhachevskiy could not have anticipated. 'He missed his goal by three or four days at the most. Speed, which had virtually ceased to exist as a strategic factor in Western Europe, was still the essence of warfare in the East'.[5]

INTER-WAR DEVELOPMENTS: THE ASSIMILATION OF AIR AND ARMOUR

In 1918 Great Britain led the world in armoured equipment and tactics. During the inter-war period this advantage was lost because of a number of factors common to all armies except the German and the Soviet. Anti-war sentiment, budgetary restrictions and the large stockpiles of 1918 vintage equipment all discouraged the adaptation of new technology to tactics and operational art. Many of the most prescient theorists were abrasive mavericks, which did not help their case. The British and French armies had to consider the contradictory requirements of continental and worldwide commitments, and the United States army expected only to have to engage second or third class opposition on the American continent. The British, astonishingly, made no concerted effort to evaluate the lessons of the Great War until 1932, when the Kirke Committee reported that: 'at present the enemy cannot be defined and this absence of a basis to the problem adds enormously to the difficulties of its solution'.[6] The committee did, however, recognize that the problem in major war would be converting the 'break-in' to the 'break-through'. Nevertheless the colonial requirement, and the spectacular success with a cavalry and armoured car breakthrough at the so-called 'Second Battle of Armageddon', against the Turks in Palestine in 1918 (very different from conditions in a European war) led to an emphasis on light tanks inadequately armed.[7]

In contrast, those who recognized a tank force's potential for deep penetration and shock action plus the need for firepower, especially Colonel Charles Broad and Fuller, emphasized pure tank formations at the expense of combined arms. Martel and Liddell Hart, two others who are famous as proponents of that non-existent abstraction 'armoured warfare', in fact had a much more balanced view of a mechanized combined arms force. The British undertook an important experiment in 1934 with Colonel Percy Hobart's first tank brigade, but this proved a disaster and rather than concentrate on independent tank formations the British adopted a compromise solution by motorizing a large proportion of the cavalry. The British therefore went into action in 1939 as a highly mobile army with insufficient firepower.

France, having taken colossal casualties in the Great War, initially adopted a cautious approach with a Napoleonic emphasis on artillery. Pétain, hero of Verdun, issued Provisional Instructions for the Employment of Larger Units in 1921 which stressed that infantry had the principal mission, in close co-operation with artillery, and helped where possible by tanks and aircraft.[8] Some Frenchmen, however, did not envisage a repetition of the infantry/artillery combination of the Great War. In 1919, General Estienne submitted to Pétain a *Study of the Missions of Tanks in the Field*. This stressed the need for armoured, tracked vehicles to carry infantry, artillery, and battlefield recovery teams alongside the tanks, and for aviation to conduct an in-depth bombardment of the enemy. This was a remarkably advanced view for the time, presaging Tukhachevskiy's ideas of the 1930s, and force structures which were not to be achieved until the end of World War Two.[9]

In 1934 Colonel Charles de Gaulle published his radical *Towards the*

Professional Army which called for a 100,000-strong armoured force. This militated against the traditional French view of the nation in arms, upset a nation which had become remarkably pacifistic by its aggressive tone, and proposed an army which was highly offensive in configuration. As a result, it probably retarded rather than assisted the progress of mechanized combined-arms doctrine. Nevertheless, the French went to war in 1940 with probably the best tank in the world – with the possible exception of the prototype Soviet T-34 – the Char B (see below).[10]

One of the most important developments in the longer term took place in the United States. This was the development between 1929 and 1941 at the US Field Artillery School of a means of concentrating any amount of artillery available on a target of opportunity. This centred on the exploitation of the new, more reliable radios instead of field telephones. More importantly, procedures were developed enabling adjustments to be recorded as if seen from the forward observer's position, instead of the battery position. Indirect fire techniques evolved before World War One and used during it enabled an enormous amount of fire to be brought to bear by a senior artillery commander. However, the system had its practical limits: fire directed at a single target had to be subject to centralized control by a senior officer. The Americans developed a system in which graphical firing tables compensated for the different locations of firing units, and a common reference point was established for all artillery in a divisional area. As a result, the fire of an entire battalion or multiple battalions could be brought down on the instructions of a single forward observer, whoever he was. This was to prove enormously important to British and American forces in World War Two and is essentially the system used today. The Russians continued to use the system whereby a senior artillery officer personally directed the fire of all his guns through the Great Patriotic War. At first, the British rather neglected the development of indirect fire technique, as artillery was required in the direct fire role to compensate for inadequacies in infantry's organic anti-tank weapons. The French retained the pre-planned bombardments of the Great War. In 1936, it took some eight minutes for French artillery to open fire on a target whose position had already been plotted and half an hour in the case of a target whose range, bearing and angle of sight still had to be assessed. This was not good enough in an era when targets might be armoured and highly mobile, and could travel a long way in half an hour.[11]

If the most enthusiastic proponents of the tank mechanization in the west actually damaged their case, the same is true of the aircraft as a component of the combined arms offensive. The disciples of air power stressed its role as an independent, strategic weapon, striking directly at the enemy's centres of power and population and independent air forces were developed with this mission. In spite of the very important lessons of the Spanish Civil War, the French devoted scant attention to co-operation between ground and air forces and to the role of the 'air arm over battle'.[12] The British were the most devoted disciples of the independent, strategic role of air power, which their island situation made particularly appropriate. Even as late as the Normandy landings in 1944, the Royal Air Foce was most reluctant to employ its bombers to give tactical support to ground forces, although it was used operationally to break up the French railway network and interdict (isolate) the intended battlefield.[13] The unified nature of the air-land battle was much better understood by Germany and the Soviet Union.

It is appropriate to consider these two countries separately from Britain, France, and the USA, first, because they did not suffer some of the limitations paradoxically imposed on the victors of the Great War, and second, because the two armies co-operated extensively in training and equipment development throughout the 1920s. The Germans had plenty of expertise, but were precluded by the Versailles treaty from training openly; the Russians were short on expertise but had vast expanses where the Germans could develop weapons and train in secret. The Germans set up tank and flying schools and a gas experimental establishment. In 1921 the first post-war German regulation *Command and Combat of the Combined Arms* envisaged not only the combined-arms assault battalion of 1918, the surgical artillery strike, but also close air support, gas warfare, and tanks.[14] In retrospect, it seems inevitable that the German infiltration tactics of 1918 plus their experience of the psychological value of the tank, close air support, decentralization of command and massing on a narrow front should lead logically to what was later called 'Blitzkrieg'. In fact, the German army did not fully accept the concept of mechanized Blitzkrieg until *after* the fall of France in 1940. Prior to this, most senior German commanders regarded mechanization as a useful but specialized tool that would not replace ordinary infantry divisions. The most influential proponent of mechanization in Germany was General Heinz Guderian who, like Percy Hobart in Great Britain, had considerable experience in the use of radio communications. He insisted on a radio in every armoured vehicle, giving the Germans a crucial advantage in command and control. This indicates how communications systems can act as a force multiplier, greatly enhancing the potency of a weapons system (the tank) which in itself may be no better than or inferior to those of the enemy. Guderian also avoided the temptation to move towards pure armoured formations which seduced many of the British tank pioneers. Only a mechanized combined-arms formation could fully exploit the potential of the tank and protect it where it was most vulnerable. According to Guderian, the idea of the mechanized combined arms formation gelled in 1929:

> In this year . . . I became convinced that tanks working on their own or in conjunction with infantry could never achieve decisive importance . . . tanks would never be able to produce their full effect until the other

> weapons on whose support they must inevitably rely were brought up to their standard of speed and cross-country performance . . . what was needed were armoured divisions which would include the supporting arms needed to allow the tanks to fight with full effect.[15]

A Panzer ('mail') division was improvised from available units in the summer of 1935, and on 15 October three were formed. Guderian met great resistance to the formation of panzer divisions, and he had to settle for tanks which were extremely lightly armed and armoured. However, the fact that these were available early and in reasonable numbers and were equipped with radios enabled the new panzer force to conduct extensive training, establish battle procedures, identify and solve problems and develop changes in organization and equipment. By 1939 the first panzer divisions were not completely ready but they had trained to fight. The Germans extracted the last drop of experience from every limited action they took: for example, the move of far from fully trained panzer troops into Austria in March 1938, when just about every gear change was analysed.[16] This advantage was denied to most of Germany's future opponents. Assimilation, training, command, control, communications were everything; raw equipment itself counted for little.[17]

The role of coincidence and accident in the development of warfare is apparent from the other component of the German air–land battle or Blitzkrieg package: the air component. A new tactical bomber, the Junkers-87 'Stuka' dive-bomber, appeared in Germany in spring 1936. Dive bombing techniques had been developed by pilots in 1917–18, but little was done to systematize the technique although it clearly offered much greater accuracy than level bombing, and this was to be crucial in the tactical use of air power. The Junkers-87 had a very modest performance by the standards of the time, and was very vulnerable to decent fighter opposition. In combination with ground forces, however, it did its job very well. Germany's limited material resources and ambitious military programme made the cost effectiveness and tactical advantages of the dive-bomber extremely attractive. Unlike heavy bombers for long-range strategic and (possibly) operational roles, Ju-87s could be mass produced quickly and relatively cheaply, and were ideal for short-range campaigns against Germany's neighbours. Destroying military targets immediately in front of advancing ground forces required maximum precision, and in tests Heinkel-111 and Dornier-17 bomber crews in level flight could place only 2 per cent of their bombs in a 100-metre diameter circle from 4,000 metres altitude. Ju-87s in a dive could put 25 per cent of their bombs in a 50 metre diameter circle (a result which makes an interesting comparison with recent and modern missile Circular Error Probable: CEP).[18]

Conversely, the Germans did not proceed with the development of long-range, four engined strategic bombers as did Britain and the United States.

Their medium bombers, the Ju-88 and Heinkel-111 were manifestations of the ideas of the first great airpower theorist, Douhet (1869–1930), who favoured fast bombers which could out-run fighters. The Ju-88 and He-111 were highly versatile aircraft. At the beginning of World War Two they were arguably the best all-round bombers in service and were extremely useful for operational bombing in concert with ground forces. Long-range bombers, like the projected Ju-89 'Uralbomber', were never favoured by the Germans.[19]

The experience of the Spanish Civil War (1936–9) suggested that the Germans were right. Large-scale co-ordination of air and ground forces in battle was first attempted in Spain. Just as the Germans had selected critical thrust points (*Schwerpunkt*) on the ground, so they selected air thrust points (*Luftschwerpunkt*). Interesting examples occurred in June 1937 at Col del Urquiola, Monte de Solluve and Lemona.[20]

The Spanish Civil War was extremely important in the development of military art, bearing much the same relation to World War Two as the Russo-Japanese war had to World War One. Its lessons were exhaustively analysed, but the conclusions drawn were not always those which now seem obvious with the benefit of hindsight. The belief that improvements in anti-tank weapons had outstripped the development of armoured forces was one example. Battles were frequently characterized by elaborate field fortifications and the emphasis on engineering was reinforced by the importance of demolition and reconstruction of bridges, embankments, and tunnels in the rugged Spanish terrain. These form a promiment part of the picture of military operations in Spain.[21] The war also provides numerous political precedents affecting the conduct of a major internal war with external intervention, Germany and the Soviet Union providing 'advisers' and more active assistance.

Some of the most far-sighted and long-term developments occurred in the Soviet Union. During the 1920s the Soviet Union concentrated on the reorganization of their armed forces, the creation of an independent industrial base (a cardinal lesson of World War One), and the development of new technologies which might circumvent the enormous material superiority of the victors of the Great War. Of particular interest was Soviet insistence on the role of the mass army, in contradistinction to the small professional armies advocated by advanced theorists in Britain, France and to some extent Germany, plus the Russian émigré Gerua (Héroys). To the Soviets mass and mobility-potency were not necessarily incompatible. Both Frunze, the founder of Soviet military thought and Triandafillov, whose enormously influential *The Character of Operations of Modern Armies* was published in four editions in 1929, 1931, 1936 and 1937, disagreed with Fuller in Britain, Zoldan in Germany and their own A. I. Verkhovsky (1886–1938). The idea of a small army of armoured 'knights', who would bring about a decision which had to be accepted by the rest of the population was flawed. It was 'naive' to think that such an army could subdue a modern state. Modern war required forces 'of high quality and in sufficient numbers'.[22] Tukhachevskiy, perhaps the most brilliant of all the Soviet thinkers of this period, expressed the same views.[23]

Another area where Soviet thinking diverged from that of western theorists was in the employment of air power. Fuller, perhaps the most strident of the proponents of deep penetration by armoured forces, had largely neglected air power as a means of imposing one's will on the enemy far behind his forward positions.[24] Spatial and geographical conditions naturally led the Russians to explore the potential of aircraft. The Russians had an impressive tradition of aeronautical record breaking going back to before the Revolution (the Il'ya Muromets heavy bomber, for example), and in the 1920s and 1930s aeronautics was seen as extremely important in opening up their vast country.[25] This has led some western commentators to assert that the Soviet military expressed interest in theories of 'strategic bombing', and to cite Soviet development of long-range bombers in support of this. However, what might be considered 'strategic' range in western Europe was only 'operational' as far as the Russians were concerned. Interdicting the battlefield, preventing reinforcement and interrupting mobilization in eastern theatres would require aircraft with this sort of range. Tukhachevskiy, for example, admitted the possibility of air attacks 'on the most important political centres and industrial regions', but does not seem to have been very interested, since these were 'not directly connected with the conduct of operations and large-scale battles, with the exception of bombing rail junctions'.[26]

Tukhachevskiy therefore saw aviation as an operational-tactical weapon and this led him to pioneer the most obvious manifestation of the unified air-land battle, the use of airborne (parachute and air-landed) troops. The first official proposal to use parachutes to deliver troops behind an obstacle probably came from American 'Billy' Mitchell in 1918, when a serious plan to attack Metz from the rear with an airborne force was drawn up, but then shelved. Italy was the first nation to create a paratroop force, with an experimental drop in 1927 and a company formed in 1928. Italy had a tradition of innovative air-power thinking and, perhaps more immediately important, a particularly good parachute.[27] The Red Army first used paratroops on active service in an internal security role in 1929. By 1931, their role in major war had been clearly formulated, 'to disorganize the army and corps rear area, delay operational transfers of troops, hinder the work of headquarters in directing combat and operations, destroy rear aerodromes and naval bases'.[28] Right from the start, Soviet parachute units were stiffened with lightweight recoilless guns and motorcycles in an early attempt to solve the key problem which has beset airborne forces ever since: they tend to be too lightly equipped to fight for long and insufficiently mobile once on the ground.

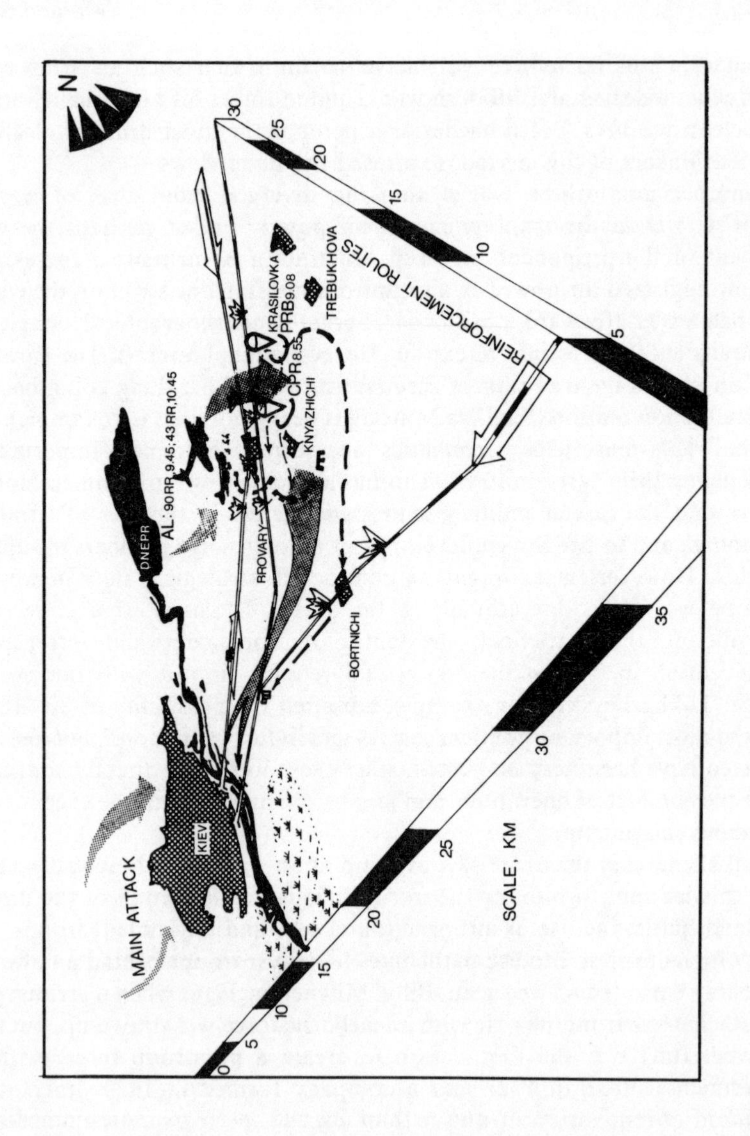

Figure 4.1 Kiev manoeuvres, 1935: large-scale employment of airborne forces

Note: Shaded arrows indicate attackers' landings and movements, white arrows defender's reinforcement and counter attack.

AL = Air Landing (*posadochny desant*) (total 1,765 men)
CPR = Composite Parachute Regiment } Parachute drop
PRB = Parachute Regiment of Brigade } (*Parashyutny desant*) (total 1,188 men)
RR = Rifle Regiment

Units' numbers are followed by time of landing on 14 September. Explosions indicate cutting of road and rail communications by *desant* force. Black and white routes are railways. Aircraft illustrated are TB-3 four engined bombers.

Source: Ramanichev, p. 75, adapted by the author.

Large-scale manoeuvres involving parachute and air-landed forces were held in the Minsk area in 1934 and 1936 and the Kiev area in 1935. The first two were made into spectacular films which attracted much interest abroad. The film of the 1935 Kiev manoeuvres illustrates the sophistication of the concept of operations for employing airborne and air-landed forces (Figure 4.1). The airborne troops were employed as part of an attack on Kiev from the west, and were landed so as to 'interdict' the objective – the city – preventing its reinforcement by forces from the east. Not all those who saw the film understood this, but it is consistent with Soviet views on the employment of airborne forces in the offensive, which almost always envisaged using them to prevent the defender plugging the breakthrough (an obvious lesson from World War One).[29]

The British General Sir John Dill, Director of Military Operations and Intelligence, who later influenced the formation of British and American airborne forces, took note of these first major trials. Tukhachevskiy visited him in London in 1936:

> Tukhachevskiy and Dill recalled various precedents from military history, and discussed what would have happened if the commander in this or that war or this or that battle had airborne forces at his disposal, and what changes this innovation would make on modern tactics and strategy.[30]

Evidently, these two forward-looking generals were in no doubt about the value of military history in clarifying and developing ideas for the future. However, in spite of their high profile, airborne forces have not had a very illustrious history when seen against the backdrop of military history as a whole. Their record supports the principle that airpower is indivisible. Only the side enjoying a pronounced superiority in the air, and with a massive air transport fleet available, is in a position to use them. This explains why the Russians, who first developed the idea, made very little use of airborne forces in World War Two, and why they were employed principally first by the Germans and then by the British and Americans. Even then, airborne operations have either been horrifically costly (Crete), or have been considered out-and-out failures (Arnhem). The first (and remarkable) published consideration of the role of paratroops in war was Miksche's book of 1943, which accurately portrayed the use of airborne forces to secure the D-day landings the following year. Luckily for the allies, the Germans took this military 'theory' to be hypothetical and not an uncanny prediction.[31]

The Red Army Field Service Regulations of 1936 are one of the most remarkable and futuristic military documents of this period. By this time, a coherent scheme had been outlined in which tanks would be used both to penetrate the enemy's forward defences (the 'tactical zone') and for the different task of exploiting the breakthrough ('converting tactical success to operational', in the sometimes excessively scientific jargon of the Red

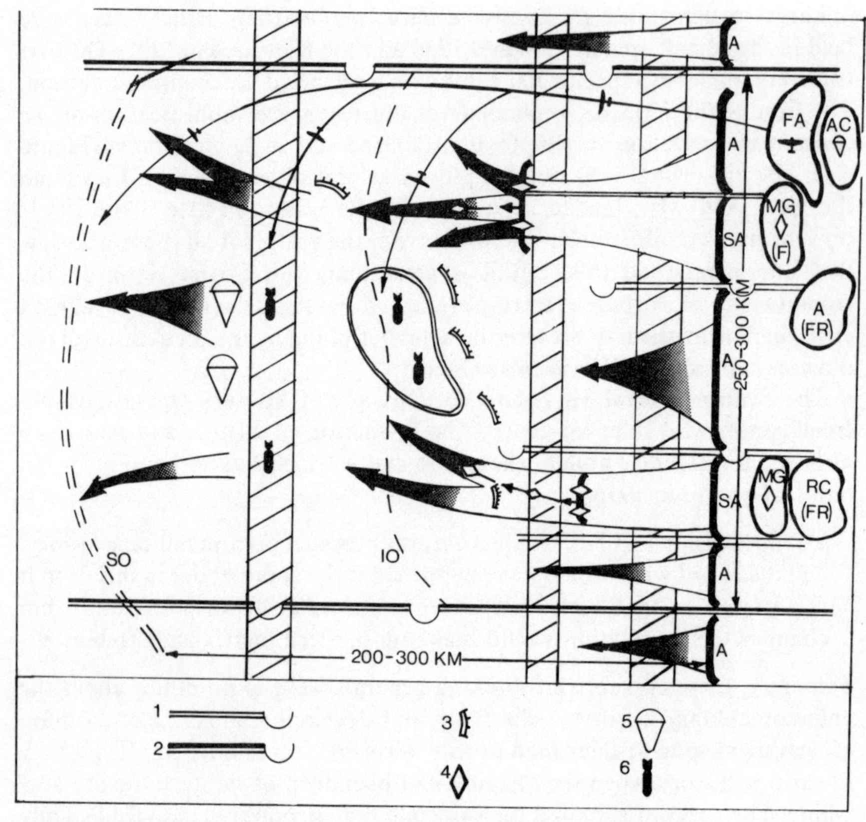

Figure 4.2 Deep operation, as envisaged in Soviet Field Service Regulations, 1936

1 = Army boundary (Soviet symbol); bulge is to include/exclude certain objectives
2 = Front boundary (Soviet symbol)
3 = Fortified area (Soviet symbol)
4 = Armour (Soviet symbol)
5 = Parachute landing (airborne brigade)
6 = Air attacks
A = Army

A(FR) = Army (Front Reserve)
AC = Airborne Corps
IO = Immediate objective
MG = Mobile Group (prototype OMG)
FA = Front Aviation
RC = Reserve Corps
SA = Shock Army
SO = Subsequent objective

Source: *Soviet Military Encyclopedia*, vol. 2, 1976, p. 577.

Army). Air power would be used both tactically, in direct support of ground forces, and operationally, to interdict, isolate, unhinge, and destroy key objectives. Airborne and air landed forces would do the same, and prevent enemy withdrawal, pinning enemy forces in place (see Chapter 1) and facilitating encirclement (Figure 4.2). Every attempt would be made to prevent a solid front forming, as it had in 1914. Mobile forces would be committed early, to keep the enemy off balance. Whereas in World War One the paradigm had been one of attempted outflanking manoeuvre, followed by positional warfare, the opposite was now true. As another extremely talented Soviet military thinker, Isserson, put it, modern manoeuvre lay, 'not in front of the enemy's forward line of own troops, as it had, in time and space, before World War One, but behind it, and in the enemy depth'.[32]

This paradigm was to prove as applicable to the Germans, the British and Americans, as it was to the Russians.

WORLD WAR TWO, 1939–45

The course of military operations in World War Two, their complex twists and turns, will not be chronicled here. They have been amply analysed elsewhere.[33] In terms of the evolution of techniques of warfare on land, it is, however, possible to break this most terrible and destructive of wars down into a number of clearly defined and relatively simple phases. In the author's view, after the opening victories by German and to a lesser extent, Japanese forces, the conduct of land warfare and ultimate victory owed very little to military skill but rather to the somewhat crude application of superior economic and human resources. The sparks of genius occur in the application of air power, the use of intelligence and strategic deception, and in technological innovations of limited effect but stunning potential. The conduct of land warfare proceeded as a logical continuation of developments noted during (and even before) World War One, and the implementation of doctrines and theories which had been amply expounded in the inter-war period.

The word 'Blitzkrieg' has done historical accuracy no service. It was not even a German military term, but appears to have been invented by an American journalist after the Polish campaign, and when Hitler subsequently heard it he said it was a 'very stupid word'.[34] One school of thought sees the initial German successes as the result of a coherent, 'Blitzkrieg' strategy (which even the linguistic evidence disproves). The theory goes that in order to maintain consumer production at a high level, Hitler planned a series of short, mobile campaigns by tanks and aircraft (themselves to a large extent spin-offs from civilian tractor and aircraft industries), with pauses between to replenish stockpiles. 'Blitzkrieg' was not a strategy at all: it describes a particular pattern of operational art, no more, and is best described as a 'fresh', rather than 'new' system of tactical combinations.[35] One must agree that:

> A close examination of the evidence does not support the theory of a Blitzkrieg strategy; rather it points to an uncertain and unclear grand strategy in which the Germans put the military pieces together at the last moment, with serious doubts and considerable haste.[36]

The striking success of the Germans at the beginning of World War Two was, arguably, 'the product of a very transitory set of advantages'.[37] The Germans had produced equipment and fielded mechanized units in the mid-

1930s, so that this equipment was still usable and the units were well organized and trained when the war began in 1939. The Germans had three principal advantages: a primitive but effective system of close air support, a command and control network which was tailored to coping with more rapid manoeuvre than any opponent could reasonably expect to achieve, and, lastly but probably most importantly, young officers and soldiers of superb quality, who had been brought up and conditioned to believe that they were bound to win, commanded by senior officers who had served in World War One, who believed that 'we was robbed' the last time, who were not going to sacrifice those superb soldiers needlessly.

The reasons for the unexpected German success of 1940 are therefore complex, but in purely air–land warfare terms it can be ascribed to classic military advantages of all arms co-operation and surprise. A notable example is the use of glider troops to seize the major Belgian fortress of

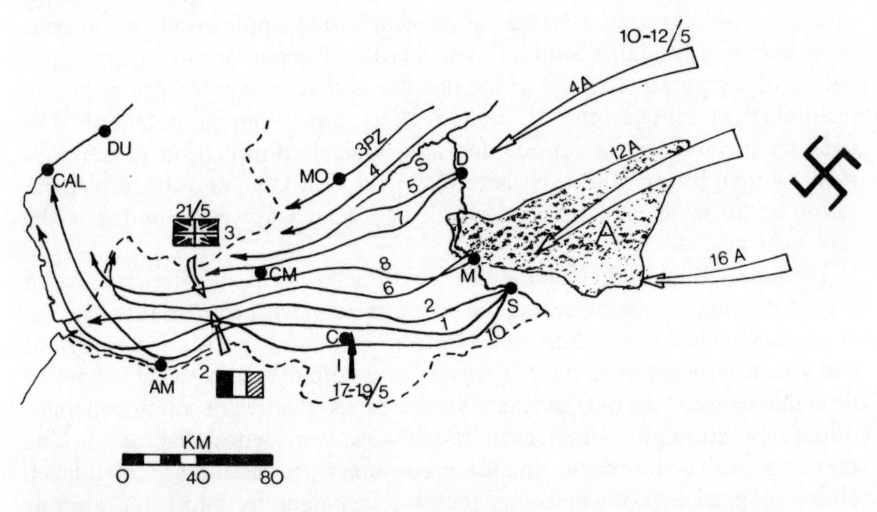

Figure 4.3 German invasion of France, 1940

Note: German armies are indicated by numerals with the suffix A. Plain numerals indicate divisions (all panzer; 3 panzer is identified as such to distinguish it from 3 below.

Large A	= Ardennes forest
AM	= Amiens
C	= Crecy
CAL	= Calais
CM	= Cambrai
D	= Dinant
DU	= Dunkirk
M	= Montherme
MO	= Mons
S	= Sedan
1	= De Gaulle's armoured counter-attack, 17–19 May
2	= Weygand's plan to link up with allied forces
3	= British attempted armoured breakthrough

Eben-Emael on 10 May 1940. In the assault on France the Germans employed the classic device of attacking from an unexpected quarter, in this case the Ardennes forest which was thought to be impassable to armour, and concentration of force: seven out of ten panzer divisions with five motorized divisions close behind advanced on a 70-kilometre front and hit the Franco-British linear defence at one of its weakest points. (See Figure 4.3).

The still commonly held view of the French Army moving ponderously into Belgium with old-fashioned equipment is not supported by the speed and precision with which the French executed the Dyle plan, moving swiftly to their positions between Namur and Houx. They committed the same error as Joffre had in attacking Alsace-Lorraine in 1914. German plans depended on the decisive commitment of French forces in a certain area, while the French failed to identify the initial German thrust. The French, furnished with more and in some cases better tanks than the Germans, fell because of inadequate command and control. As a revisionist article published in 1970 pointed out:

> Given the powerful French armoured structure organized into six armoured divisions on 10 May 1940, joined by two additional armoured divisions several days later and supported by more than 700 infantry support tanks and several thousand anti-tank cannons, the French Army had to do little more than clarify the strength and directions of the major German thrusts in order to determine the major one and smother it.[38]

It sounds simple, but of course, nothing in war is ever simple. Clearly, however, the German success was a very near-run thing.

The 1941 campaign against Russia saw the acme of the so-called Blitzkrieg method of war and especially the encirclement battle. Where resistance was weak, the German could lead with armoured units, but otherwise they preferred to penetrate with a conventional combined arms force. As the war lengthened, this became more and more difficult. Once a penetration or flanking manoeuvre had succeeded, the Germans sought to encircle the enemy in a two-layered pincer: an inner one to hold the surrounded force and reduce it, and an outer one to deflect attempts to relieve those caught in the ring (see Figure 4.4 and Chapter 1).

Germans usually used a combined arms battle group of battalion or regimental size to spearhead each jaw of the pincer. These forces were extremely weak for the task they were given and in practice Soviet soldiers and even entire units were able to work their way out of the thinly drawn encirclement.[39] The initial German encirclements cut off Soviet formations of disproportionate strength, and had a psychological effect which recalls that of relatively small numbers of mongol troopers. However, you cannot go on fooling people forever and once the Russians had managed to stabilize their defence they began to out-encircle the encirclements. The classic example of

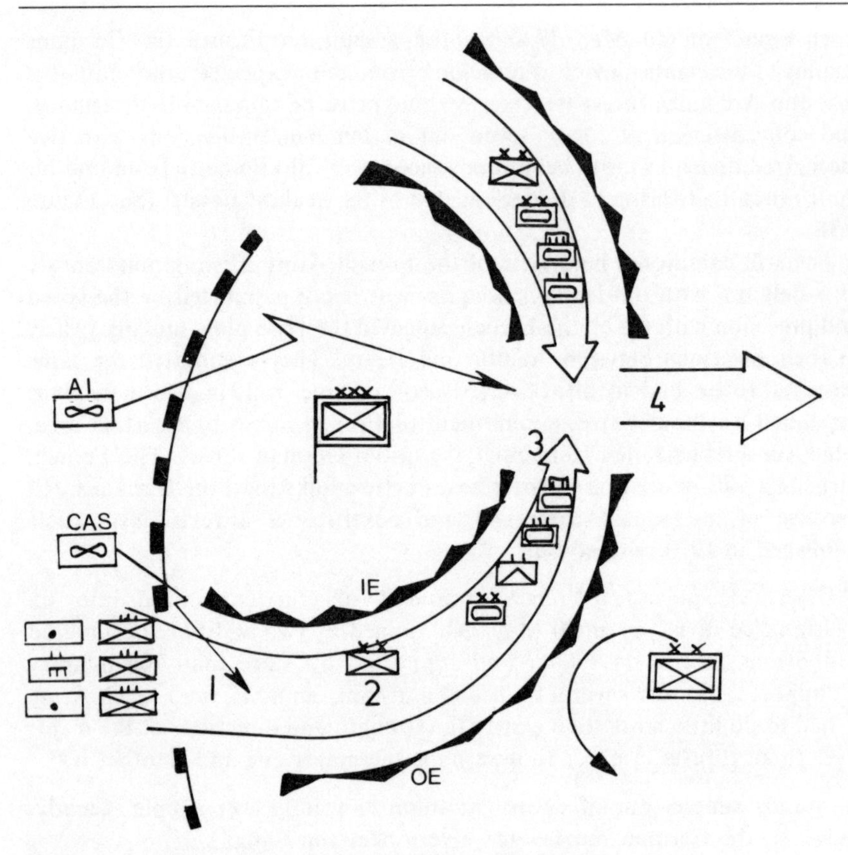

Figure 4.4 Blitzkrieg encirclement 1941–2, schematic

AI = Air Interdiction
CAS = Close Air Support
IE = Inner edge of encirclement (holding enemy in)
OE = Outer edge of encirclement (deflecting attacks to relieve encircled force)
1 = Phase 1: penetration
2 = Infantry or motorized infantry division
3 = Phase 2: encirclement
4 = Phase 3: continued exploitation
Symbols used are standard NATO

this is the very large-scale operational envelopment at Stalingrad in November 1942. The Soviets were clearly thinking bigger than the enemy, and in this great battle they also managed to dislodge the psychological advantage which the Germans had formerly enjoyed. It was the turning point of the war (see Chapter 5, Figure 5.11).

The British and French defeat in 1940 and the Soviet disasters in 1941–2 can all be ascribed to poor deployment, training, command, and control. The British readjusted organization and training, and created armoured divisions which in fact comprised a combined arms mix of infantry, armour, artillery and anti-tank forces. The British had considerable difficulty integrating the various arms at low level, owing to slow command and control procedures and poor inter-arm co-operation. However, Montgomery was able to overcome this by carefully pre-arranged 'set-piece' battles such as that at El Alamein in 1942. The Russians also adjusted their organization and procedures, notably through the 'artillery offensive' and the carefully targeted committal of armoured spearheads (Chapter 5).[40]

From late 1942 onwards, the conduct of major land warfare begins to exhibit many of the characteristics of World War One and recalls the preoccupations of the inter-war theorists and planners. Huge fortified belts, like that in the Kursk salient in summer, 1943, and the positions successively occupied by the withdrawing Germans on the eastern front thereafter, are a close parallel with the defences of World War One. They were much more oriented towards anti-tank defence than those of 1914–18, but the solution evolved by the attackers was the same: artillery bombardment. Mines were extensively used in North Africa and Italy, but less so on the eastern front, where the size of the theatre and the difficulties of supply made large-scale mining impracticable.[41] The task force by the British and Americans in re-establishing themselves on the European mainland was similar. The first and most difficult task was to break through a fortified belt, in this case, the Atlantic wall. If we recall the precedent of World War One, it is still totally relevant. The problem, as Fuller observed, was maintaining momentum once the hard crust was cracked. In 1944, the British and Americans possessed the means of land mobility to do this, but they were dependent on passing supplies (principally fuel, in this case) across the area over which they had assaulted, something which in World War One had been very difficult. The use of a submarine pipeline to pass fuel to the Anglo-American forces once they were ashore was a logical development of this old dilemma. We thus see the fulfilment of Isserson's paradigm, the breakthrough, and manoeuvre behind the enemy 'front line'. Subsequent British and American operations (the Siegfried line, the Rhine crossing) were manifestations of the same problem.

The British and American experience from 1943, when they again secured a foothold on the European mainland, illustrates the expanding scope of operations. The landings in Italy and France were immensely complex combined (in British parlance) or joint (in American parlance) operations, involving the most careful co-ordination of land, sea, and air power. This had been even more apparent in the Pacific theatre, where the land operations were of relatively limited scope and everything depended on creating favourable conditions for them through achieving dominance in the air and at sea. Operations in Italy, north-west Europe, and Asia were all heavily dependent on sea and air cover and support. To talk about 'land warfare' in isolation was hardly possible.

Whether assaulting over land or sea, the immensely strong and thick crust of the German defences, and subsequent lines of defence in depth could only be overcome by one means: firepower. Once again, the parallel with World War One is more apparent than some analysts would admit. British, Americans, and Russians all relied heavily on massive concentrations of firepower to stun and physically annihilate opposition, at least in breaking through the first strong defensive belt. Co-ordinating this mass of firepower was no easy matter. In the west, the application of firepower to facilitate the D-day landings was a masterpiece of planning, intelligence and target acquistion, and skilful application. Whether in the form of the rocket batteries fired from the sea, or the use of heavy bombers to interdict the landing area, it was a classic example of the achievement of fire superiority. Similarly, in the east, the Soviet artillery was largely responsible for cracking open German defences. The Russians deployed huge numbers of guns and rocket launchers, 9,500 artillery pieces, excluding 'assault guns', which at this time were categorized along with armour, in the Belorussian operation in June 1944; 25,000 in the East Prussian operation in January 1945; 33,500 in the concurrent Vistula-Oder operation; 42,000 in the Berlin operation. The use of this amount of ordnance presented huge problems of organization and supply. The deployment of guns and ammunition was carried out with skilful deception measures, and Clausewitz's maxim that one should be very strong, first generally, and then at the decisive point, was followed. With artillery, as with ground forces generally, secondary sectors were rigorously denuded in order to achieve massive concentrations of forces and equipment on the chosen axes of advance. In the Stalingrad counter-offensive, for example, Soviet and German-Italian-Rumanian forces had been roughly equal in strength, but the Russians adjusted their 'force ratios' (the modern military jargon) to achieve decisive superiority on chosen axes. The same principles applied to the allocation of firepower (notably in the Yassy-Kishinëv operation). This required staff work, security and deception, of the highest order.[42]

The British and Americans also relied heavily on air power to deliver the requisite dose of fire superiority. The US Army Air Force was given a primary mission of independent strategic bombing against Germany and Japan. Operational-tactical air support comprised three missions: first, air superiority; second, isolation of the battlefield (air interdiction, in modern parlance); third, attacks on ground targets 'in the zone of contact' between ground armies (close air support).[43] The latter was a 'priority three' mission. This contrasted with the Soviet attitude, resulting from an intensive debate in 1938, which made tactical air support the first priority complete with a special aircraft, the *Shturmovik*, which had been expressly designed as a flying tank.[44] The Americans did not develop a formal doctrine and training procedures for air ground co-operation until late in the war. Neither did the British, but in spite of the dominant stress on the strategic value of air power, Air Vice Marshal Leigh Mallory did manage to build up the British Second Tactical Air Force as a support for the Normandy invasion.[45]

One obvious question is why this diligent application of firepower did not have the same effect as it had in World War One, churning up the battlefield and making it impassable. First, although the air and artillery bombardments of World War Two were more intense than those of World War One, they did not last for so long, and so the dreadful combination of weeks of shellfire and weeks of rain was less likely to occur. Second, they were generally targeted more precisely, aimed at key locations spread out in a force's intended area of advance, rather than churning mechanically through a given area. Third, ammunition fragmented more efficiently, into smaller pieces, making it more lethal towards people but less so towards the landscape. Even so, the allied air bombardments certainly had a similar effect on the landscape, as photographs taken after the bombardment of Monte Cassino in Italy show clearly.

Where a heavy World War Two bombardment was combined with thawing and rain-sodden ground, moreover, its effect was similar to World War One. In operation Veritable – the British clearing of the area between the Meuse and the Rhine – the German defences to be overcome comprised three defensive lines, one of which was the northern extension of the Siegfried line. It was not dissimilar to the defences to be overcome by the Russians in the east. The solution adopted was similar: firepower. Although the British did not use as many guns as the Russians, they probably used them even more efficiently: 576 field, 280 medium and 122 heavy and superheavy guns, plus 72 heavy anti-aircraft guns were fired in a 'pepperpot' barrage (one in which every available weapon was used). The effect on the thawing and rain-soaked ground was not dissimilar to a World War One bombardment. However, out of 55 Germans artillery pieces disposed against the British, 28 were destroyed by the initial artillery attack. The British took over 1,100 prisoners for a loss of 350 assaulting troops (a ratio of 3.2:1). Given that a three to one superiority, at least, is normally considered necessary for a successful attack, this was a complete reversal of the normal ratio. It was a vindication of the role of firepower.[46] Similar considerations applied in the Rhine crossing, where every gun (3,500 on the British 2nd Army front alone) was fired in support of the assault. This included suppression of German anti-aircraft guns which threatened the airborne assault. Such examples, and the use of anti-aircraft guns against ground targets either in direct or indirect fire modes (see Chapter 2) underlines the complex interrelationship of land and air operations.

Although the initial German and Japanese victories in World War Two were strikingly successful and accomplished a disproportionate amount for the effort expended, the conduct of major land warfare rapidly acquired the same characteristics as 1914–18. The approach to breaking through thick, strongly defended belts, on all fronts, was not essentially different. The main advantage enjoyed by World War Two armies appears to have been that they possessed better means to exploit the breakthrough (the internal combustion engine and the track) if it was achieved.

While there were shortcomings in the use of air-delivered firepower, air transport was used originally and effectively, especially in Asia. Vast distances, poor road and rail networks, and the inscrutable oriental manoeuvrability of the Japanese led British Empire commanders to exploit their advantages in air mobility. Thus, in March 1944 General Slim correctly predicted a Japanese offensive against his main base around Imphal. Using large numbers of RAF and USAAF transport aircraft, Slim parachuted or air-landed supplies for all his bypassed forces, thus enabling them to fight on independent of their threatened lines of communication. Slim also air landed most of the 5th Indian Division around Imphal, enabling them to engage the infiltrating Japanese immediately. In 1945, the advance of the 14th Army into Burma was facilitated by the combination of air and surface mobility. As each objective was taken, the conquering land force constructed an air field for resupply. The third brigade in each division was equipped with air-mobile transport and artillery which could be air-lifted to the newly constructed airstrips. The British therefore created a concept of integrated armour, motorized infantry and air-mobile forces, often operating interposed between those of the enemy, which anticipates the modern air–land battle.[48] The last great operations of the war reveal an astonishing complexity and scale. If genius in war is an infinite capacity for taking pains, these operations were works of collective military genius. Every component part, from the way supplies were loaded so that they could be unloaded in the right order, to the zeroing of thousands of artillery pieces, soil sampling and weather forecasting, had to be got just right. Once the vast oiled machine of such an operation was put in motion, very little could be done to adjust its course.

Most complex of all was the largest combined land, sea, and air operation ever, the Normandy invasion – D-day. The assault on fortress Europe would begin with the seizure of a bridgehead which by day 12 would be 100 kilometres wide and 100 to 110 deep. The assault would be conducted by 21st Army Group (1st American, 2nd British and 1st Canadian Armies): 32 divisions and 12 independent brigades, plus 11,000 military aircraft and 6,939 ships. Overall, the expeditionary force comprised 2,876,000 men.

At the time of the expedition, the bulk of German forces – 179 divisions – were fighting the Russians. In France, Belgium, and Holland there were 58 divisions. Most of the German divisions in France were concentrated in the Pas de Calais area, where Hitler's astrological insight and allied strategic deception suggested that the invasion would take place. Only six German divisions, including one panzer, were actually in the invasion area.[49]

The allied strategic deception, involving 'phantom' armies as far north as Scotland (suggesting a possible attack on Norway) continued to influence the war even after the allied landing and breakout from the beachheads. Hitler still believed that the real invasion would fall on the Pas de Calais, and this hindered the mobilization of reserves and delivery of a counter stroke. The strategic deception for D-day is a pre-eminent example of *maskirovka*, a concept which goes beyond deceiving the enemy, towards inducing him to do what you want. If, as Clausewitz said, 'war is an act of violence intended to compel the enemy to do our will', and if, as Sun Tzu realized, 'the acme of skill is to subdue the enemy without fighting', then if you make the enemy do what you want, you've won.[50] This is known as 'reflexive control'. It goes beyond deception, becoming manipulation, much as a conjurer manipulates his audience. In the modern military world of high technology, communications and data processing, 'reflexive control' is of critical and potentially decisive importance.

Once ashore, the allied plan of campaign involved a British advance on Caen (operation Goodwood), which would draw in most of the German forces enabling the allied right (the Americans) to envelop the enemy in a wheeling movement south and east. The battle went broadly according to plan, Patton's 3rd Army breaking through the Avranches gap on 1 August. By mid-August, British, Canadian and US forces in the north and Patton's forces sweeping round to the south had encircled the German forces between Falaise and Argentan, the 'Falaise Gap' (a 'golden bridge': see

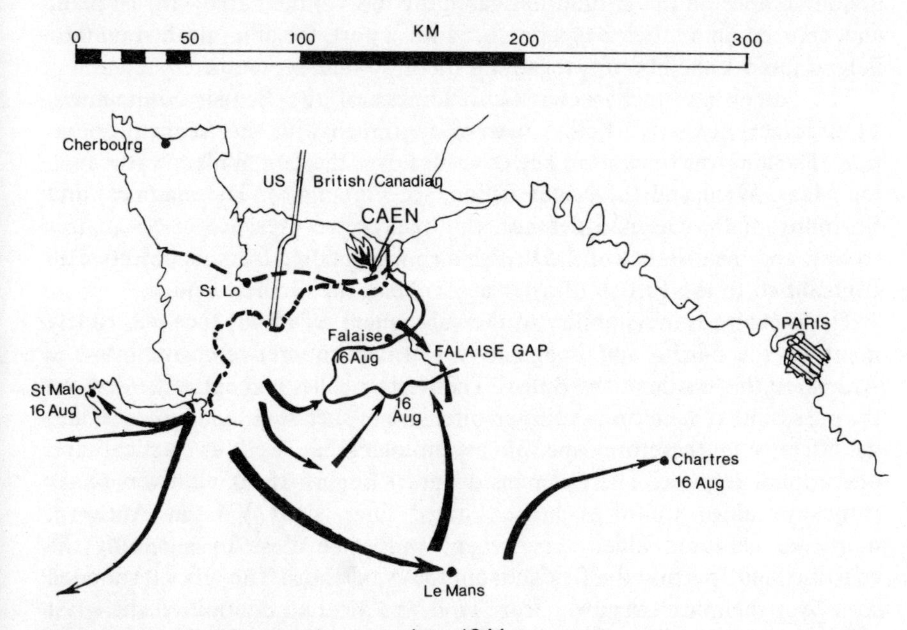

Figure 4.5 The Normandy campaign, 1944

Note: The Germans are drawn in to the 'anvil' at Caen, permitting the Americans to break through on the right flank. Arrows indicate Allied movements.

Chapter 1) being sealed on 20 August. The Normandy campaign thus presents a series of classic military manoeuvres (see Figure 4.5).

The British subsequently made their main effort along the coast, mindful of the need to capture the great port of Antwerp to provide the armies with a shorter supply route into Germany. Patton's successful pursuit stopped at the end of August because of a shortage of fuel. Logistics decreed that Antwerp should be the objective, and logistics constrained the Anglo-American advance, as it was impossible to supply the entire Anglo-American force simultaneously. The latter gave rise to the dispute over the 'broad front/narrow front' alternative.

The problem now facing the western allies was that of pursuit, a question which western armies have arguably neglected when compared with their eastern counterparts. Pursuit by all four allied armies was impossible, because of the resupply requirements and the already critical fuel situation. A strong push by part of the allied force, encircling the Ruhr might pay dividends. The British commander, Montgomery, and American General, Bradley, both favoured this plan. The US supreme commander Eisenhower preferred a slow advance on a broad front. He feared that a 'pencil-like thrust', with exposed flanks, might easily be chopped off. The choice between racing ahead with spearheads (the narrow front) and advance on a broad front is shown in Figure 4.6. The allies' air superiority and their great preponderance on the ground mitigated the risks of the narrow thrust plan, and, spurred on by their desperate need for a port, the allies pushed on into Belgium and Luxembourg, regaining them by mid September.[51]

The somewhat uncharacteristic readiness of the British commander, Montgomery, to strike boldy forward continued with the Arnhem operation. The aim was to capture key crossings over the Zuit Willemsvart canal, the Maas, Waal and the Neder Rhine (see Figure 4.6). The readiness and flexibility of the German defence, the unexpected presence of SS panzer troops, and the distance of the British airborne landing from its objective all contributed to the British disaster at Arnhem, the furthest bridge.[52]

The ponderous inevitability of the subsequent allied advance was rudely upset by the daring and imaginative German counter-offensive into the Ardennes: the 'Battle of the Bulge'. The western allies had entirely ruled out the possibility of a German counter-offensive of such scale and violence, and its effect was therefore one of psychological as well as tactical and operational surprise. The Germans dropped English-speaking diversionary troops in allied uniforms behind allied lines, fired V-1s at Antwerp, murdered captured allied servicemen, and came close to regaining the coastline and splitting the British from the Americans. The allies responded with overwhelming firepower from land and air, and committed their last reserves, the airborne divisions resting after the Arnhem operation (see Chapter 2 and Figure 3.2). With the benefit of hindsight, it seems unlikely that the Germans could have severely dislocated the allied plans (they perhaps delayed the westerns allies' advance by six weeks). Yet it was a savage lesson in the arduous art of war: even when you think you are certain to win, you get kicked hard and have to pick yourself up. Ultimately, the cost to the Germans was higher: the loss of the only relatively fresh forces that might have checked the concurrent and subsequent Russian offensive.[53]

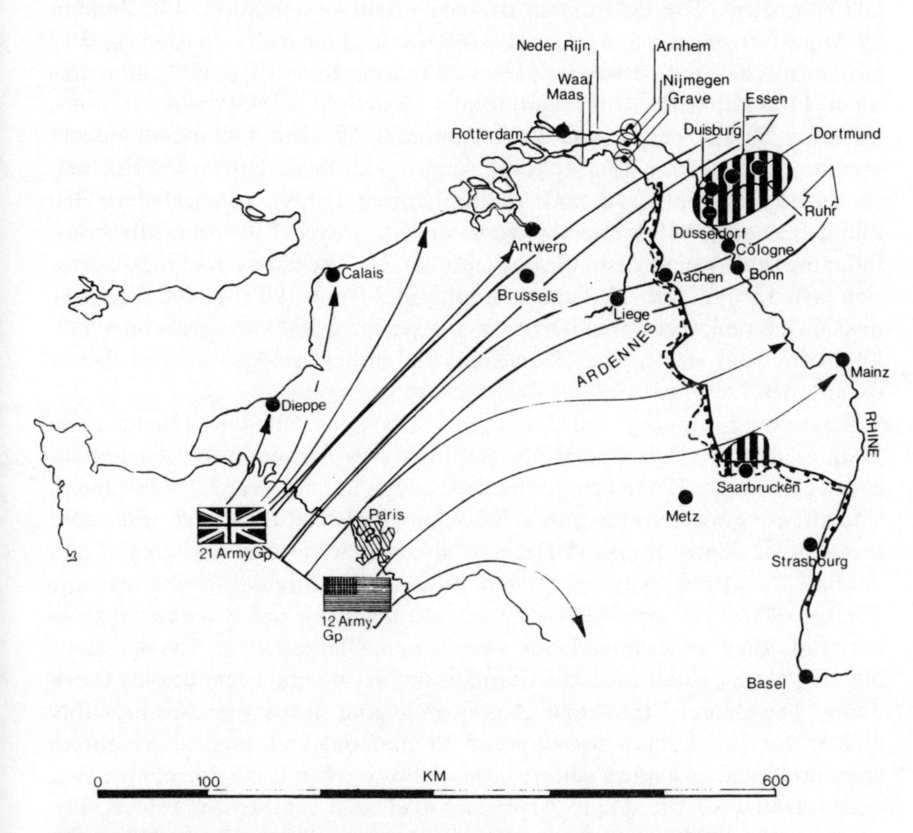

Figure 4.6 Broad and narrow fronts, north-west Europe, 1945

Note: The big white arrows show the 'narrow front' option, spearheads probing to either side of the Ruhr. The small black arrows indicate the general course of the broad front, moving relatively uniformly through France. The three key bridges involved in the Arnhem operation, giving the option of out-flanking the Siegfried line to the north, are also shown.

Major industrial areas/conurbations, the Ruhr and Saar

Siegfried Line (West Wall)

Bridges in Arnhem operation (begun 17 September 1944)

Note that the Ruhr is an area considerably larger than greater London, a variegated patchwork of cities, suburbs, industry, and open country. 'Occupying the Ruhr', then or now, would be a vast undertaking, which would easily swallow an entire army group.

The western allies' operations were relatively efficient and cost-effective. In three months, between June and September 1944, they suffered some 40,000 fatal casualties but inflicted some 700,000 losses on German forces. The Russians had tied down more German troops, but had paid dearly for so doing. During summer 1944 they launched three overlapping offensive operations on a total frontage of over 2,000 kilometres, to a depth of 500 to 600 kilometres. The Belorussian strategic offensive operation of 23 June to 29 August involved 4 Soviet fronts (166 rifle and 6 cavalry divisions and 12 tank or mechanized corps). It penetrated to a depth of 550 to 600 kilometres on a 1100-kilometre front and totally destroyed 17 German divisions, inflicting 50 per cent casualties on another 50. The Lwów-Sandomierz operation lasted from 13 July to 31 August, employed 1 front (80 rifle and six cavalry divisions, 10 tank or mechanized corps), penetrated to 350 kilometres on a 440-kilometre front, and destroyed 8 German divisions, inflicting 50 per cent casualties on another 32. The Yassy-Kishinëv operation lasted from 20 to 31 August, employed 2 fronts (90 rifle and 6 cavalry divisions, 6 tank or mechanized corps), penetrated 300 kilometres on a 500-kilometre front, destroying 18 divisions and inflicting 50 per cent casualties on another 7.

Between 12 January and 3 February 1945, the Russians executed the Vistula-Oder strategic operation, employing two fronts (109 rifle and 12 cavalry divisions, 17 tank or mechanized corps, in total over 2,200,000 men). The offensive was executed on a 500-kilometre broad front and penetrated to a similar depth. Some 35 German divisions were totally destroyed and another 25 suffered between 50 and 70 per cent casualties.[54] Although the number of troops employed on the eastern front was vastly greater than in the west, they were spread out over a much larger area. The Russians inflicted more casualties on the Germans in toto, but paid very heavily for so doing. The effect of the Anglo-American assault in the west was arguably greater for the human (as opposed to material and logistic) resources committed and casualties suffered, the more so when it is remembered that relatively few of the Anglo-Americans had seen real action before. The western allies' offensive was a striking example of the effect of scientific selection procedures and scientific training, a precedent for an age when forces involved in major conflict are less and less likely to have real combat experience. Money, firepower, grey matter and sweat saved much blood, but even then, the commitment was colossal, the consequences of failure, catastrophic and unthinkable.

The final major land operation by any participant in World War Two is of extreme interest. For a long time it was neglected in western historiography, but recently American scholars with a keen professional interest have made strenuous efforts to publicize it.[55] The Soviet operation against the Japanese Kwantung Army in Manchuria (9 to 16 August 1945) involved the co-ordinated action of three fronts on a 4,400-kilometre arc, each striking at objectives from 400 to 900 kilometres deep. Geography had a major effect on Soviet plans: in variegated terrain the three fronts could not even expect to keep in contact with each other, separated as they were by mountain ranges, desert, rivers, lakes, and marshes. The whole theatre was so remote from Moscow that instead of the fronts being controlled directly by the supreme headquarters (Stavka), a special Far East command under Marshal Vasilevskiy (a former war-time officer in the Tsarist army, to put the thing in historical perspective) was created.[56] This can be seen as a sort of prototype of the modern Theatre of Strategic Military Action (TVD; see also Chapter 5), although in the author's view this is not totally accurate. The Far Eastern theatre was framed by geographical and spatial considerations: a TVD is more of an electronic and intellectual abstraction. Nevertheless, the Far Eastern theatre is an example of the further expansion of the scope of a single, linked series of military actions.

> Unbroken strategic direction of the armed forces in a huge *Theatre of Military Operations* required the creation of the Far East Command. The deployment of an *intermediate operational-strategic level of command* was also brought about by the distance of the Theatre of Military Operations, the gigantic extent of the front over which operations were planned and the depth of operations, the diversity of operational and strategic objectives and the great number of arms and services involved.[57]

If this theatre of military operations was regarded as operational-strategic, a modern one, certainly in Europe, would be purely strategic.

The strategic and operational deception employed in the Manchurian offensive also parallel those that would apply in a modern Theatre Strategic Operation. The Japanese assessed that the scale of the deployment precluded the Russians attacking before the end of 1945 or early 1946; speed itself creates surprise. The Soviet attack was probably fortunately timed (Stalin had promised to attack three months after the end of war in Europe) and not precipitated by the explosion of the Hiroshima bomb on 6 August. On 8 August, the Japanese ambassador in Moscow was informed that a state of war would exist between the two countries from the following day.[58] This might appear to prejudice surprise, and may have been necessitated by the need to get in on the act diplomatically in case the Japanese surrendered to the other allies, although that was not considered likely. However, the Japanese had no idea that the morning would bring an attack of such ferocity and swiftness.

The main move to the Far East was controlled by a very small circle of staff officers, and telephone conversation and even written correspondence concerning the plan was forbidden. Many high-ranking officers moved into the theatre wearing junior officer rank and under assumed names. Within the theatre itself, formations occupied concentration areas far from the border: 5th Army, for example, arrived from East Prussia at concentration areas 100 to 120 kilometres from the Manchurian frontier. Its final movement to assault positions occurred between 1 and 6 August. At front headquarters, the plan and objectives were known only to the front commander, his political officer, chief of staff, and chief of operations. Movement only took place at night, and army-level artillery units, one of the major indicators, often moved considerable distances into their waiting areas in single nocturnal bounds. Strict radio silence was maintained except for normal border guards traffic, and elaborate engineering works provided physical screening for the move forward. Engineers erected masking walls and camouflaged overhead covers along roads under Japanese observation. Fifth Army's zone alone contained 18 kilometres of screening walls. Normal peacetime routine was ostensibly maintained, with troops being allowed leave and border troops carried on normal duties, including harvesting crops. The Russians were therefore able to conceal not only the general scale of the deployment, but also the detailed timing and axes of their attack.

The effect of this was compounded when the Russians launched their attack in the dark, in many cases through heavy August rain, and across terrain which the Japanese considered impassable. Many of the eleven armies constituting the three fronts moved across terrain which, like the Ardennes in 1940, was considered a barrier, especially to armour, and was therefore left undefended. Leading with mobile groups or forward detachments at every level, the Soviets bypassed many of the strong points but, aware of the damage that these could do in disrupting communications, employed follow-on forces to isolate and contain them. Finally, many of the Soviet formations attacked without the customary artillery and air preparation, advanced infantry battalions taking the Japanese completely by surprise. 'Japanese miscalculation, combined with the Soviets' ability to achieve strategic surprise and to use imaginative operational and tactical techniques, produced the rapid and utter defeat of the Kwantung Army.'[59]

The more varied nature of World War Two operations and the diversity of its theatres has given greater prominence to some of its commanders and the role of inspired leadership. Seen against the back-cloth of the war as a whole, the importance of certain individuals is liable to be overrated. There can be little doubt, for example, that Rommel was a far more talented, indeed, brilliant, commander than Montgomery, but in the end he lost. Against overwhelming superiority in men and materiel, there is only so much that you can do. However, World War Two does underline one important fact about military leadership: that in order to get the best out of the fighting men of a given nation, a commander has to understand and utilize that nation's characteristics.

Rommel's style of leadership encapsulated many of the traits of the German soldier: highly professional, studious, and academic, but also tough and daring. Montgomery similarly appealed to many of the instincts of the British. A certain rugged individualism, and also a certain stolidity. One of the most charismatic and successful Army commanders was the American George S. Patton. Patton appealed to numerous American instincts: a certain theatricality, glitz, and razzamatazz, combined with a thinly disguised wild west penchant for violence. Patton was a highly sensitive, artistic man, but he concealed this under a most successful veneer. He released some of the underlying characteristics of the men under his command. 'The only way you can win in a war is to attack and keep on attacking, and after you have done that, keep attacking some more.'[60] He was probably the best western general of the war. 'The speed and brilliancy of your achievements is unsurpassed in military history.'[61]

The Russians come over as being very like the British. The similarity between Russian and British soldiers is, indeed, pronounced throughout recent military history. In spite of the political propaganda, the Russian soldier seems to have been as cynical about fighting for ideals as his British counterpart. A humour as dry as the British was, however, reinforced by burning hatred for the Germans because of what they had done. Russian generals, particularly Zhukov, seem to have understood this very well. It is perhaps no coincidence that the British and the Russians are the last two surviving imperial armies.

Just as the heroic warrior-commander seems to have largely disappeared, so too has the heroic subordinate, distinguished both by courage and individual skill at arms. The emergence of elite, specialized forces with clearly defined roles can be seen both as a function of the collectivization, democratization, and mechanization of most destruction in war, and of a human need to individualize and personalize this process. The elite units (paratroops, commandos, rangers, the British Special Air Service, the Waffen-SS) were not popular with the other professionals in World War Two, however. As Michael Howard observed,

> They were believed to divert useful manpower to activities that were intermittent, usually marginal and invariably overpublicized. Their wartime operations were often seen as luxuries sustained by the patient and unspectacular efforts of their more self-effacing colleagues, and their activities did not appear to make much contribution to the massive and collective destruction of which war now inevitably consisted.[62]

DEVELOPMENTS IN MAJOR LAND WARFARE SINCE WORLD WAR TWO

Since World War Two ended in 1945, major land conflict and large-scale military operations have been confined to Asia, a fact which underlines the pertinence of studying Asia's military past (Chapter 6). With the exception of the Iran–Iraq war, which began in 1980, all the conflicts have been the result of dissatisfaction at the settlements reached after the world war: the creation of the state of Israel, which led to three major wars (1956, 1967, 1973), the division of Korea after the Japanese were defeated, the division of Vietnam, the situation in Malaya, the division of India and Pakistan, and so on. The Iran–Iraq war has been the only 'new' major war.

The most striking technological innovation in weaponry, the nuclear warhead or bomb, has not been used since 1945. It has influenced the conduct of one major conflict, arguably the largest traditional land conflict, the Korean war. Korea underlines traditional military principles, notably the efficacy of envelopment as spectacularly demonstrated by MacArthur's landing at Inchon. South Korean and UN troops had been pushed back to a small perimeter around Pusan in the south east of the peninsula, when a major landing took place on the west coast, cutting North Korean lines of communication and threatening their rear. The result was a precipitate North Korean withdrawal almost to the Chinese border. This, in turn, led to Chinese intervention, and UN forces were once again pushed back to the thirty-eighth parallel. Korea is a classic example both of principles of the art of war and of the conduct of limited war, which almost became unlimited. The Chinese intervention was a major point of escalation, and UN forces kept the possibility of using nuclear weapons very much in mind, and used it as a threat to press the north to accept terms. Operation Hudson Harbour, in which single American B-29 bombers would approach North Korean installations and release single, very large conventional bombs, the nature of which was not apparent until they exploded, was a remarkable exercise in controlled escalation, and a lesson in the difficulties of 'verification' – of telling whether a given weapons system has a conventional or nuclear warhead. Korea is probably the nearest precedent we have to a possible armed confrontation between east and west.[63]

DEVELOPMENTS IN MILITARY ORGANIZATION SINCE WORLD WAR TWO

After the Korean war, the US army had to develop doctrine and organization which would allow it to operate on battlefields ranging from low-intensity to nuclear. This led to the development of the pentomic division. The need for dispersion on the nuclear battlefield led to five subordinate units being substituted for the familiar three, hence the name: penta – five, -omic – implying its ambiguous nuclear or non nuclear role (although it was more suited to the latter). Each of the five subunits was therefore larger than a battalion but smaller than a regiment.[64] The pentomic division was subsequently replaced by the ROAD (Reorganization Objectives Army Division), from the mid 1960s. This was designed to fight across the whole spectrum of war from guerrilla operations to nuclear, and reflected the Kennedy administration's new concept of flexible response. The ROAD division was able to 'task organize' units at any level. Armoured, mechanized, infantry, airborne and, later, air-mobile divisions could be created from the basic building blocks, depending on the expected threat. In practice, of course, this organization suffered from the same weakness as the World War Two US divisions: commanders tended to become attached to the specialized units temporarily attached to them, and tried to keep them once they had got used to operating together. Nevertheless, the ROAD division gave the US army a mixture of control and flexibility which the pentomic structure had lacked.[65]

Armies have continued to experiment with different organizations ceaselessly to the present day. The differing requirements of nuclear, heavy conventional war, in Europe, and more diffuse operations elsewhere have contributed to the continued dissatisfaction with any one force mix. As Freytag-Loringhoven pointed out, 'no organization can possibly cover all the possible contingencies of war, and . . . therefore it is of the first importance to make it as elastic and adaptable as possible'.[66] According to Freytag, in World War One divisions in the German Army became 'strategic units' (he would probably later have used the word 'operational'); army corps became 'army groups' and the number of their divisions underwent constant fluctuation. Napoleon, similarly, had never hesitated to alter the number of divisions in his army corps.[67] In the 1970s, the British experimented with abolishing brigades, splitting the division instead into variable 'task forces', but then reintroduced brigades. The structure of the 'light' divisions required for possible intervention outside Europe and the 'heavy' divisions needed for major mechanized war has been the subject of constant debate in the American military press.

Soviet military organization underwent constant evolution in World War Two: tank and mechanized corps were reintroduced; they were not always big enough for independent missions (see Chapter 5) so tank armies were created; rifle corps were reintroduced and acquired greater permanence. After the war, corps and brigades were done away with (though not finally until 1953), and a fairly rigid structure of armies, divisions, and regiments (the Radzievskiy model) was adopted. This was ideal for nuclear war and, since the operational and tactical organization was the same as the logistic and administrative organization, also suited a peacetime army of occupation. By the early 1980s there was evidence that experiments with a revived corps-brigade structure were taking place. Changed doctrine – the need to be able to switch rapidly from the defensive to the offensive and back again, and the complex and unpredictable nature of protracted conventional operations, made a model more like the World War Two pattern appropriate. The new structure did not of itself overcome all the Russians' problems, but it arguably provided a framework better able to assimilate key new pieces of equipment: C^3I, helicopters and armoured vehicles, as they became available. There were clear lessons to be learned from the Soviet mechanized corps of June 1941, which had inadequate command and control, having only the same number of radio sets as the pre-1939 corps of half their size, and nothing like their statutory complement of the new T-34 and KV tanks.[68]

Similar experiments have continued at lower levels of organization and in all armies, the trend, as throughout this century, being the progressive integration of different arms of service at lower and lower levels.

HELICOPTERS (THE 'ROTARY-WING REVOLUTION')[69]

In terms of the mechanics of land warfare, the most significant innovation since World War Two has been the helicopter. It has overtaken the fixed-wing aircraft for battlefield transportation and close air support, and has given new meaning to the term 'air-land battle'. Heliborne air assault teams, rather than cumbrous and vulnerable fleets of transport aircraft, gliders and dangling parachutists, are the modern manifestation of the 1930s visionaries' ideas. The helicoper fits uneasily into independent air forces' structures, but has proved an outstanding army weapons platform and transport system. However air forces have tended to oppose any expansion of army aviation as a challenge to their own position.

Helicopters had been used to a limited extent by the Americans in Korea and during the late 1950s the US Marine Corps experimented with helicopters, primarily for transport. In 1962, the US Secretary of Defense, McNamara directed the US Army to study the use of aviation to improve the tactical mobility of ground forces. General Howze recommended the formation of air assault divisions, air cavalry brigades for screening and covering force roles, and air transport brigades to lift conventional divisions. An air assault division could manoeuvre to attack a conventionally equipped enemy from various directions. Practice lagged far behind theory, but from 1963 to 1965 the 11th Air Assault Division at Fort Benning conducted experiments with air assault troops. Artillery, aviation, and infantry had to co-operate closely in an assault. Initial suppression was effected by artillery and air force aircraft, and then helicopter gunships would take over with more accurate direct-fire suppression while other helicopters landed air-mobile troops. At the same time, the US began to employ helicopter units to support South Vietnamese forces, initially as transport aircraft but by 1964 for fire support as well. In 1965 the 11th Air Assault Division became the 1st Cavalry Division (Airmobile) and was deployed to Vietnam.

The key to the use of helicopters was close co-operation and training within the same unit. Using helicopters to lift or give fire support to another unit was vastly less efficient, another reminder of the integrated nature of combined-arms warfare. The helicopter also brought into focus the nature of operations without a fixed front line, which was particularly marked in Korea and Vietnam, and may be very relevant to the nature of future war. In addition to ground artillery firing from a more secure rear area and direct fire support from helicopters, artillery pieces were also lifted to new locations further forward, again emphasizing the integrated air-land battle with its widely spaced 'focal' positions. The Soviet Union and western European nations all took note of the Americans' use of helicopters in Vietnam. By the early 1980s assault helicopters (primarily anti-tank) were a key part of the ground forces' structures of all major land powers. For example, the West German III corps committed a regiment of anti-tank helicopters in close formation as part of an exercise in 1983.[70]

THREE EXAMPLES OF AIR-LAND BATTLE

Lam Son 719 (1971)

This operation was executed by US and South Vietnamese forces against the growingly confident and ingenious forces of the north under the cautious and cunning Giap (Chapter 6). The fact that some have considered it a failure does not make it any less instructive with regard to the problems of conducting the air-land battle and the merging of new technology (helicopters) with the clogging terrain and timeless verities of war. The sweating mountain forests on the Vietnam-Laos border imposed a unique character on operations, like any distinctive terrain. General Dave Palmer described the area on the Vietnam-Laos border as a 'forbidding verdant fastness over which the communists had wisely and energetically superimposed a superb, in depth defensive system'.[71] The aim of the operation (see Figure 4.7) was to destroy the North Vietnamese base area in Laos, especially the numerous logistic installations around Muang Xepon (Tchepone), thus forestalling any North Vietnamese offensive to conquer the northern provinces of the Republic of Vietnam. It was a spoiling attack in four phases: first, US forces would seize the approaches inside South Vietnam leading to the Laotian border. Then, the 1st South Vietnamese Corps would attack along highway 9 to Xepon in a series of leap-frogging air assaults and armoured advances. Third, the South Vietnamese forces would carry out search and destroy missions in area 604. Depending on opportunity, the fourth phase was either a withdrawal along highway 9 or further destructive missions in area 611. The operation was named after the defeat of the Chinese invasion in 1427 (see Chapter 6). The operation has been described as one of mid-intensity war, perhaps in recognition of the heavy hand of political restraint which constantly impinged on all the Vietnam fighting. For the first time, South Vietnamese forces were employed on a large scale without American ground forces, who were forbidden to enter Laos, or even any American advisers. On the other hand, the Americans dominated the skies and flew thousands of missions in support. In terms of numbers engaged and casualties, this was undoubtedly

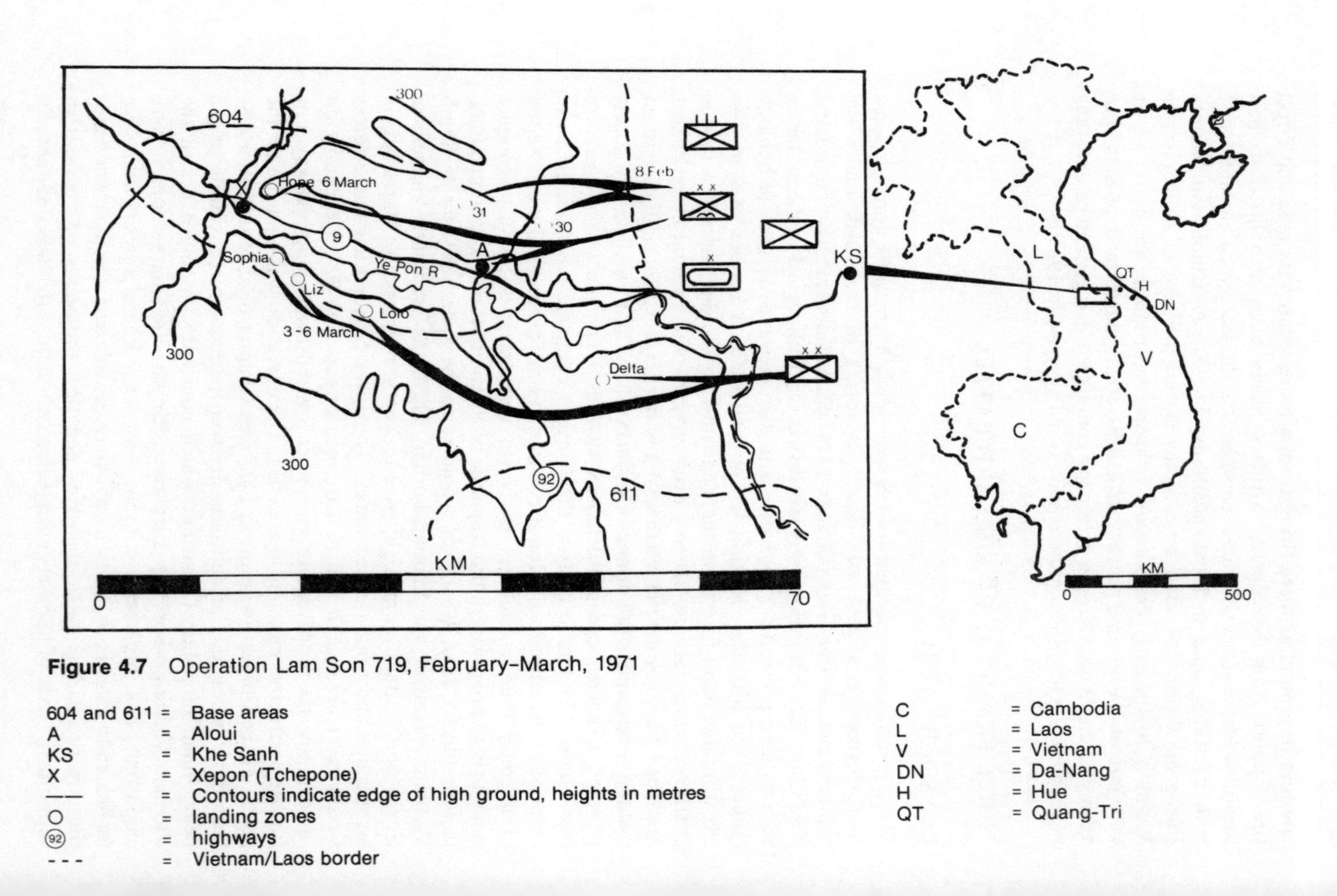

Figure 4.7 Operation Lam Son 719, February–March, 1971

604 and 611	= Base areas	C	= Cambodia
A	= Aloui	L	= Laos
KS	= Khe Sanh	V	= Vietnam
X	= Xepon (Tchepone)	DN	= Da-Nang
——	= Contours indicate edge of high ground, heights in metres	H	= Hue
○	= landing zones	QT	= Quang-Tri
(92)	= highways		
- - -	= Vietnam/Laos border		

an operation of major war, and is particularly interesting as an air-land battle.

The operation began at midnight on 30 January 1971, the Americans firing heavy artillery concentrations from positions inside Vietnam. Meanwhile, military engineers began to make bases abandoned after the 1968 Khe Sanh campaign fit to mount air sorties, and Khe Sanh itself, the American Verdun, was re-opened as a forward base. It was estimated that this would take four days: in fact, it took over a week longer because of weather, mines and unexploded shells. Meanwhile, the operation had to go on and the South Vietnamese ground forces moved into Laos on 8 February.

The shape of the land inevitably channels an approach to Xepon along the Ye Pon river valley. The valley was so narrow that there was hardly room for the leading three armoured squadrons to manoeuvre. Meanwhile, cloud swathed the high ground on either side of the valley, naturally funnelled helicopters into it, and the silver gleam of the river was the most obvious navigation landmark. The North Vietnamese knew this well, and the nineteen anti-aircraft artillery battalions in the area were sited with this in mind, and also around the few obvious helicopter landing sites, which were self-evident amidst the dizzy vista of jungle and mountain. Artillery fire on all likely helicopter landing zones was pre-planned. As a result the Americans came up against much stiffer fire from the ground than they had expected, and helicopter gunships had to be allocated to escort even casualty evacuation helicopters in order to suppress ground fire, reducing their availability for other tasks. When the South Vietnamese forces landed, they were savagely counter-attacked, and one landing zone, 31, was overrun by North Vietnamese T-34 tanks which, seldom having faced heavy ordnance on the other side before, the air-landed forces were ill-equipped to counter. The American Army Aviation commander for the operation concluded that more emphasis needed to be placed on the anti-tank helicopter.

After several weeks of limited success, the South Vietnamese commander abandoned plans for a straight ground advance west of Aloui, and instead set up helicopter bases for an air assault towards Xepon. On 6 March two battalions carried out a *desant* (a Russian term which summarizes the type of manoeuvre graphically) into landing zone Hope, and in spite of enemy anti-air defences lost only one helicopter out of 120 employed. This and other air assaults were carefully planned, co-ordinating various layers of air elements: strategic bombers raining bombs from miles high, tactical bombers swathing anti-aircraft positions with fire, and helicopters and air-delivered smokescreens to guard the air-mobile infantry as they actually landed.

Meanwhile, general Giap threw everything he could into an attempt to destroy the raiding force: 36,000 North Vietnamese troops including two armoured regiments with T-34 tanks attacked the penetration. In order to escape American bombing, the North Vietnamese hugged the South Vietnamese positions, accepting terrible casualties from their ground fire instead: it is estimated that 10,000 to 15,000 were killed. The ARVN forces accomplished their mission, destroying the support facilities around Xepon before withdrawing. They thus delayed a major North Vietnamese offensive for a year, but two crack divisions, the 1st Division and the Airborne Division, had been terribly mauled in the process. Expert opinion is therefore sharply divided on the success of the operation: one authority considered that the South Vietnamese forces were 'routed': another that they broadly succeeded in their missions and had a major effect in delaying further North Vietnamese advances.[72]

No US troops were involved on the ground, but the Americans provided lavish air support for Lam Son 719. They flew 160,000 air sorties, losing 107 helicopters to enemy fire. There were over 10,000 strikes by tactical fixed-wing aircraft, and the B-52 heavy bombers dropped 46,000 tons of bombs.

Many observers have cited Lam Son 719 as proof that air-mobile operations are too vulnerable to enemy air defence and counter-attack on the ground and could not be carried out in mechanized wars. Yet the loss of 107 helicopters, although high, must be seen in context. Out of 160,000 sorties over nearly two months (the last South Vietnamese troops fell back over the border on 24 March), that is not unacceptable, bearing in mind that war is a horribly expensive, bloody business. The terrain neutralized many of the advantages of the air-mobile force, allowing the defenders to concentrate on known axes of advance. This and low cloud over the land forced the helicopters up, flying at perhaps 1000 metres and more. It is ironic that although in some ways helicopters free military activity from the constraints of terrain, in others they are more dependent on it. Lam Son 719 did not prove conclusively that air-mobile operations are impossible. The operation has many lessons for future war, especially as the North Vietnamese were well provided with organic air defences and artillery, and both the Americans and Russians have studied it in developing their own doctrine for employing the air elements of ground forces. It was the first great air-land battle using helicopters in major war.[73]

The Israeli Counter-Attack Across the Suez Canal, 1973

Egyptian forces launched a surprise attack across the Suez Canal into Israeli-occupied Sinai on Saturday 6 October 1973. Simultaneously, Syrian forces attacked Israel in the north. By dusk on 6 October, the Egyptians had ten bridges across the canal thanks in part to the ingenious idea of using high-pressure water jets to hose out slots in the steep canal bank. During the following night some 500 Egyptian tanks crossed the canal. In spite of total surprise and great success in repelling the Israeli armoured counter-attack, the Egyptians did not press on far into Sinai, but sat tight, their air defence missiles shielding them against Israeli air power. While Israeli forces in the north battled desperately to bring the Syrians to a halt, thoughts turned to a major counter-attack (in the precise sense of a move to recover lost ground vital to the defence) in Sinai.[74]

The Egyptians continued to build up their forces prior to the next stage of their advance. In contrast to the near linear attrition battle raging in the north, in Sinai there might still be room for a classic operational-level revolving-door movement, swinging round behind the Egyptians and knocking them off balance. The plan was the brainchild of the dynamic Major-General Arik Sharon, and had been prepared long before the war. It involved crossing the Suez canal, thus getting behind the Egyptians and also exploiting the flat, brown terrain on the Egyptian side which was ideal for fast-moving armoured operations favoured by the Israelis. The plan as it eventually crystallized also hinged on overrunning Egyptian surface-to-air missile (SAM) sites thus creating a corridor which opened the way to the Israeli air force and restored freedom of the sky. This is a good example of how ground and air systems and fighting interact in modern air-land warfare.

Sharon had selected a crossing point about 20 kilometres south of Ismailia in the vicinity of Deversoir (see Figure 4.8). Here, there are two side roads off the main north–south canal-side road which runs between Lake Timsah and the Bitter Lakes, enabling vehicles to get right up to the canal. At this point the walls of the canal bank had been weakened, and the section marked with red bricks. A parking area, protected by sand revetments had also been constructed so that vehicles could be marshalled in safety prior to a crossing. In fact, this is not far from a place where the Egyptians had had difficulty flattening the canal bank with their high-pressure water jets, and if they had found the weakened section their progress there would have been faster, and Sharon's plan doomed. Once again, ground was absolutely central to the conduct of operations.

Sharon was given three weak armoured brigades, totalling some 200 tanks, plus a force of engineers with earth-moving vehicles, assault boats, and bridging equipment. The way to the crossing place was blocked by the Egyptian 21st Armoured Division, and Sharon had to get round behind them. He planned to use one of his armoured brigades to seize the main north–south road and another to carry out a feint from the east. The engineers would then pass through and secure a crossing. Finally, the 3rd Armoured Brigade would cross the canal, cut the Egyptian 3rd Army's lines of communications and destroy the SAM sites, enabling air cover to be brought in and assisting air operations over Sinai generally.

As the autumn night began to fall at 1800 on 14 October operation Gazelle began. Sharon's force slipped in along the boundary of the 2nd and 3rd Egyptian armies, and drove north along the lake shore, meeting no resistance. However, at the Chinese Farm, an irrigation project so-named

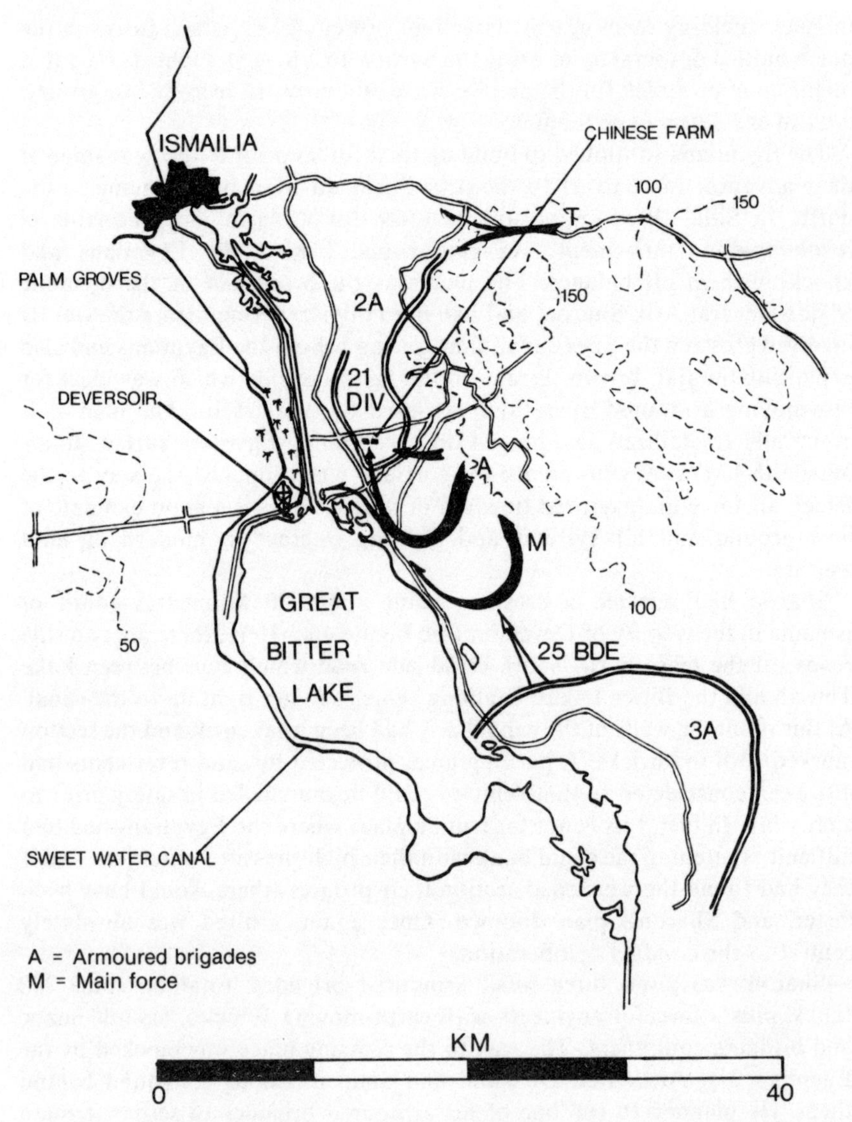

Figure 4.8 The Israeli counter-attack towards the Suez Canal, October 1973

Note: Israeli thrusts in black, Egyptian formations indicated by double lines.

because much of the equipment was of Japanese origin, the 14th Armoured Brigade ran into heavy Egyptian opposition and fought a savage battle to try to secure the road junction area. At this point the seconds were ticking away and it looked as if the plan might fail. A task force did break through to the canal, and the first Israelis, including Sharon, crossed the canal in rubber boats at about 0100. The main crossing force had to bypass the fighting at Chinese Farm, rather than driving straight down the road as planned, so they did not begin to arrive until 0300, and the barges to ferry the tanks across the canal did not get there until 0500, dangerously behind schedule. Dawn brought Egyptian artillery fire, which was directed at the crossing point, and two ferries were hit. In the meantime, the two armoured brigades to the east were fighting like mad to keep open the corridor to the crossing while Sharon was on the other side of the canal with a small force at the end of a tenuous lifeline. Had the Egyptians launched a counter-attack on the west bank of the canal at this time, there is no doubt that the forward detachment would have been wiped out but, fortunately for the Israelis, the Egyptians had been taken completely by surprise and it was not until late morning that General Ismail, in charge of the Egyptian land operations, was told of the crossing. Even then, the Egyptians under-estimated the size and importance of the crossing force. The situation remained precarious as the Egyptians attacked from the east in the evening, blocking routes to the crossing point but, on the other hand, Egyptian resistance at Chinese Farm began to crumble. On the next day Israeli engineers were able to start work on the first of three bridges across the canal. Meanwhile, rather than dig in on the far side of the canal, making himself a sitting target, Sharon split up his small force into raiding groups which ranged far and wide searching for SAM sites, fuel and ammunition dumps (see Figure 4.9). The destruction of four SAM sites was critically important in parting the sky for Israeli Phantoms. Being oriented skywards, such positions are particularly vulnerable to ground attack, although in one case the Egyptians fired a big SA4 anti-aircraft missile straight at Israeli ground forces over open sights. The effect of Sharon's armour swinging north and Bren's and Magen's south was disproportionate, forcing Ismail to consider pulling back some of the 500 tanks he had just put across the canal with such skill, and cutting of the Third Army's communications. It also caused the Egyptians' Soviet mentors considerable concern as to the way the war was going, in spite of Egyptian attempts to play down its significance. It was a classic eccentric enveloping movement (see Chapter 1). Nevertheless, even when apparently encircled and cut off, the Egyptians fought with unex-pected determination and tenacity, notably as the Israelis attacked Suez, a reminder of how troops who keep their heads and fight on can disrupt plans for swift encircling moves severely, and possibly critically. Meanwhile, the Egyptians pounded the Israeli bridgehead across the canal with artillery and also because of their weakness in aircraft, big Soviet FROG surface-to-surface rockets.[75]

The Israeli counter-attack on the Egyptian side of the canal thus highlights many of the factors which have been prominent throughout the history of war. It illustrates the criticality of ground, and of time; the integrated and mutually supportive nature of ground and air operations; the

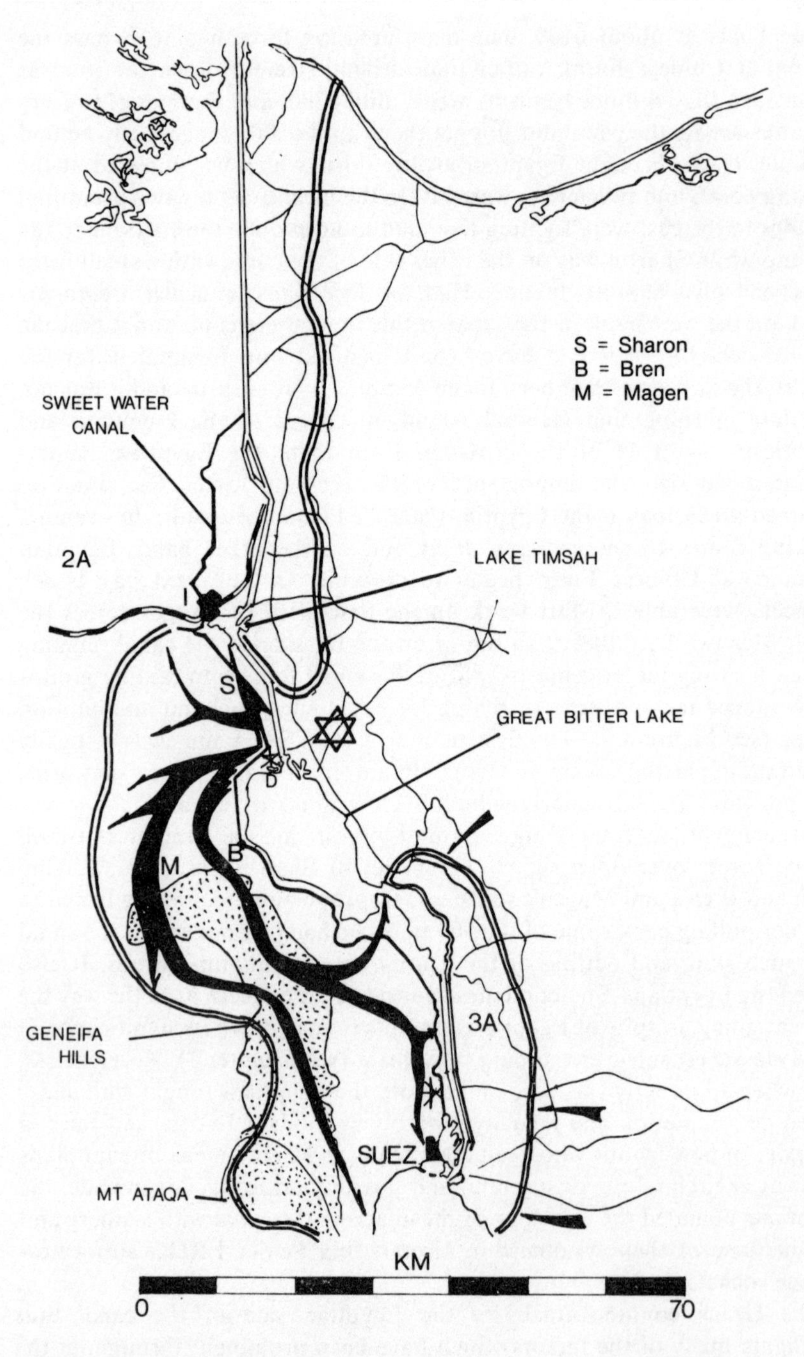

Figure 4.9 The Israeli breakout on the African side of the canal

Note: Double line is ceasfire line

problem of slipping a forward detachment behind the forward enemy troops and the use of formation boundaries to help. It underlines the timeless importance of the military engineer and the problems of overcoming dry and, in particular, wet gaps. It highlights the position of a forward detachment in the enemy depth and the difficulty of resupplying it, the particular value of movement and sustained aggressiveness and initiative in maximizing such a force's potential and, indeed, assuring its survival. Sharon's forward detachment, initially of three brigades, was tasked to seize particular objectives, especially those which tend to be located in the enemy depth such as SAM sites, rather than to destroy enemy forces *per se*: indeed, it sought to avoid contact wherever possible. A major water obstacle was involved, as it probably would be in Europe. Finally, it exercised influence disproportionate to its size in tipping the enemy off balance operationally and, by appearing on his home territory, upsetting him politically. It also underlines the need for the high command to react very quickly indeed to the insertion of such a force.

Russian Roulette: the Jijiga Operation, Ogaden, 1977

Different again, bare and starved and also mountainous, is the tortured land of Ethiopia. Here in 1977 Somalian forces were pushing Soviet-backed Ethiopian forces back to the north. The Russians sent large quantities of modern weapons and equipment, Cuban forces, and Soviet advisers. One of the latter was General Vasiliy Ivanovich Petrov, a dynamic fireball who later briefly commanded the Soviet ground forces. Another, who arrived in mid-1977, was General Barisov, who was to command the Ethiopian counter-offensive. The Soviet equipment began to arrive early in September along with 6,000 Cuban troops. Meanwhile, Cuban artillery began to arrive from South Yemen and the Ethiopian air force attained control of the skies. It took a couple of months for the Ethiopians to get used to the equipment and for the Cubans to acclimatize, but by early 1978 they were ready to launch a counter-attack to recapture the strategic base of Jijiga, and destroy the Somalis there having cut off their escape route to the Ogaden plain (see Figure 4.10). The plan was elaborate and made extensive use of air mobility, and was most probably the work of the Soviet general, Barisov. It is therefore extremely interesting as an example of a classic enveloping movement using air-mobile forces *à la russe*. The plan was:

- To persuade the enemy that their main thrust would come from the west down the Kara Mandah pass.
- To make a sweeping outflanking move with a reinforced Cuban tank brigade north of the mountains.
- To establish a staging post north-east of the mountains to facilitate helicopter resupply of the outflanking force.

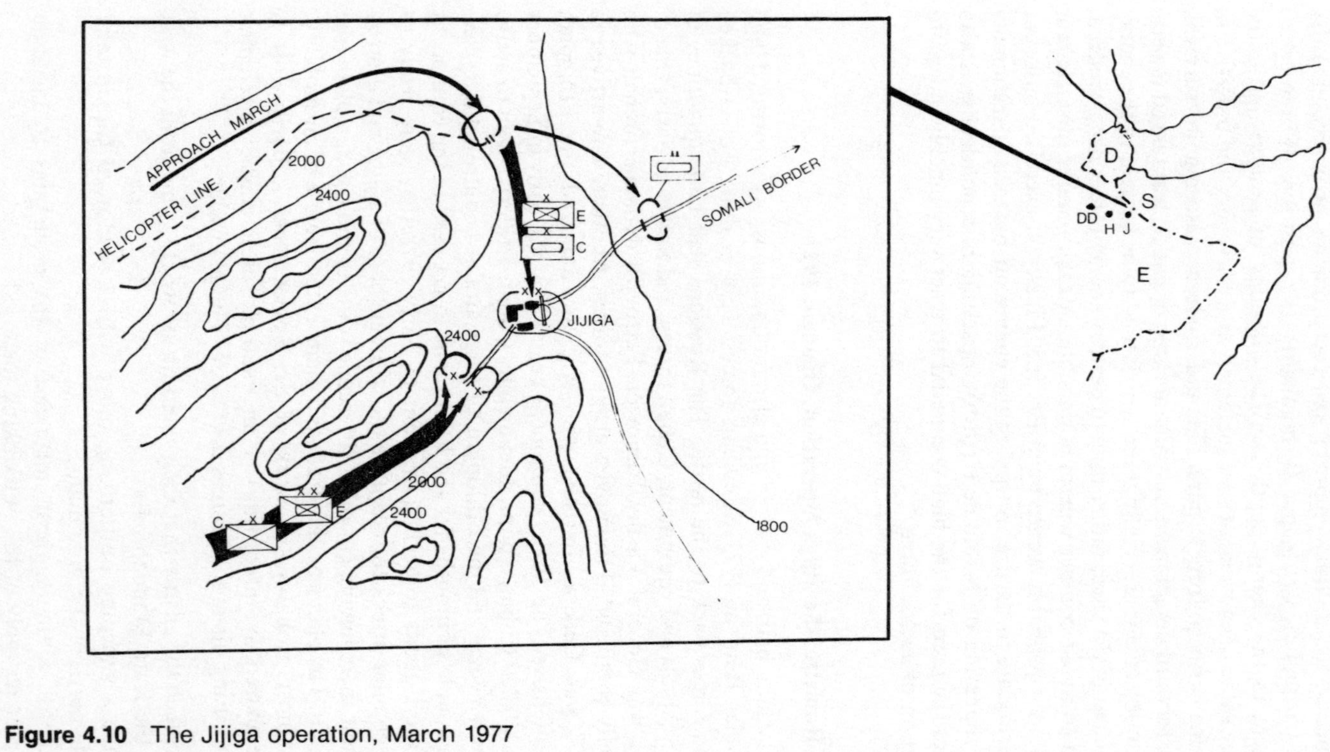

Figure 4.10 The Jijiga operation, March 1977

C = Cuban
E = Ethiopian
D = Djibouti
DD = Diredawa
E = Ethiopia
H = Hara
J = Jijiga
S = Somalia

- To make a strong diversionary attack down the pass with the Ethiopian 3rd Infantry Division supported by a Cuban mechanized brigade.
- From the staging point protected by two companies of PT-76 air-mobile tanks, a heliborne battalion would cut off Somali withdrawal routes.
- To destroy the Somali forces in a pincer movement from north and west.

Everything went according to plan. The 3rd Ethiopian Division reached its assault position on 1 March and began skirmishing to keep the Somalis looking west while the Cubans made their 100-kilometre air-assisted flanking march and reached the staging point 30 kilometres north of Jijiga during the night of 2–3 March. On 4 March the outflanking force attacked Jijiga from the north while the Ethiopian 3rd Division stormed the pass. Of 6 Somali brigades in the area, 4 were completely destroyed.[76] The operation reaffirms the classic truth that the best way of taking a bridge or a mountain pass is both ends at once; the danger of setting up defence in mountain passes thus creating a trap for oneself; the value of air-mobile forces in executing envelopments; and the use of air-landed forces to establish staging posts to accelerate a ground forces advance (a device used by the Americans in Vietnam and the Russians in the Panjsher valley in Afghanistan). It also shows the ability of Russian generals to execute the elaborate encircling moves which their military art stresses, even with disparate and, one suspects, relatively poorly trained forces. Vasiliy Petrov no doubt took note. On the other hand, both this operation and Lam Son 719 were carried out with complete air superiority, in the case of Jijiga against an opponent without strong anti-air defences. They also took place in a situation where the rest of the battlefront was relatively passive and all resources could be concentrated on one operation. There was plenty of time to plan and co-ordinate the complex interaction and leap-frogging of land and air-mobile forces. On a European battlefield, if not an Asian one, erupting in a thousand interacting combats, such carefully prepared operations would surely be very rare and confined to the opening phases, or after a major pause.

FROM PAST TO FUTURE

We thus reach the point where we stand too close to recent military operations to get a clear historical picture: where events of the recent past are intertwined with the politics and technologies of the present. The Arab-Israeli war of 1973 is still used as a database for predicting the nature of future conflict (as, indeed, the Russians use World War Two), but even that is fast receding. We shall return to the question of historical time in the concluding chapter. Before that, however, we focus, as with the 'mouse' of a computer, on two particular areas of recent and current concern where military history provides vivid illumination of general and particular points.

In the first case study, great detail is necessary to prove the points and to reveal just how extensive a database is available to the researcher who is prepared to do some homework, and how it can be used. In the second, somewhat broader strokes portray the evolution of warfare in particular ways in the special circumstances of Asia.

Chapter five

Case study one: *Corps volant* to OMG: The practical utility of military history

WESTERN PERCEPTION OF THE OMG THREAT, 1982–9

The early to mid-1980s saw a general resurgence of interest in the 'conventional option' in the disastrous and unthinkable event of any future major conflict between east and west. This interest was fuelled, perhaps, by the Iran–Iraq war and Russia's 1979–88 war in Afghanistan. Both western (particularly American) and Soviet strategies for major war acquired high profiles with well publicized schemes for defeating the other side. Naturally, these doctrines interacted. A cardinal component was the west's identification of the Soviet Operational Manoeuvre Group (OMG). This has proved to be such a central touchstone and catalyst for modern military thinking that it is worthy of special consideration. In fact, formations such as this have occupied a prominent place in the military theory and practice of all nations, not just Russia, for a very long time, and are ultimately only manifestations of classic principles of war. The OMG provides an outstanding illustration of the practical usefulness, indeed, the essentiality, of military history.

By the beginning of the 1980s it had become clear to western analysts that the Soviet High Command was intent on acquiring the ability to execute a swift offensive in the Western Theatre of Military Operations (TVD) and adjoining TVDs without using nuclear weapons and before NATO had a chance to use theirs. Something significant was happening within Soviet theatre ground and air forces, but it was not clear what. With the benefit of hindsight, it seems that the Russians had decided that the Radzievskiy army-division-regiment structure, designed for nuclear war, was too cumbersome and inflexible for complex and possibly prolonged conventional operations, and were experimenting with task-organized offshoots which had much in common with the 'mobile groups' or 'cavalry-mechanized groups' of World War Two. That was only a partial solution: since then, they have moved towards organizing all their theatre land-air forces into more flexible task-oriented groups.[1] However, the sensational snapshot obscured the plot of the film.

During 1981, American analysts including John Hines noticed that a number of Polish articles were expressing particular interest in the Soviet mobile groups of World War Two. Between February and April, 1982, an American team carried out extensive briefings and shared data with other NATO nations under the auspices of Supreme Headquarters, Allied Powers Europe, at 'SHAPEX-82'.[2] The idea was transmitted to the British analyst, Christopher Donnelly, Head of the Soviet Studies Research centre at Sandhurst, who was able to present it in a form not drawing on classified sources and thus suitable for widespread dissemination. Chris Donnelly's article was originally entitled 'Recent changes in Soviet operational strategy' (an oxymoron as far as Soviet military vocabulary is concerned, but Mr Donnelly was not addressing a Soviet military audience), and was scheduled for publication in the Geneva-based *International Defense Review*. The magazine's editors, unknown to Mr Donnelly, altered the title to 'The Soviet Operational Manoeuvre Group: a New Challenge for NATO', and disseminated copies at the beginning of October 1982, prior to publication.[3] The evolution of this appraisal of the revived mobile group concept, which was far from new but had antecedents throughout military history, was very much a joint American and Britist effort, and owed much to John Hines and other analysts. The story had media impact: the typescript leaked to Associated Press prior to publication was reported in western newspapers and by the Soviet news agency TASS (for abroad) on 4 October. Western newspapers reported the assessment accurately enough, stressing the need to collapse NATO defence at high speed and describing the OMGs as 'Blitzkrieg' units.[4] The Soviet media responded with remarkable alacrity, suggesting that Chris Donnelly had indeed touched a sensitive nerve. They linked his assertions with remarks made by General Bernard Rogers, then Supreme Allied Commander Europe, the previous week. Rogers had demanded strengthening of NATO conventional defences and the destruction of Warsaw Pact 'follow-on forces': the 'new' (although, again, it has innumerable precedents throughout military history) NATO policy of 'follow-on forces' attack (FOFA)!. This, it was discovered, interacted profoundly with the OMG concept (not least since the OMG might result in the removal of 'follow-on forces' for FOFA to attack, and thus ruin the Americans' entire day).[5]

Donnelly's article was published shortly afterwards, the first of many to explore different aspects of OMGs. Others sought to stress the strategic and operational context, of which the OMG is only a part; to probe the practical difficulties of implementing the idea, and in a few cases to examine historical antecedents. A proper examination of the latter would have shown that there was nothing new or particularly special about OMGs, would have highlighted the fact that they were part of an evolving strategy, and would have indicated their composition, capabilities, and how to counter them. Debate continued for over five years but by 1988 it had become apparent that the OMG had been neither a complete nor a permanent solution and that further reorganization was underway. In February 1989 two highly placed Soviet officers admitted publicly for the first time that OMGs existed, but that all were to be eliminated by 1990. Thus, this study is based on history and not, even in part, on current speculation, however well-informed.[6]

Returning to 1982, the Warsaw Pact sources cited as evidence for this revived nostrum were quite explicit about its historical pedigree as they explained its contemporary purpose:

> 'It is now necessary to seek new methods of employing forces so as to break up the cohesion of the enemy deployment over its whole depth. This might be accomplished, for example, by destroying objectives or elements of the grouping which are essential to the viability and combat readiness of enemy forces. These requirements, on the basis of analyses of past wars, have led to the re-appearance in modern battle of detachments engaged in raiding activities and also operational marching groups (*operacyjnych grup marszowych*). The latter derive from the so-called high speed groups formed from armoured troops and widely employed by the Soviet Army during the Great Patriotic war.[7]

The term operational manoeuvre group was taken from Polish although all analysts agree that there is a corresponding term in Russian.[8] It was assessed that both Fronts (Army Groups) and their subordinate armies would launch OMGs, of Army and division size, respectively, in the first day or so of hostilities, into gaps or weak spots, existing or created in NATO forces' deployment. These would break out into the 'operational depth' as fast as possible, front OMGs moving over 150 kilometres ahead of the main forces and army OMGs up to 70 kilometres ahead. They would splinter the command, control and cohesion of NATO armies, blocking withdrawal routes and thus facilitating encirclement, and preventing reinforcement. Soviet OMGs would thus exemplify the 'eccentric movement' or 'encirclement from within' described in Chapters 1 and 3, and exploit the force-multiplying effect of attack on the flanks and rear and destruction or paralysis of the brain and nerve system familiar to all students of the history of war. The illustration of OMG penetrations used at the 1982 SHAPEX briefing is shown in Figures 5.1 and 5.2. Modern OMGs would enjoy the additional advantage that in an environment where nuclear weapons were a constant threat, rapid penetration would facilitate the seizure or paralysis of NATO tactical nuclear weapons. At the same time, OMGs' presence close to and among NATO forces and centres of population would deter retaliation with nuclear weapons against OMGs themselves. In fact, this was not new, either; back in 1964, Marshal Rotmistrov noted how 'one can inhibit enemy nuclear retaliation in the theatre by carrying an offensive rapidly into the enemy's rear, thereby displacing or capturing his nuclear delivery means'.[9]

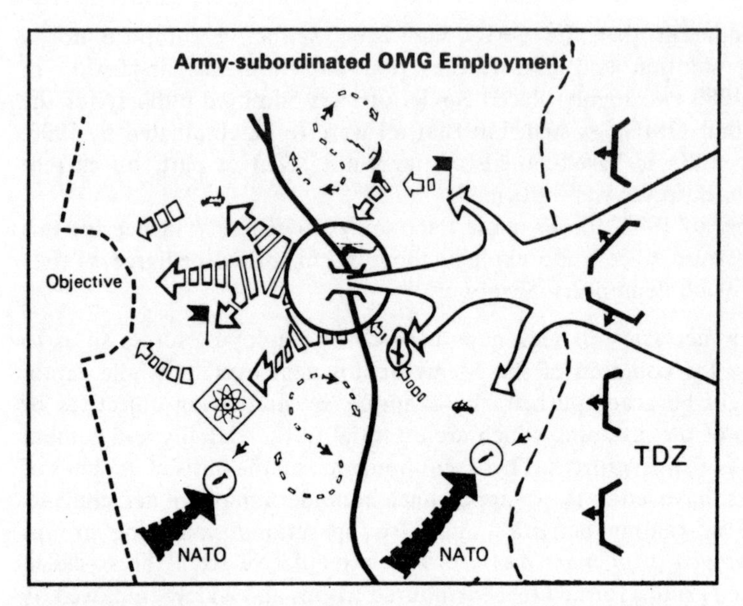

Figure 5.1 Army OMG operations, as presented to SHAPEX briefing, 1982

TDZ = NATO tactical defence zone

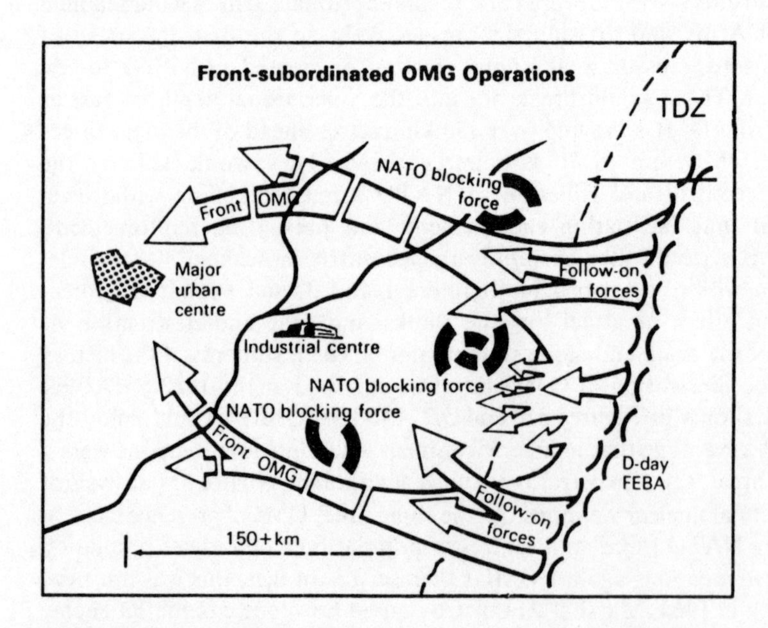

Figure 5.2 Front OMG operations, as presented to SHAPEX briefing, 1982

TDZ = NATO tactical defence zone

The larger, longer-range front OMGs might even aim to grab political or very deep logistic centres (Bonn, the main reinforcement routes running through Holland), and thus have a 'strategic', rather than strictly operational, role. The possibility of surface forces linking up with air-dropped or air-landed (particularly heliborne) troops far in the NATO depth compounded the possible synergistic effects. It was assessed that such formations might be made more mobile, more heavily armed and better provided with fuel, ammunition and other essentials, than comparable formations remaining as part of the main body, because they would be acting 'severed from the main forces' for an uncertain period.[10] All these points were thrashed around by disputatious analysts who seem to have been oblivious to the vast body of information available from military history. The vastness and complexity of the subject makes a case study approach the only practical one: the examples below lead to answers to the analysts' dilemmas, and violent challenges to some of their assumptions.

The Russians use the historical term mobile group to refer to OMGs in open source literature, which immediately focuses attention on the large, front subordinated formation. In fact the OMG has three possible antecedents in military history, defined in Russian as:

The mobile group. This is defined as an element of an army or front's forces designed to exploit success in the offensive. Soviet historians consider the prototype of the front mobile group to have been the cavalry army in the Russian Civil War. An army would similarly have a cavalry division available for such missions. This concept was further developed in the inter-war years (see below). The mobile group was supposed to enter combat after the first echelon had accomplished a breakthrough of the enemy tactical defensive zone. Mobile groups began to appear as a regular component of Red Army formations in 1943: the first front mobile group of the Great Patriotic war appeared in the Kursk counter-offensive in August. Army-level mobile groups would be committed on the first day of operations and front-level groups on the second or third.[11]

After the war the mobile group disappeared temporarily from writings on contemporary operations. According to the relevant volume of the *Soviet Military Encyclopedia*, published in 1978, the task of exploiting success would be undertaken by the second echelon of the army or front.[12] This may be one reason why western observers were confused by the appearance of a new element in the composition of armies and fronts in the late 1970s, and some thought that the Russians might be giving the second echelon another name. However, it was later believed that there might be up to three separate elements behind the 'first echelon' (if there is one): the 'second echelon', the OMG, and special reserves.[13]

The forward detachment. This is defined as a reinforced unit or sub-unit earmarked for independent tasks in the course of a battle or operation. In the offensive, the forward detachment is ordered to penetrate quickly into the depth of the enemy defence, seize important objectives, and execute other special tasks. In the last war, tank or mechanized corps usually had brigade-strength forward detachments operating 30 to 35 kilometres ahead of the main body.[14]

The distinction between mobile group and forward detachment is most important. The former were larger and provided the element of operational manoeuvre when the main forces following were constrained to a step-by-step advance. Forward detachments were primarily intended to accelerate the advance of the following forces although, unlike the latter's advance guard, were able to operate independently of them. In addition to executing tactical manoeuvre, forward detachments were able to initiate operational manoeuvre, by paving the way for mobile groups, leading them through the fragmented tactical defence and into the operation depth.[15] The aims and characteristics of mobile groups and forward detachments overlap to some extent, but a formation designated an OMG, exploiting success and manoeuvring independently of other forces in order to trap the enemy, was perhaps closer to a Great Patriotic War mobile group.

The raid. This is now defined as a determined penetration, leading to combat action in the enemy depth, by highly mobile (airborne, tank, mechanized, amphibitious or, formerly, cavalry) formations and units, or by partisan forces. A raid usually takes place along a previously determined route and aims to destroy personnel and equipment; seize important objectives; disorganize command, control, and communications, sow panic, divert enemy forces from their objectives and, sometimes, assist partisan operations. There were early examples of raids in the Patriotic War of 1812, with further development in the Civil War. Whereas in western parlance a 'raid' usually means a small commando or diversionary action, a Soviet 'raid' may well comprise forces as large as a corps. The terms 'raiding force' or 'raiding action' were often used as euphemisms for OMGs.[16]

The OMG therefore owed something to the forward detachment, the mobile group and the raid, had aspects of all their characters, and studying all these antecedents can be instructive. The Russians do so exhaustively although, as the influential and formidable Soviet military commentator General M. M. Kir'yan, Doctor of Military Science, pointed out in 1987, some have taken a selective view of military history and glossed over the problems and failings of such formations. When they were first employed, they frequently failed to achieve their objectives.[17] Kir'yan stressed that there have been fundamental changes in technology and methods of conducting operations, and that these should be borne in mind. Even so, 'the experience of the past war not only has not lost its relevance, but continues to influence the development of tactics, operational art and strategy to a certain extent'.[18]

IMPERIAL ANTECEDENTS

The imperatives of steppe warfare (see Chapter 6) and the military institutions of the Russian Empire combined to produce an emphasis on deep penetration operations by mobile (cavalry) forces. In the Cossacks, the Russians possessed a force which, contrary to all the preconceptions and clichés about Russian methods of command and control, was ideally suited for independent operations on the flanks and in the rear of enemy armies. The Cossacks were widely employed in this way in the Napoleonic wars, harrying French forces and attacking from all sides.[19] The Cossacks were increasingly regularized at the end of the nineteenth century, which enabled them to be more precisely controlled as a component of large-scale operations, but did not reduce their traditional suitability for raiding. Mishchenko's raid in the 1904–5 Russo-Japanese War and the widespread employment of Cossacks by both sides in raiding actions during the Russian Civil War affirm the Cossacks' continuing pre-eminence in the operational manoeuvre role. This coincided with the change in cavalry's prime role from a shock arm on the battlefield to the operational manoeuvre force. The Cossacks, in particular, would 'spread beyond the tight limits of the field of battle . . . to fulfil potent tasks in the Theatre of Military Operations (TVD)'.[20]

There was also a tradition of deliberate planning for such operations. Examples are the raid on Berlin in 1760, during the Seven Years' War and on Cassel in 1813. The former cannot claim an exclusively Russian pedigree: the Austrians had raided Berlin in 1757, and the 1760 Russian raid was apparently suggested by the French liaison officer at Russian headquarters, Count René de Montalambert. The Russians did, however, respond to the suggestion with enthusiasm, and earmarked a raiding force of some 20,000 men (a 'corps', in modern terms), under General Z. G. Chernyshev (1722–84). The forward detachment of this force 4,000 to 5,000 men, predominantly cavalry, was commanded by Major General Totleben. The aim of the raid was to divert resources from the simultaneous action in Pomerania, to seize politico-economic objectives (the capital: Berlin), and to destroy enemy warmaking capacity. Chernyshev's 'corps', as it is referred to in modern Soviet accounts, therefore equated with a modern Front OMG with Chernyshev's brigade as forward detachment. The Russians entered Berlin on 28 September OS (9 October, NS), wrecked the royal cannon foundry and other installations, and liberated a large amount of money from the treasury before withdrawing.[21]

The raid on Cassel in 1813 was commanded by another Chernyshev, General A I Chernyshev (1785–1857). It was but one of a number of such raids carried out by Russian light cavalry forces in 1812 and 1813, which French authorities, including Marshal St Cyr, considered a principal contribution to Napoleon's defeat. Chernyshev's force was known as an 'independent cavalry detachment' or 'Army partisan detachment' and comprised 3,000 horse – a brigade, essentially. The force was detached from the army of the north and reached Cassel at the end of September, forcing Napoleon's brother, Jerome, to flee the capital of his Kingdom of Westphalia, and contributing to the collapse of French imperial power.[22] It was therefore more a strategic deep-penetration force than an operational one, but an indicator of a strong Russian predilection for this type of action and a well developed appreciation of what it could achieve.

As we saw in Chapter 3, the American Civil War was cardinal in the evolution of deep-raiding operations. European commentators noted the low density of population in the American theatre compared with western Europe. This was an indicator of the different conditions which made the former particularly ripe for raiding actions by cavalry.[23] Conditions in Russia were not dissimilar to America and the Russians took the American experience to heart 'alone of all European powers'.[24] As early as June, 1863 the Russian journal *Military Collection* published a report on the significance of light cavalry in modern war, drawing lessons from the American conflict. It noted the far more pronounced 'material element' in modern war, the increased dependence on the telegraph for command, control and communications, and on the railway for munitions. Conversely, depriving the enemy of the same facilities had attained 'exceptional importance', and cavalry was now to be used 'primarily in the rear, against communications'.[25] Such detailed analysis so soon after the event is signficant. Stuart's campaigns were studied in detail in the 1860s and 1870s, notably by the cavalry officer Vladimir Sukhomlinov (1848–1926), later to be Russian War Minister at the outbreak of World War One.[26]

Russian interest in raids and wide-ranging cavalry action within a TVD was reinforced by two other factors. First, the use of cavalry in raiding operations were intimately connected with their use as mounted infantry. Many Russians advocated the employment of cavalry in this way, as evinced by the award of the Tsar's prize for the best history of cavalry to the Canadian Colonel George Denison. Denison analysed the American Civil War and concluded that cavalry's future lay as mounted infantry. He also examined the Franco-Prussian War (1870–71) and advocated strategic raids.[27] In 1884 a former British military Attaché to Russia confirmed that 'an influential section of officers in the Russian Army' believed that cavalry would in future be employed as in the American Civil War.[28] Cavalry raids had not been prominent in recent European conflicts but the Russians' use of them in the Napoleonic Wars was stressed.

The second factor in favour of the development of the deep raid in Russia was the need to practise the concept. Theory is all very well, but unless armies can practise it with some semblance of realism, the chances of it succeeding when put into practice against a real enemy are slim. To practise the movement of a cavalry corps across hundreds of kilometres in 'any civilized country' is not easy. However, in Russia:

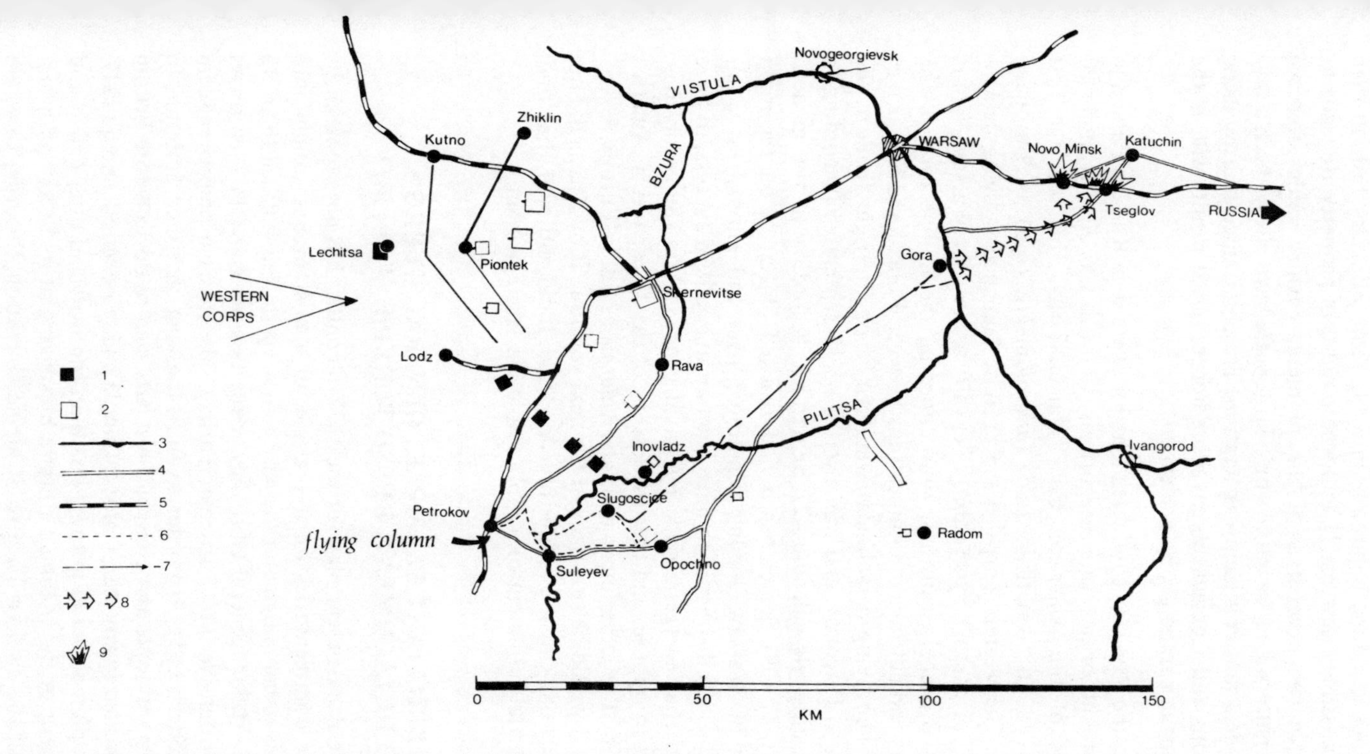

Figure 5.3 The 'American raid' beyond the Vistula: Russian manoeuvres in Poland, 1876

1 Invading (western) cavalry 2 Defending (eastern) cavalry 3 Rivers 4 Roads 5 Railways 6 Movement of 'flying column', day 1, 15 September 7 Movement of flying column, 16 September 8 Movement of flying column, 17 September 9 Final objectives of flying column

> 'where things are done habitually in a far more arbitrary fashion than would be possible elsewhere[!], and where rights of property are made to give way to the necessities of the moment, cavalry raids after the American pattern and sometimes on a large scale have always been a very favourite idea, and confident expectations are entertained that by using the dragoons and Cossacks as invading hordes, great results will, in the next European campaign, be achieved.[29]

True to form, the Russians conducted a major exercise in Russian Poland in 1876. Russian authorities state categorically that this was directly modelled on the raids of the American Civil War and that this was the first employment of an 'American raid' (*amerikanskiy reyd*) in European manoeuvres. The western corps in the exercise launched a 'flying detachment' (the historic term *corps volant*) of cavalry against the Warsaw-Vienna and Warsaw-Bromberg railways, thus impeding the mobilization of the eastern forces and cutting communications between Poland and Russia (see Figure 5.3). The flying detachment was formed on the right flank of the western force at Petrokov, and was inserted on 15 September (according to the Russian old-style calendar, 27 September according to the modern calendar). By 17 (29) September it had reached objectives on the main railway East of Warsaw, having penetrated over 150 kilometres. Commentators considered the raid to have been an unqualified success, to have demonstrated that energetic raids by 'cavalry masses' on the enemy rear were difficult to oppose, and that the lessons of the 4-year war in America were upheld by this experiment with an 'American raid' in Europe.[30]

After this, it would be surprising if the next war fought by Russia did not furnish at least one classic example of a deep raid. It did, although that 'raid' was but a partial fulfilment of a more grandoise strategic plan.

'A SINGLE BRILLIANT EXAMPLE': GENERAL GURKO'S FORWARD DETACHMENT, RUSSO-TURKISH WAR, 1877

Although the Russian war plan for the 1877–8 conflict has not, apparently, survived, the commentaries on it provide an almost unique insight into Russian plans of campaign and the way actual operations fulfilled them. In March 1877, Major General Obruchev wrote two notes on the then secret plan for an offensive into European Turkey. The Russian objectives seem strikingly modern today. In order to fulfil the long-cherished ambition to capture Constantinople and control the straits, and to do so before Britain became embroiled in any war, Constantinople had to be seized, and quickly. The Russian plan aimed at nothing less than the swift and utter defeat and dismemberment of the Ottoman Empire by striking at its heart and brain, the essence of the modern Theatre Strategic Operation: 'In order to attain decisive results, the aim of our strategic action must, more than ever before, be Constantinople itself.'[31] Obruchev argued that this objective was attainable, provided that Russian military action was determined and swift:

> We must exploit time and the speed of advance. . . . It is essential, that an army, moving into Turkey, should immediately deploy beyond the Balkans, not a weak detachment, but fully adequate forces for the rapid seizure of Constantinople. In other words, we now need to deploy not one, but two armies, as it were, of which one could concern itself with battle in Bulgaria on this side of the Danube and the other – immediately after crossing, move directly on Constantinople, which would only be some 500 versts [kilometres] before it, a distance which they would endeavour to cover in five or, maybe, four weeks, not being distracted from this aim by any secondary operations, by guarding their rear, by the attack of fortresses, or even by operational battles on the flanks.[32]

In order to guarantee the seizure of Constantinople, even if the British deployed land forces, Obruchev envisaged that an army of 100,000 to 120,000 men, fully equipped for the energetic attack of a strongly fortified position, should be hurled through the Balkans. A second army was required to fight on the near side of the Balkans and to protect the forward army's lines of communications. Obruchev therefore recommended pushing one army through another and, disregarding all other missions, having it seize the ultimate strategic objective. The importance of speed was reiterated by General Staff Colonel Artamonov in his notes on the best means of acting against the Turks, which comprised the most 'unwavering, decisive and, as far as possible, rapid action'. There can be few clearer precedents for the modern OMG. Artamonov analysed the Russian experiences in a previous war against the Turks in 1828–9 extensively, and concluded that it was necessary to get a strong detachment to the other side of the Balkans as soon as possible in order to conduct essential reconaissance.[33]

Obruchev's grand design was not fully implemented, but strategic reconnaissance was one of the missions of General Gurko's force. Contemporary English language sources refer to Gurko's command as an 'Advance Guard', but in Russian it was a *peredovoy otryad*, the same name as the 1941–5 forward detachment.[34] Lieutenant General Yu V. Romeyko-Gurko was commanding the 2nd Guards Cavalry Division at the outbreak of the Russo-Turkish War. He was sent to the front on the Tsar's personal orders, leaving his division behind in St Petersburg, and was appointed to command the forward detachment on 24 June (6 July) 1877.[35] This was an ad hoc force created six days before, comprising 5,800 infantry, 5,000 cavalry and 40 guns, or about 11,000 men in all.[36] Like its modern OMG counterpart, the Forward Detachment had extra engineer support to help fulfil its mission to prepare routes for the main army: a special mounted engineer squadron of 200 Caucasian, Don and Ural Cossacks. The total strength of the Russian Army in the Balkan Theatre at this time was some

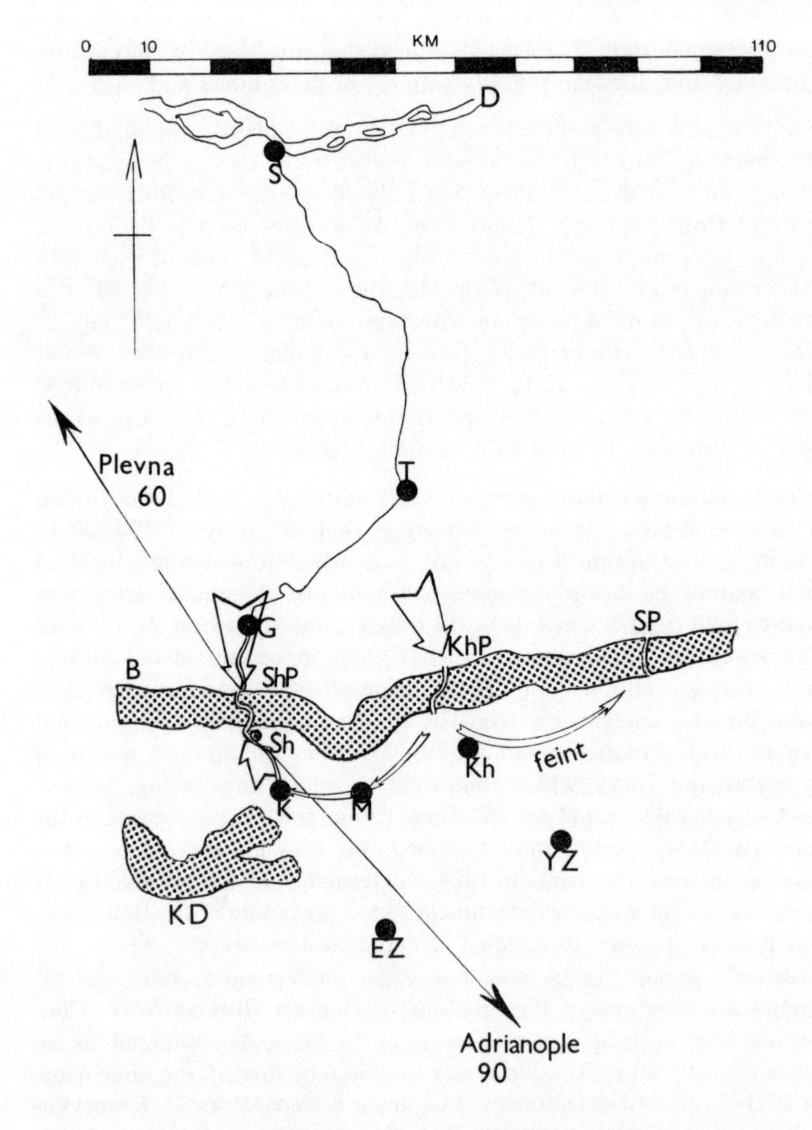

Figure 5.4 Operations of General Gurko's forward detachment in 1877

B = Balkans
D = Danube
EZ = Eski Zagra
G = Gabrovo
K = Kazanlik
KD = Karadj-Dagh
Kh = Khankioi (village)
KhP = Khankioi Pass
M = Maglish
S = Sistova (Main Russian bridgehead across Danube
Sh = Shipka (village)
ShP = Shipka Pass
SP = Slivno Pass
T = Tirnovo
YZ = Yeni-Zagra

185,000 men.[37] Gurko's force therefore equated roughly to a modern division in size and to a Front OMG in function, albeit, at under 10 per cent of the total, a rather weak one. The detachment's weakness to undertake its allotted function was commented on at the time.

Preliminary orders had been given to Gurko's second-in-command before Gurko's arrival, by the theatre commander. These were to reconnoitre in the direction of Tirnovo and Selvi (see Figure 5.4), to collect information about the passes over the Balkans, and to prepare to move into the mountains, although the movement was not to be executed without direct orders from the theatre commander. On receipt of the latter, the second phase would begin: to seize the passes, send cavalry south of them to raise the populace in revolt, scatter those Turkish detachments which were small enough to be dealt with easily, and protect the passes while they were made ready for the passage of the main army.[36]

The mission, composition, and control of this Forward Detachment, like all the examples in this study, have clear similarities with those associated with the idea of the OMG. Seizing objectives enabling the main body to advance, intelligence gathering, and avoiding engaging enemy forces are all similar. The formation's crystallisation from a division comprising assorted Cossack regiments, regiments from two regular divisions, horse artillery batteries, specially trained engineers and Bulgarian volunteers is instructive. The whole, 'under the name of Forward Detachment [was] placed under a general who had not reached the theatre of war . . . all the higher officers entered on their duties simultaneously with the formation'.[39] The forward detachment was only to be launched into the enemy depth on the express orders of the Commander in Chief and at a moment chosen by him. Once launched, however, its commander had carte blanche to do as he thought fit within the broad guidelines he had been given.

Security was of great importance. The composition of the detachment was kept secret initially and no-one apart from Gurko knew its destination. The orders stressed the need for the utmost exertion in pursuit of the object in view, and that there must be no stragglers, talking, singing or smoking. Subsequent analysis reiterated the point. 'The fact that a raid is contemplated as well as its destination should be made known to no-one except the commander, until the force is well on its way.'[40]

Ammunition, fuel and food are a problem for any such force. In order to preserve mobility, ammunition carts were not taken. The men carried as much as possible in their pockets and the artillery only had its regular *unit of fire*. For food, Gurko intended to live off the country as there were plenty of cattle in the area and grazing for the horses, but five days' worth of biscuit was taken as emergency rations for the men and three days' worth of barley for the horses.

The best route across the Balkans for the main army to use was the Shipka pass, Gurko's first objective. It was a very strong defensive position held by Turkish forces of unknown strength. The best way to capture a mountain pass, like a bridge, is both ends at once. The most difficult pass across the Balkans was the Khankioi, and a local superstition warned that 'ill luck awaits him who crosses the Khankioi pass'.[41] Gurko surmised, correctly, that he would meet least resistance here. His plan was therefore to take the forward detachment through the Khankioi and attack the Shipka pass from the south simultaneously with an assault by the main forces from the north (Gabrovo). Thus the forward detachment was acting in concert with the main forces to achieve encirclement. In the days before radio there was no way for Gurko to communicate with the main force to synchronize their attacks and the Russians used the Mongol method: specify a time in advance: 5(17) July. The passage of the Khankioi (1–2(13–14)July), securing its exit and the subsequent move into position at the southern end of the Shipka via Kazanlik took longer than expected and Gurko had to push his men to the limit:

> late in the evening of the 4th [16th] the whole of General Gurko's force was assembled at Maglish . . . they were still some seven miles from Kazanlik and the men were exhausted by a march of more than 13 miles performed under the burning sun and three or four hours' hot fighting. . . . The force was still 20 miles from the Shipka pass and haste was necessary in order to open up communications with the Gabrovo force.[42]

The pass was attacked from the north as planned on 5(17) July but without success. Gurko, having been delayed, attacked from the south on 6(18) July, again without success, but the Turks now realized they were caught between two pincers and melted away into the mountains on either side. On 7(19) July General Skobelev advanced from the north and Gurko from the south to find the pass devoid of the living. Thus Gurko re-established contact with the main army. The Russian achievement was impressive; the force had crossed the Danube on 21 June (3 July) and by 7 (19) July had taken the Shipka pass.

At this point Gurko's forces were exhausted. Ideally, such a force would then be withdrawn but Gurko's force was the only one suitable for deep penetration in 1877. Therefore, after a brief respite, it was pushed forward again to accomplish its remaining tasks: to cover engineering work on the passes, rouse the populace to revolt and cut Turkish communications. Gurko had wanted to push forward as far as Adrianople but the delay in capturing the fortress of Plevna and the consequent check to the entire Russian Army obliged him to restrict himself to a more limited advance to Eski-Zagra and Yeni-Zagra.[43] It was only when Plevna had fallen and the strength of the Russian forces nearly doubled that it was possible to advance on Adrianople. As a British observer remarked: 'the plan was too ambitious for the numbers available'.[44] If Gurko had the equivalent of a Front OMG, which his mission warranted (20 per cent of forces in the theatre), that would have been over twice his actual strength.

Although delayed, the forward detachment's move into the Tundja plain, south of the Balkans boded well for similar operational-strategic penetrations in the future:

> The moral effect of our appearing in that plain was immense. A crowd of despatches came flying from Constantinople with anxious enquiries and hasty orders. It was resolved to replace the Minister of war . . . and the Commander in chief. . . . 'In consequence of the ground covered by the enemy, the Empire is between life and death', ran a despatch. . . . 'The capture of Adrianpole would reduce the Turkish empire to the level of the Khanate of Bokhara' . . . 'The existence of the Empire hangs on a hair'.[45]

Subsequent Russian analysis of the operation is revealing. An article of 1896 considered the main tactical lesson of the operation to be the need for all-arms co-operation. Strategically, the forward detachment was the thin end of the wedge formed by the whole army, which clove the Turkish deployment in two.[46] There was confusion as to the exact purpose of the detachment; on 29 June (11 July) the Russian HQ ordered Gurko to confine himself to the capture of the passes, and a week elapsed between the capture of the Shipka on 7(19) July and Gurko receiving orders to move on.[47] The Russian command initially envisaged an operational-level action – to seize the passes in concert with the main army, but this became confused with independent, strategic action. Had the Imperial Army developed the concept of operational art, as distinct from strategy, they would have been able to clarify the detachment's mission in their minds.

The clearest appreciation of the role and significance of Gurko's forward detachment was penned by Captain Greene of the US Corps of Engineers:

> This expedition of Gurko's was more than a mere cavalry raid: it was an admirably conducted movement of an advance guard composed of all arms . . . it had in less than a month gained possession of one of the principal passes of the Balkans . . . which they finally used in January for the passage of a large portion of their army; it had carried a panic through the whole of Turkey between the Balkans and Constantinople; and its scouting parties had penetrated to within 70 miles of Adrianople, the second city of the Empire, and had destroyed the railroad and telegraph on two principal lines; finally, it had gathered accurate information concerning the strength and positions of the large Turkish force advancing towards the Balkans.[48]

Greene correctly appreciated the role of the forward detachment as an integral and essential part of the general offensive. It was a product of peacetime operational experiment and strategic planning, modified in accordance with the circumstances of real war.

MISHCHENKO'S RAID, RUSSO-JAPANESE WAR, 1904-5

The success of Gurko's forward detachment led to considerable optimism about the potential of operational manoeuvre by cavalry during the long period of peace between 1878 and 1904. General Kuropatkin, who commanded the Russian forces in Manchuria in 1904–5 was, however, cynical about the concept, and in his opening orders effectively quelched the idea, which had apparently been highly popular in the 1880s:

> One must avoid being carried away with the strategical [operational manoeuvre] role of cavalry . . . it is true that now, during manoeuvres, we no longer have those quite improbable movements of cavalry masses on each side in rear of the enemy as was the case 15 to 20 years ago.[49]

The war did however witness one interesting Russian raiding action. Mishchenko's raid, in December 1904 and January 1905, was an independent, commando-style action deliberately launched during a lull in main forces' operations, and was not designed to accomplish operational manoeuvre. The reasons for its failure and some detailed aspects are, however, instructive. Mishchenko's force was the forward detachment of the Manchurian Army, a weak division, 7,500 strong. It is possible that at the end of 1904 Mishchenko knew of the planned offensive against Sandepu, executed from 12(25) to 15(28) January, 1905, and had instructions to be on hand for these operations. However, on 20 December 1904 (4 January 1905) the Russian stronghold of Port Arthur fell. The railroad from Liao-Yang to Dal'ny immediately became the key artery along which Nogi's investing army could be brought to the front. A cavalry raid against this line of communications would have made operational sense, but the orders of General Kuropatkin, on 22 December (4 January) made Inkou the main objective with incidental damage to the railroad.[50] Furthermore, Inkou was an ice-locked harbour and the Russians should have known that Dal'ny would be the main base for the Japanese. The Russian orders reveal a strange preoccupation with the destruction of stores and 'capital', especially the nine million roubles' worth of stores accumulated at Inkou.[51]

In addition to the wrong objectives, the timing of the raid limited its effect. The exponential value of a deep penetration in conjunction with a general objective was well appreciated:

> The psychological effect produced by the appearance of cavalry in the rear is incomparably greater when the army is engaged, when psychological and physical exhaustion lead to such tension that a single spark is sufficient to bring about a panic among even the most disciplined troops.[52]

The Russian army was in no position to undertake a general offensive at this time, however, and it was necessary to counter the effect of the fall of Port

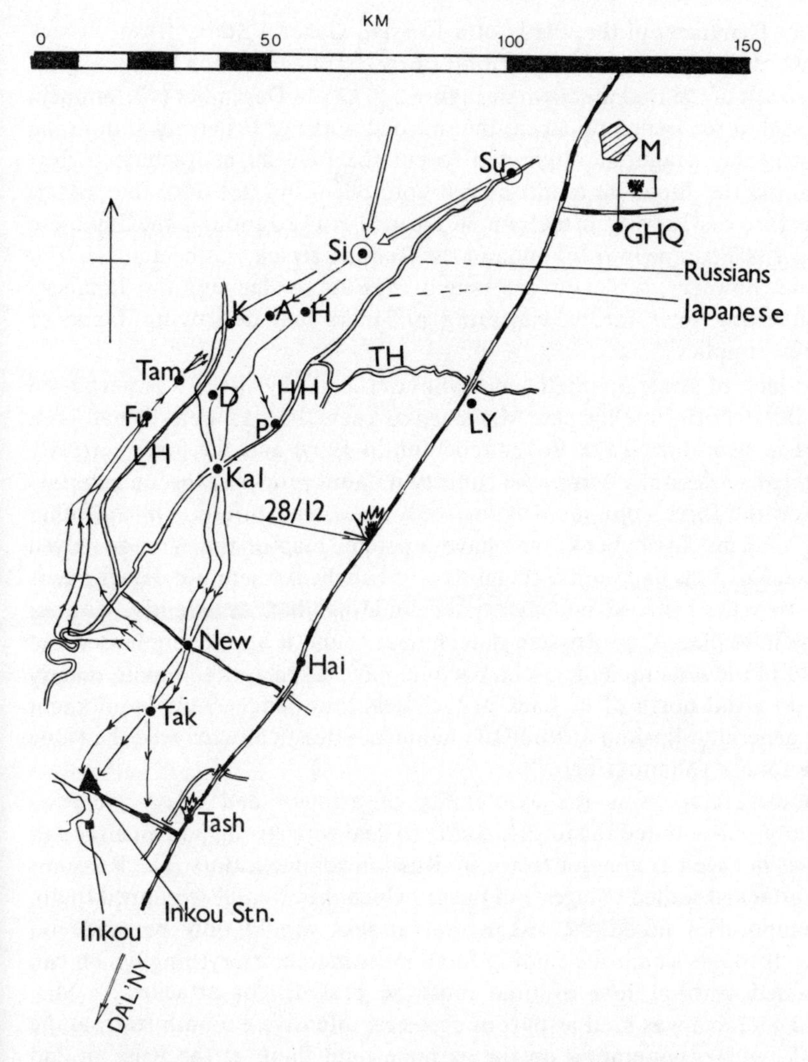

Figure 5.5 Mishchenko's raid, December 1904–January 1905

A = Ashenyula
D = Davan
GHQ = Main Russian Headquarters (Kuropatkin)
Fu = Futsyanzuan
H = Huhuandi
Hai = Haicheng
HH = Hun-Ho river
K = Kalyama
Kal = Kalikhe
LH = Liao-Ho river
Ly = Liao-Yang
M = Mukden
New = Newchuan (Newchwang)
P = Peidagou
Si = Sifantai (concentration of all Mishchenko's force)
Su = Suhudyapu (concentration of most of Mishchenko's force)
Tak = Takauzhen
Tam = Tamienpuz
Tash = Tashichao
TH = Taitzu river (Taitzu-Ho)

Arthur. Therefore, in the words of a Russian General Staff officer, 'it was decided to throw in one of our trump cards – strategic action by cavalry'.[53]

The path of the raid is shown in Figure 5.5. On 28 December (10 January) a Cossack force peeled off to cut the railroad north of Haicheng and on the following day dragoons attempted to cut the railroad at Tashichao thus preventing the Japanese reinforcing Inkou, but it did not have that effect. Just before the Russian attack on 30 December (12 January) the Japanese moved 1,000 troops into Inkou, and the Russian attack was beaten off. The raid did, however, achieve some tangible results in alarming the Japanese and diverting their forces, capturing prisoners and destroying Japanese essential supplies.[54]

The lack of strategic intelligence about crucial objectives was paralleled by faulty detailed intelligence. Mishchenko knew the area well (he had been stationed here during the Boxer rebellion in 1900) and the force certainly navigated successfully across the country without proper maps, and contact between the three columns was lost only once, and briefly. On the other hand, why did Mishchenko not have a proper map of the railroad, given that the Russians had built it themselves? Mishchenko sent out detachments to destroy the railroad and associated buildings but, apparently, without any definite plan. One Russian detachment found a bridge unguarded but instead of blowing up a pier, which would have taken ages to repair, merely blew up a rail north of it. Lack of technical knowledge, faulty equipment and a generally slipshod attitude to intelligence drastically reduced the value of the force's valiant efforts.[55]

Another factor was the availability of artillery and heavy weapons generally. This limited the force's ability to deal with strong points, although this was not seen as a major factor by Russian commentators. The Russians often attacked walled villages and farms, which they were ill-equipped to do. Contemporaries noted the lesson that attacks should only be made on defiles 'through which the raiding force must march; everything which can be passed without loss of time must be passed, not attacked':[56] Mishchenko's force was used as part of a general offensive a month later, at the end of January, operating on the extreme right flank of the Russian 2nd Army and attempting to envelop the Japanese left.[57]

One vital lesson of Gurko's forward detachment and Mishchenko's raid is liable to be overlooked. Both Generals were exceptionally able, energetic and charismatic commanders. In both cases their formations were welded together in the enemy depth in the most difficult conditions (scorching heat, freezing cold), primarily by the personality and will of the commander.

OPERATIONAL MANOEUVRE BY CAVALRY WITHIN THE THEATRE OF STRATEGIC MILITARY ACTION (TVD)

A Russian study into the role of cavalry in future war in 1898 considered the operational roles of cavalry in covering their own forces' mobilization and concentration, seizing important strategic objectives in the frontier area at the start of the war, conducting reconaissance in front of the main army, impeding the mobilization and concentration of the enemy and the synchronization of his armies within the TSMA.[58] Using cavalry to seize important strategic objectives before the enemy was fully deployed in a series of pre-emptive strikes has parallels with the 'advance guard echelon' of the 1930s and the possible employment of an OMG in the single echelon of an offensive against a partially prepared defence today. The main objectives would be rail junctions and lines, state and private stables, telegraph lines, mobilization and administrative centres. The capture of senior military and civilian administrators and their staff was a legitimate objective, indicating that these initial cavalry raids had some of the characteristics of OMGs and Spetsnaz.

The forces which would conduct these raids were often referred to as 'partisans', (as Chernyshev's force had been in 1813), a reminder that in the Russian mind the distinction between partisans in the accepted sense and regular deep-penetration forces is far from clear-cut. Whereas in the war of 1812—13 partisans had remained in the enemy rear for long periods, it was considered more likely that disruption in future war would be achieved by periodic raids, as in the American Civil War. Practical considerations and the need for mobility had brought about evolution from 'partisan' forces in the usual sense to mobile deep-raiding forces. These raids would be conducted by 'partisan' detachments from regimental to brigade strength (1,000 to 2,000 men), and would penetrate up to about 200 kilometres. The latter shows a remarkable consistency with later developments, and is also a reflection of Russian concepts of scale. In western Europe, a 200-kilometre penetration might well knock a state out of the war.

Following this, the period of main operations would occur, divided into two phases. The first was the approach of the two sides, the period of the 'march-manoeuvre' during this phase, the primary task of mobile forces (cavalry) would be to cut the links between the advancing enemy 'masses' or formations, preventing lateral reinforcement and communication. Furthermore, a powerful cavalry force should be sent into enemy rear to cut communications, railroads, bridges, and so on to prevent his withdrawal. This 'partisan' action might not only have a favourable influence on main forces' operations, but confound the enemy's plans by itself.

Then came the period of concentration and decisive battle. Even before the Russo-Japanese war, it was recognized that the large forces available could not be concentrated on the battlefield in a single day, and battles lasting two to three days were forecast (see also chapter 3). Cavalry would be involved in interdicting the arrival of forces on the battlefield over the course of 3 to 4 days and between 100 and 150 kilometres from the concentration area. This foreshadows the later employment of mechanized manoeuvre and airborne forces to prevent enemy reinforcement.

Lastly came the period of strategic withdrawal and pursuit. Strategic pursuit by cavalry, reinforced by a strong, fresh advance guard, was of greater importance than ever due to the difficulty of conducting effective tactical pursuit on the modern battlefield. Furthermore, this pursuit should ideally be conducted in parallel with the retreating enemy on one or, preferably, both flanks, so as to constantly threaten to envelop him and thus prevent him establishing new lines of defence. To pursue a retreating enemy and threaten envelopment required speed, above all, and the ability to cover great distances. Cavalry was the arm best able to save time and space.[59]

Given the Imperial Russian Army's tradition of deep raiding, its numerous and excellent cavalry forces and the amount of theoretical work on the employment of cavalry in the TSMA, it is not surprising that such preemptive strikes and raids were envisaged in the event of a major European war. Such plans appear throughout the 1890s. In a discussion with the French Military attaché in 1895, General Dragomirov, for example, mentioned launching, at the very outbreak of war, 'a Cossack division from Kamenets-Podolsk into Bukovina . . . going round all the obstacles which it could not break and avoiding engaging the enemy as much as possible'.[60] Such ideas appear to have flourished up to 1914. On the eve of World War One the Russians enjoyed a 2 to 1 advantage in cavalry over their potential adversaries, and, according to General Yuri Danilov, a massed raid into enemy territory was considered. Figure 5.6 shows the railway networks in Russian Poland and the German territory of East Prussia in 1914. The much denser German network gave them an advantage in concentrating and moving forces quickly, which they used with devastating effect in the sequential engagement of Samsonov's 2nd Army and Rennenkampf's 1st Army (see also chapter three). Had the commander of the Russian North-West Front, General Zhilinskiy, controlling these two armies, launched an operational pre-emptive raid as envisioned in the 1898 article, it could have dislocated the German railway network and paralysed the German command's response. Sukhomlinov, the Russian War Minister, had studied Stuart's raids and was aware of the potential of such raids as a force multiplier.

According to Danilov, the Russians examined the option of a raid from the middle Vistula along the left bank, deep into Germany (see Figure 5.6). However, such an operation required the side launching it to hold the initiative. Russia, in Danilov's view stood on the strategic defensive and the initiative would always lie with the enemy. The Russian cavalry had numerous other tasks to perform in the opening phases of war, and was

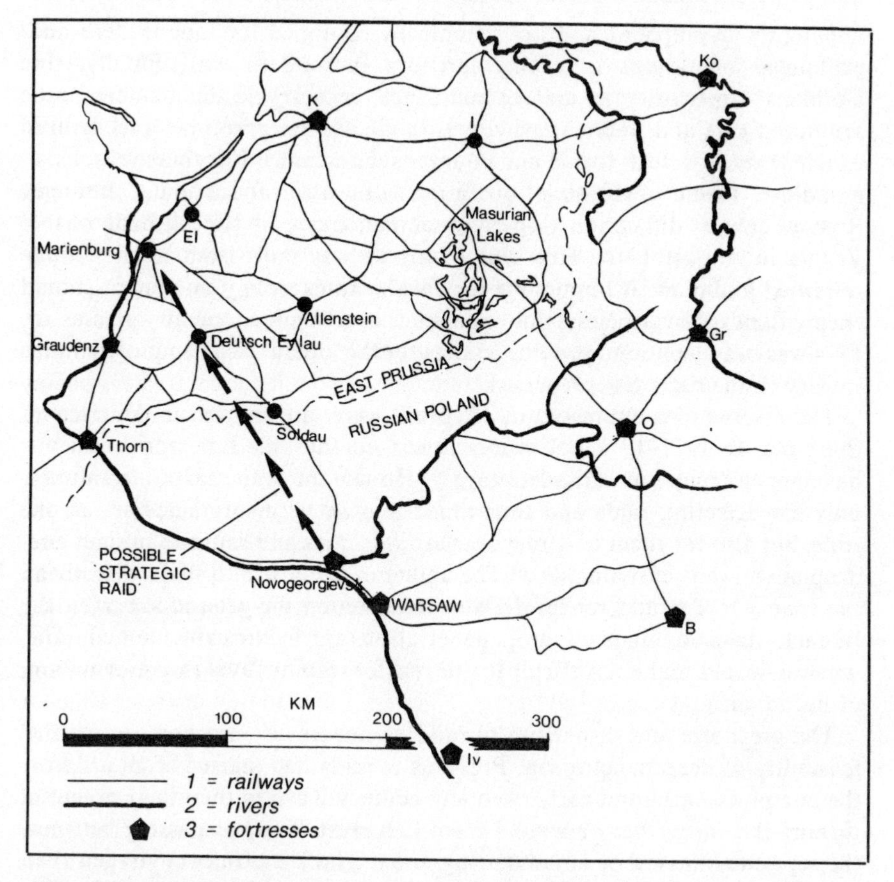

Figure 5.6 East Prussia, 1914: Russian and German rail networks, and possible 'strategic raid'

Note: The possible raid emanated from the 'middle Vistula'. The author assumes the Russians would have avoided the great fortress at Thorn (Torun), and have followed the shortest route to cut the main lines carrying troops and supplies into East Prussia. After crossing the border into East Prussia, 50 kilometres would have taken the Russians to Deutsch Eylau, enabling them to cut two lines (one near the border, one through Deutsch-Eylau). A further 70 kilometres to Marienburg (Malbork) would have enabled them to cut the third line. Such an operation would have been very difficult to execute and the chances of cutting these key lines for very long probably slim. For this and other reasons, the Russians did not attempt such a 'strategic raid'.

Other cities/fortresses

B = Brest (Brest-Litovsk)
I = Insterburg (Chernyakovsk)
El = Elbing (Elblag)
Iv = Ivangorod
Gr = Grodno
K = Königsberg (Kaliningrad)
Ko = Kovno
O = Osovets (Bialystok)

apparently ill-prepared and not technically equipped for independent and prolonged operations in enemy territory in modern war. Finally, the Germans' superiority in air reconaissance, motorcycle detachments and armoured cars and their extensive telephone and telegraph network would enable them to detect, locate and counter such a raid.[61] For these reasons, a grandiose Front OMG-style operation was not carried out, although Russian cavalry did conduct operational manoeuvre on the left bank of the Vistula in August 1914. This type of action was short-lived as the fronts solidified and made it impossible for cavalry to exert any influence against enemy flanks. Nevertheless, this experience of employing 'cavalry masses' in 1914 was drawn upon to provide lessons for the 'operational' employment of cavalry in the early Soviet period.[62]

The reservations expressed by Russians were also current in America at this time. In 1911–12 a US officer remarked that 'modern improvements have put the raider at a disadvantage'.[63] He saw the potential of aircraft not only for detecting raids and thus rendering the necessary surprise impossible, but also for them to 'circle over armies, cities and railroad bridges and drop at will tons of dynamite'.[64] The ability of air power to strike directly at the enemy rear would render deep penetration on the ground superfluous. In fact, the two kinds of deep penetration are inextricably linked. The wireless would make it difficult for the raider to out-travel the information of his advance.

The great size and dispersion of modern armies also worked against the feasibility of deep penetration. Previously, raids had started from a flank: the late nineteenth- and early twentieth century Russian theorists envisaged turning the enemy before a solid front had crystallized or passing between the separate 'masses' of an advancing army. The US officer envisaged that raids might still be possible and put forward specific proposals for the composition of deep-penetration forces. He stressed the importance of engineer support and the need for 'expert telegraphers and probably a wireless section'. Finally, success depended on the good judgment, nerve and luck of the leader.

Most interestingly, the American anticipated Danilov in considering that the automobile and motor cycle could be used to carry infantry to guard choke points in order to cut off raiding parties.[65] Everybody, in Russia and America, saw the new technological developments working in favour of the strategic defender and against the deep-penetrating attacker. Nobody seems to have thought that, far from paralysing the latter, the aeroplane and the petrol engine would give the deep raid still greater potential, and a new and more violent lease of life.

REALMS OF DIS: THE RUSSIAN CIVIL WAR

'A thin, cold, insubstantial conflict, in the realms of Dis'.[66] So Winston

Churchill encapsulated the essence of the Russian Civil War. This explains in large measure the reappearance of manoeuvre after the stalemate which, with some exceptions, characterized western and eastern fronts during much of World War One. The Russian Civil War took place in a vast theatre within which the available forces were thinly spread. Although a few armoured vehicles were available, the main means of manoeuvre were the horse and the railroad, which give the war its particular nature, at once old-fashioned and yet, simultaneously, futuristic.

According to the authoritative *Soviet Military Encyclopedia*, the civil war provided the prototypes of Mobile Groups, which were used initially within armies (cavalry divisions) and, later, within fronts (cavalry armies) as a 'means of developing the offensive'.[67] It also saw the first practical realization of the *cavalry-mechanized group* (*KMG*), which aimed to 'combine the mobility of cavalry with the high degree of shock action and protection possessed by tank and mechanized forces'.[68]

The Russian Civil War began in December 1917 when Cossacks under Kornilov revolted against the new Bolshevik regime. Those who opposed the Bolshevik (Soviet) government were known as 'Whites', to distinguish them from the 'Reds'. By the end of 1920 the Bolshevik government had crushed the main concentrations of internal revolt and foreign intervention, although strong opposition remained in Siberia until at least 1922 and 'banditism' continued in some areas throughout the 1920s.[69] The vast empty spaces and thinly held fronts made deep-raiding operations both possible and highly profitable. As in Imperial times, Cossacks, who fought on both sides, were prominent in deep-raiding operations.

MAMONTOV'S RAID, AUGUST-SEPTEMBER 1919

At the end of June 1919 a renewed Red offensive was ordered southwards in two directions. The main thrust (10th Army and Budënny's cavalry corps) pushed towards the lower Don; the subsidiary one towards Kharkov.[70] The former was halted by Wrangel's Caucasian army at the beginning of September, the latter by White General Kutepov. At this point, the Whites put Mamontov's raid (see Figure 5.7) into operation.

At the end of July the Whites had worked out a plan envisaging the formation of a large cavalry group on the left wing of the Don Army in the Uryupino area comprising two cavalry corps under Generals Mamontov and Konovalov. These corps were to thrust north towards Kozlov to destroy the railroad tracks behind the Red front in that area. This was to take place in conjunction with a combined advance of the Volunteer and Don armies: an army-sized OMG in concert with a front offensive, and fully realizing the value of a deep penetration synchronized with a main forces' action. However, Konovalov's corps was heavily engaged elsewhere and the Red assault on Kharkov prevented the White army conducting a general

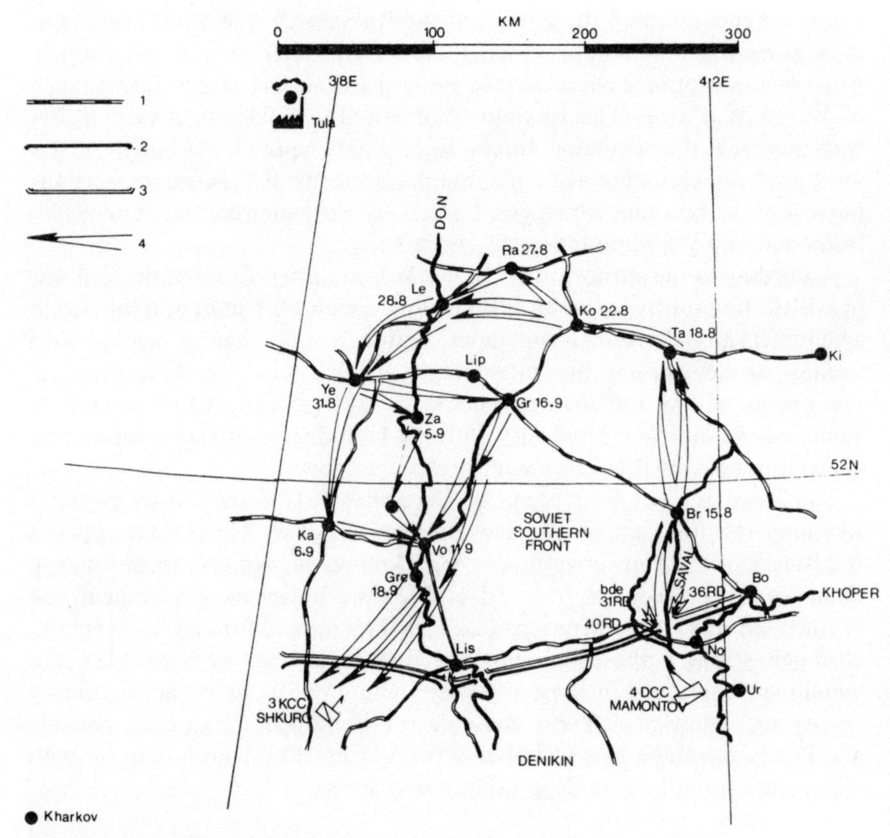

Figure 5.7 Mamontov's raid, August–September 1919

1 Railways in area of raid
2 Main White Front
3 Main Red Front
4 Advance of Mamontov's force

Lines of latitude and longitude are shown as general indicator of position, and Tula and Kharkov as frame of reference

DCC = Don Cossack Corps	No = Novokhopersk	Ka = Kastornoye
Le = Lebedyan	Br = Bratki (Burnak)	Vo = Voronezh
KCC = Kuban Cossack Corps	Ra = Ranenburg	Ki = Kirsanov
Lip = Lipetsk	Gr = Gryazi	Ye = Yelets
RD = Rifle Division	Ta = Tambov	Ko = Kozlov
Lis = Liski	Gre = Gremyache	Za = Zadonsk
Bo = Borisoglebsk	Ur = Uryupinsk	

offensive. Like Mishchenko's raid, therefore, Mamontov's was constrained by other events and not conducted in ideal circumstances. In response to the shifting and slippery fortunes of war, Mamontov's corps was now ordered to conduct a shallow penetration in the region of Voronezh and Liski in co-operation with Kutepov's corps.[71]

The raid had clear operational aims: to disrupt the Soviet rear and preparations for the Soviet southern Front's offensive.[72] Mamontov's 'corps' comprised 6,000 Cossacks in three 'divisions', 3,000 infantry, 12 guns and a number of armoured cars.[73] In size it was therefore consistent with Tsarist experience of raids although small for a front OMG.[74]

Mamontov veiled his concentration by spreading rumours that the White cavalry was completely exhausted by the recent fighting. After a forced march of three days from the safety of the deeper rear to the point of penetration, Mamontov's corps forded the river Khoper on 10 August and suddenly broke through on a weakly held 40-kilometre-wide sector at the junction between Red 8th and 9th Armies.[75]

Mamontov's corps pushed towards Tambov and also outwards against the Soviet 36th and 40th divisions which were simultaneously engaged by the main forces and fell back north west and north east, respectively. Mamontov then ordered the bulk of Postovskiy's division, the 'forward detachment' of the mobile group – equating roughly to a modern Soviet regiment in size – forward to Bratki. Meanwhile the Red Commander ordered 36th and 40th divisions back against Mamontov's flanks, and brought 31st division up from reserve. However, the Reds were unable to co-ordinate these forces' actions and the White force slipped on into the operational depth. The forward detachment reached Tambov on 18 August, the main force two days later.

The second phase of the operation, the advance westward towards Kozlov, began on 20 August. To deceive the Reds, Mamontov sent a cavalry regiment and an infantry detachment in 300 motor vehicles south. Strong reconaissance detachments were also sent north, while the main body advanced in two columns.

Up to this time the Red command did not appreciate the raid's importance or its real objectives. They called up Budënny's corps from Tsaritsyn to cut off Mamontov's withdrawal, but were mistaken about the route so Mamontov's force, fortunately for them, never encountered Budënny.[76] At this point the Revolutionary Council in Moscow decided to take special measures to deal with the raid, indicating that it had higher strategic and political as well as operational impact. Six provinces were put on a war footing and major towns and railroad stations formed military committees. The southern front formed an 'inner front' of about divisional strength with armoured train and air support, underlining the effect of an OMG-type penetration in creating such a front.[77]

Mamontov's forces were approaching the deepest point of the penetration (see Figure 5.7). The Reds were taken in by his clever deception measures and assumed him to be driving for Tula, an important armaments-manufacturing centre.[78] As a strategic and economic objective it could have been most important but this had to be weighed against the probability of success and failure and more pressing operational objectives. The Reds, thinking in economic terms, attributed their views to the enemy and were deceived.

Mamontov's escape was assisted by General Shkuro's cavalry corps which engaged the Red 8th and 13th Armies from the South. While the latter were frontally engaged, Mamontov struck them from the rear, and on 17 September forded the Don. On the next day his reconnaissance detachments once again made contact with the main body of White troops.

The original intention had been to conduct a major raid with two corps in conjunction with a Front offensive. However, this had been modified into a more limited raid to take pressure off the Don Army. However, Mamontov went beyond the mission given him. All Red reserves had to be called on to counter the raid, prejudicing regular operations in progress. In an area of over 2,000 square miles considerable damage was done to railway rolling stock, communications, depots and stores. In fact, the disorganized and fragmented nature of the civil war mitigated the effect of Mamontov's raid, as the Soviet forces at the front were largely self-sufficient: in a modern operation, where armies are vastly more dependent on supplies from the rear, the effect of a comparable penetration would be far greater.[79]

The configuration of Mamontov's force is of interest. Mamontov made good use of screening forces, with a whole 'division' (modern regiment sized) forward detachment operating over 60 kilometres ahead. He also seems to have employed the Soviet device of march security elements – stronger than mere reconaissance patrols, which ensured security for the main body on the move. The force generally advanced in three columns, a common configuration for deep-penetration forces, which shortens the length of the formation overall but offers maximum elasticity, freedom of manoeuvre and ease of control.[80]

It has to be said, however, that the Red response was inept. When it is compared with the way that the White General Wrangel dealt with a penetration by a Red cavalry corps in 1920, and the Polish encirclement of Budënny near Zamość, the Bolsheviks were indecisive and ill-informed.[81]

The Reds did put the salutary experience to good use, while recognizing the limitations of such raids. A Soviet work on the lessons of the Civil War of 1921 considered the most exemplary raids to have been Mamontov's operation and that of 5th Kuban Cossack Cavalry division against Wrangel in October 1920. However, such raids would, in the analyst's opinion, never attain their ultimate aim of completely collapsing the other side's will to resist as both fought with fanatical determination. Also, Mamonov's raid took such a toll of men and horses' strength that the corps was 'non-combat-effective' for well over a month after its conclusion. Furthermore, the raid required such a large number of horsemen (at least 5,000 – it may have been 6,000) that the rest of the front was severely weakened.[82]

CRIMSON COSSACKS

The Reds had already begun to use their own Cossack forces in raiding actions. In January 1918 the outstanding cavalry commander, 20-year-old Vitaliy Primakov had led his 'Crimson Cossacks' across partly frozen ice in a daring manoeuvre to encircle Kiev and break through its defences.[83] In July, 1919, the Crimson Cossacks, as they came to be known, had executed a raid on Kigichevka station to destroy the railway lines from Konstantinovgrad to Lozovaya.[84] Immediately after Mamontov's raid, the southern front employed a 'shock group', comprising a Lettish (Latvian) division, an infantry brigade and a Red Cossack brigade against Denikin's forces. On 3 November, the Lettish division broke through the White front and the Red Cossacks under Primakov were pushed into the gap where they operated for four days. As a result, the Whites were pushed back 120 kilometres.[85]

On 14 November 1919 Primakov's cavalry began a more ambitious raid on the key railroad junction of L'gov, some 50 kilometres east of Kursk and a principal base for the supply of Denikin's forces. The decision to launch the raid was taken by the commander of 14th Army, Uborevich, later to be one of the key figures in the development of deep-operation theory and executed along with Tukhachevskiy. The raid was entrusted to 8th Cavalry division, which reached L'gov on 17 November under cover of a heavy snowstorm. Engineer teams blew up rail points while 8th Division engaged enemy forces. After a brief but bloody fight, the Whites withdraw from the town. The L'gov raid brought the Reds 1,700 prisoners, killed 500 White troops and induced the Whites to withdraw 150 kilometres.[86]

On 5 December, a front mobile group comprising 8th Cavalry division and a regiment of the Lettish division was created. Its mission was to launch a raid towards Kharkov to cut off the enemy's line of retreat towards Zmiev, Lozovaya and, by 12 December, to recapture Kharkov from the west. The force ruptured the White front on 7 December and pushed on into the enemy rear. On 11 December they fought a brief and violent meeting engagement with a large force of Denikin's cavalry in which 4 Red cavalry regiments and 40 machine gun *tachankas* participated.[87] The latter were brought in to deliver a hail of fire at the decisive moment. The Reds realized the need to stiffen the firepower of cavalry units and as a result of a series of victories over Denikin's forces were ultimately able to deploy up to 100 machine guns per cavalry regiment.[88] Stripped of the inconsequentials of early twentieth-century technology, these meeting engagements between fast-moving forces deep inside the territory of one, with victory going to the one which first uses manoeuvre to concentrate attrition, foreshadow the combat of tank and mechanized forces in the Great Patriotic War and an image of future war which still has some currency in the 1990s.

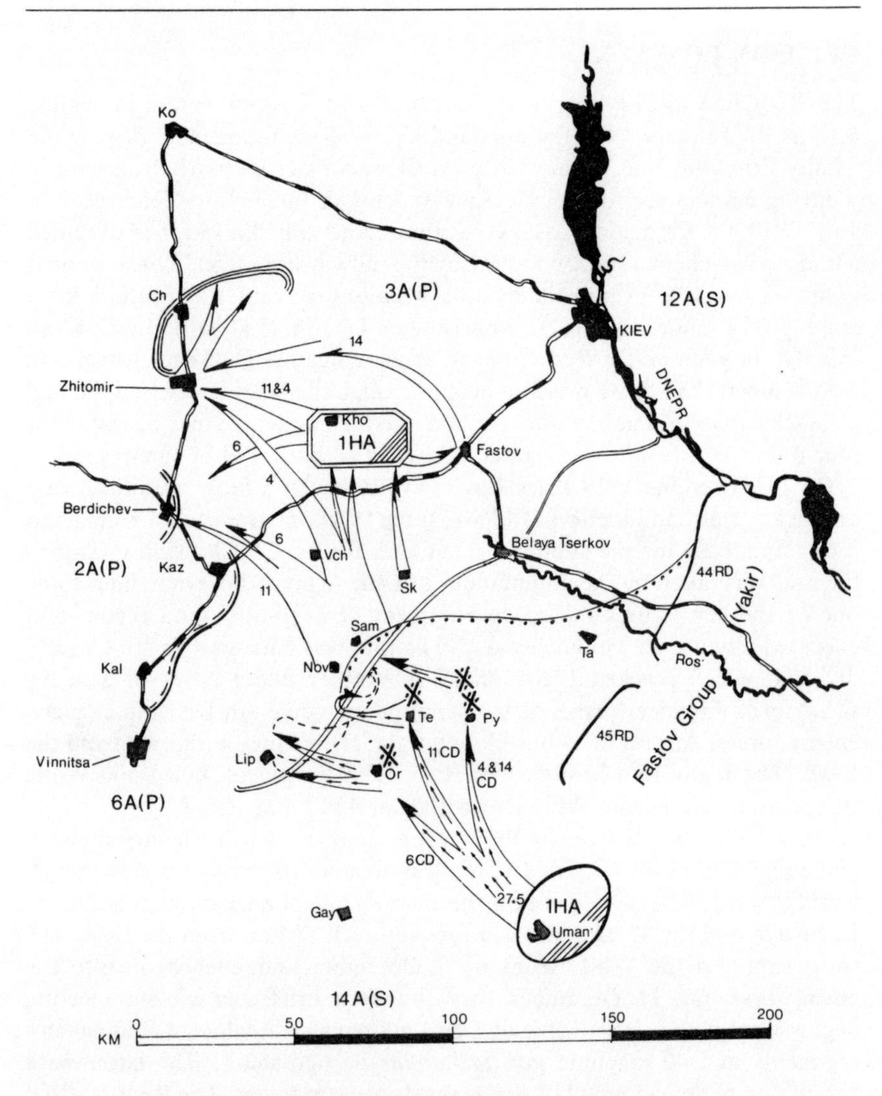

Figure 5.8 The Kiev Operation, Soviet-Polish War, 1920

Note: For simplicity, operations of the Fastov grouping, attacking towards Tarasha, and other Soviet formations, acting in co-ordination with 1st Horse Army, are not shown.

A = Army
(S) = Soviet
(P) = Polish
RD = Rifle Division
CD = Cavalry Division
numbers indicate cavalry divisions of 1st Horse Army
Ch = Chernyakov

Gay = Gaysin
Kal = Kalinovka
Kaz = Kazatin
Kho = Khodokov
Ko = Korosten
Lip = Lipovets
Nov = Novofastov
Or = Oratavo

Py = Pyatigori
Sam = Samgorodok
Sk = Skvira
Ta = Tarasha
Te = Tetiyev
Vch = Vcheraische

FIRST HORSE ARMY IN THE KIEV OPERATION, RUSSO-POLISH WAR, 1920

This operation is of particular importance as 1st Horse Army's action was more closely integrated with that of other forces than in the more readily distinguishable 'raids' of previous years; because of the greater use of armoured vehicles, though still on a small scale, the importance of radio both as a means of command and control and a source of intelligence, the subsequent influence of its commander, Budënny, and the psychology and reactions of Budënny and his opponent, the Polish Commander in Chief, Pilsudski.

Following their invasion of Soviet territory on 25 April, the Poles had captured Kiev on 7 May and reached the line Belaya-Tserkov-Lipovets-Gaysin, where they had been halted by the Soviet South-West Front (A. I. Yegorov). Soviet forces on this front comprised 12th Army, 14th Army (Uborevich) and 1st Horse Army (Budënny), plus the independent 'Fastov Grouping' (Yakir) (two divisions plus elements of the Dnepr flotilla). Against them were the Polish 2nd, 3rd and 6th Armies plus White Guards and Ukrainian nationalists. Overall, 78,000 Poles faced 46,000 Soviet troops. The main enemy grouping was around Kiev and to the south-east of Vinnitsa. The Soviet command aimed to encircle and destroy the Kiev groups (3rd Army) and then destroy 6th Army around Vinnitsa (see Figure 5.8).[89]

The Soviet orders of 23 May envisaged 12th and 14th Armies commencing operations on 23 May, Budënny's 1st Horse Army a day later. First Horse Army comprised four divisions (4th, 11th, 6th, 14th), plus a cavalry

Figure 5.8 *Notes continued*

- rivers
- railways
- Polish front, 25 May 1920
- initial Soviet positions
- Polish front, 4 June
- Movements of 1st Horse Army, 25 May–4 June, 1920
- Movements of 1st Horse Army into breakthrough sector
- Movements of 1st Horse Army, 5–16 June
- Limit of 1st Horse Army advance, 16 June 1920
- Polish positions, 16 June
- Concentration area of 1st Horse Army, before 27 May
- Concentration area of 1st Horse Army before 10 June
- ⁂ Irregular White bands, destroyed by 1st Horse Army on approach

regiment for 'special duties', totalling 16,700 men, about 300 machine guns and about 50 artillery pieces.[90] It therefore equated roughly with a modern division or unified corps, which might form a front OMG although, at 36 per cent of the total Soviet Front forces it was proportionally larger.

By 25 May Budënny's Army had reached Uman', having travelled 1,000 kilometres from Rostov in five weeks. Movement of the force by rail was out of the question after five years of international and civil war: only lines vital to the supply of the front and economy were usable. Budënny made good use of the approach march and skirmishes with irregular bands to raise the standard of training and combat readiness of his troops.[91]

The main offensive had begun on 26 May but made little headway until the beginning of June. Between 27 May and 4 June, Horse Army travelled a further 100 kilometres, reaching a line just south-east of Novofastov-Samgorodok by 4 June. Meanwhile, the Soviet High Command became increasingly dissatisfied with lack of success and on 2 and 3 June, S. S. Kamenev, commander in chief of all the Soviet Republic's armed forces issued orders allocating 1st Horse Army a new direction of advance. Whereas formerly the intention had been for the cavalry to break through in a westerly direction towards Kasatin, it was now to advance in a near northward direction and 'by means of a continuous advance envelop the area around Fastov and smash the enemy grouping with strikes against its rear'.[92] The sector on the former main axis, Pustovari-Lipoviets, should only be held with weak forces. Although the army group's orders directed him past Fastov to the west of Kiev, Budënny remained attracted towards Kasatin. His first two days' objectives, to seize the areas around Karapcheyev and Vcheraishe, diverged somewhat from his assigned mission; but his third day's objective – an enveloping attack towards Kasatin and a strike on Berdichev – depart from it quite clearly.

Budënny envisaged concentration of maximum force at the decisive point. He planned to break through on a sector only 15 kilometres wide, with three divisions (14th, 4th and 11th cavalry, the latter minus 3rd Brigade) in the first echelon and 6th Cavalry Division, behind the 11th, comprising the second echelon. Very careful and detailed orders were to be given envisaging, in the event of a successful breakthrough, destruction of railway lines and road bridges. To protect the mobile group against attack from the south-west, 3rd Brigade of 11th Cavalry Division was to feint against the sector Plistov-Lipoviets. It was not to rejoin the cavalry army, but, when ordered to do so by Army Group Headquarters, connect with 14th army, a good example of deception and flexibility. Concurrently with this, the Soviets had intercepted a Polish army order envisaging a concentric attack by 3rd and 6th Armies against 1st Horse Army on 4 June. For this reason, Budënny immediately ordered the attack for the next day. The Army therefore stood in a high state of readiness for 24 hours, taking strict measures to conceal itself against aircraft. However, fairly contemporary sources consider that it was the rain and low cloud that concealed the precise position and aims of the Cavalry Army from the Polish High Command, rather than the air-defence alert. The same bad weather conditions caused the Poles to postpone their concentric attack until 6 June, by which time the Horse Army had bolted. The relative passivity of the Poles and the absence of significant engagements on this sector from 2 to 4 June also helped veil the waiting Horse Army. This is a reminder of the role of chance, luck, and weather in all operations of war and above all of the close interaction of air power with the fate of mobile groups throughout history; had the skies been clear and the ground dry, the outcome could have been very different.

First Horse Army's breakthrough, therefore, took place on 5 June. The Soviets had also found a particularly favourable sector; the junction between the Polish 2nd and 3rd Armies. At this point the deepest salient cut into the Polish deployment, and a penetrating force would have the shortest distance to cover to reach the rear of the Polish Army Group 'Kiev'. Furthermore the Polish position here was weak, with only two battalions of the 19th Infantry Regiment inserted on the night of 3–4 June. The choice of timing and place seems to have been a result of good intelligence and skilled military judgment, in a fluid situation. There was no real obstacle to the penetration of the cavalry mass.

The Soviet assault began at daybreak on 5 June, the Russian artillery belonging to the mobile group concentrating its fire on a very narrow sector. In this, it differs from the likely form of a modern OMG breakthrough. However, there were no elements of main forces anywhere near enough to provide fire support. Even if Budënny had considered waiting for them to arrive, which is unlikely, he correctly assessed that the key to success was to seize the transient 'window of opportunity' presented to him. It has been suggested that the basic question is 'whether gaps should be opened by force or sought as coincidentally occurring opportunities'.[93] Budënny's breakthrough highlights the artificiality of this distinction; an opportunity appears, is seized, the gap prised open by fire, and exploited, it all depends on chance, luck and timing.

In concert with the brisk artillery bombardment, three cavalry divisions and all the armoured cars overran the Polish positions, although the Poles put up a stiff fight, particularly the battalion defending Samgorodok and surrounding area. The Poles fought with all the courage of despair, but by midday the Soviets had unlocked the defensive system and fire points further out were falling one by one. The Soviets had broken through on a sector 10 to 12 kilometres wide, and were through to open country. The Polish 3rd Cavalry Brigade mounted a *counterstroke*, but against the overwhelming local numerical superiority of the Soviets this was just nugatory self-sacrifice. The brigade was rapidly attacked from all sides and repulsed with heavy casualties. Budënny had undisputedly achieved his first

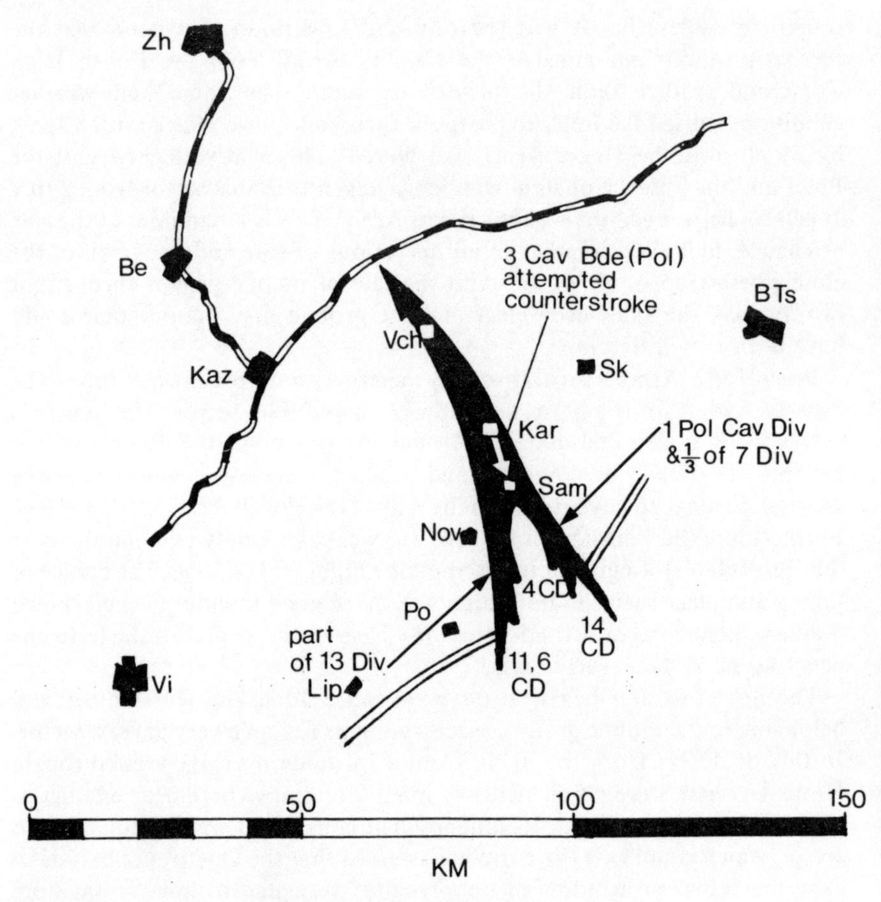

Figure 5.9 Detail of the Zhitomir breakthrough

Kar = Karapcheyev Po = Pogrebishche

Other locations see Figure 5.8.

day's objective: breakthrough of the enemy tactical zone (see Figure 5.9).

The Polish counter-attack came the next morning. Part of the Polish 13th Division attacked from the Polish right from Pogrebishche in towards Samgorodok, while 1st Polish Cavalry Division, plus a third of 7th Division assaulted from the left from Belaya Tserkov. However, this was a blow at thin air: the Soviet force had already passed through the area. The Polish report confirmed that 'as water closes behind a stone which sinks to the bottom, so our front closed behind the Red Cavalry'.[94] Budënny, heading north-west, was not unduly troubled by this. He was well aware that he would soon be cut off from his own forces anyway. His columns carried considerable supplies of munitions with them, while food and forage would have to be found in territory formerly occupied by the Poles. His position, cut off from his own forces, was indeed grave, but so was that of the enemy in whose rear he was, a fact recognized by German analysts of the raid in 1931.[95] Success or failure depended on the commander's nerves and on the fighting qualities of the troops. Only troops with very high morale and the will to win can hope to operate in the enemy rear. Budënny managed to maintain this in spite of fruitless but weakening attacks by the Poles, which does him great credit as a commander.

By the evening of 6 June Budënny had successfully completed his second day's objective. On the following day Soviet forces advanced on Zhitomir, in the latter case cutting deep into the rearward communications of the Polish Army Group 'Kiev'. They took both cities, the Headquarters of Polish Army Group 'Ukraine' only escaping from Zhitomir at the last minute. The Russians continued to slash railroad lines, highways, and telephone lines. The planned assault on the railroad junction at Kasatin was not undertaken, because of the enemy's strength and the fact that the lines through there had been cut in other places. Budënny studiously avoided unnecessary contact with enemy forces. On the following day, 8 June, the Polish 1st Cavalry division was beaten off by the Soviet 6th which pushed on towards the Berdichev-Zhitomir area. It is clear that Budënny was not planning to move towards Kiev as ordered, and perhaps create a Cannae type encirclement in co-operation with 12th Army and the Yakir Group. Instead, on 9 June he moved towards Fastov, which he reached on the following day. Here he met the Soviet 45th Division and a Cavalry Brigade of the Yakir Group, while the Polish Kiev Group was free to break out and escape towards Fastov and the north west.

The Soviet High Command meanwhile planned to encircle 3rd Army, a group from the Soviet 12th Army moving on the line Ivankov-Dymer, with one brigade far out to the right, reaching the station at Teterev and cutting the Kiev-Malin railroad. The Cavalry Army meanwhile devoted its resources to meeting an enemy offensive eastwards from Berdichev or south-east from Zhitomir. The fact that the Polish 3rd Army, which was still intact in the Kiev area was able to break out of the nearly closed Soviet encirclement, albeit with heavy losses, was probably due to the Cavalry Army's preoccupation with its front, rather than driving into the back of the enemy. Whether Budënny decided to concentrate on piling resources against the enemy front at the expense of attacking in the rear elsewhere, or was forced to by the High Command cannot be divined at this stage. It may be that the Army Group command was unsure of its own intentions and, perhaps, could not therefore express its aims sufficiently clearly. Either way, the result is the same: the Polish 3rd Army escaped. At this stage the Horse Army's intelligence, hitherto good, also seems to have broken down: it identified a non-existent 'enemy' at Fastov, while apparently unaware of the withdrawal of substantial enemy forces from the vicinity of Kiev to the north west.

Scarcely had Budënny reached Fastov than he received a radio message from the Army Group pulling him back onto the original axis: 'Engagements to the east unnecessary. Move west and seize area Zhitomir-Kasatin. Take 45 Division if need be.'[96] The whole Polish deployment around and south of Kiev was shifting. First Horse Army's breakthrough, whatever its limitations and errors, had worked.[97]

Polish sources confirm the detail of this account, but are especially revealing about the psychological effect of such a penetration. The Polish Supreme Commander, Pilsudski, was not unduly concerned about the immediate military effect of the breakthrough, but his cool appreciation of the situation was shaken as the impact on the deeper rear became apparent. The first Soviet attempts to penetrate the Polish front east of Kasatin had been repulsed, as we have seen. In view of the extended front, Pilsudski was not, apparently, surprised to see it broken, but:

> supposed that we should find it comparatively easy, by the combined used of infantry and cavalry, to beat Budënny's cavalry in detail and to compel it to retreat; and in view of the fact that I was not concerned to be obstinate about holding this or that point of occupied territory, I decided to keep my freedom of manoeuvre, and not to tie my activities to the occupation of one strip of land rather than another.[98]

The destruction of enemy forces was the first priority. However, Pilsudski admitted that:

> I was slightly alarmed by the violent panic which broke out in the rear, but up till then I had not thought that there would be any serious moral effect on our troops at the front. Thus, when at the end of our counter-offensive in the north I took stock of the situation, and as a result of my survey came to conclusions which took too little account of Budënny's activity, I resolved to refrain for the moment from seeking a decision on the northern front, but to endeavour to dispose of Budënny's cavalry as soon as possible.[99]

Pilsudski had been informed of Budënny's 1st Horse Army's approach from the neighbourhood of Rostov-on-Don, but had believed the reports of its strength to be exaggerated. World War One experience had derogated the value of cavalry, and Pilsudski, understandably enough, was very sceptical about the idea of 'employing cavalry almost in the manner of nomadic peoples', a manner recalling the Tartar expeditions: 'Cavalry covering great distances, almost without organising its rear'.[100] The initial collision, with Budënny's cavalry did little to shake Pilsudski's confidence; their successes could be attributed to weak resistance. It was only later that the unhinging impact of the deep penetration became apparent. Pilsudski's orders to his forces immediately after Budënny's breakthrough were to carry out a counterstroke while it was still concentrating in and around Zhitomir. He ordered General Rydz-Smygly, commanding the Polish 3rd Army, to retire from Kiev along the Kiev-Zhitomir road and thus to back into Budënny's Mobile Group around Zhitomir and crush it. For reasons which Pilsudski could never explain, the order never reached 3rd Army and Rydz-Smygly withdrew north west along the Kiev-Korosten road 'as if he were endeavouring to avoid all contact with Budënny's cavalry'.[101] After this, the Poles tried several times to attack Budënny's force from various sides at once, all of which failed because of the impossibility of co-ordinating the movements of the different units in a concentric manoeuvre. However, Pilsudski confirms that the majority of the troops engaged preserved their combat readiness and morale intact and returned to the attack again and again, completely indifferent to any checks received. However, a sinister development recalled Fuller's observation on the March 1918 German offensive, where the British deployment seemed to collapse from the rear forward.[102] According to Pilsudski:

> 'Worst of all, however, the influence of these events became perceptible not at the front itself, but beyond it, in the [deep] rear. Panic set in time and again in areas even hundreds of kilometres in the rear, many a time in the Headquarters of higher formations, and repeatedly spread deeper and deeper. Even the State Government began to crack, and within it a certain precarious, unsteady pulse could be felt. Close by [me], unjustified accusations gave rise to uncontrolled fear and nervous impulses. I observed this constanly around myself. This new instrument of warfare, this raid (*jazda*) of Budënny's, was unknown and strange to our unprepared forces, and acquired the character of a legendary, insurmountable force. Furthermore, one can say that the further from the front, the more irresistable and stronger this suggestion, not subject to reason, became. In this way, for me there began to form the more perilous of fronts: the *inner front*[103]

The term 'inner front' is of particular interest. It is fairly certain that Pilsudski was here referring to the deeper, politico-strategic rear, or interior, embracing political will and civilian morale, as well as the operational (military) rear. The modern Polish Armed Forces are organized into 'external' and 'internal' fronts, the former comprising conventional forces and the latter, troops to defend lines of communications, notably against parachute and other deep-penetration attack, and internal security forces. The Polish preoccupation with the 'inner front' is doubtless a result of Poland's strategic situation, surrounded by potential enemies, and of the country's shape. A few years later, the Soviet military theorist Vladimir Triandafillov seized on Pilsudski's statement to illustrate the potential overlap between operational deep penetration and politico-strategic results.[104]

Polish sources admit that Budënny's breakthrough was one of the key reasons why the Poles found themselves pinned against the gates of

Warsaw. It also had a profound effect on Polish morale.[105]

The Reds kept up the pressure. Between 3 and 9 July 8th Red Cossack division from Uborevich's 14th Army raided the Polish 6th Army's rear as far as Proskurov for five days. They caused considerable casualties forcing 6th Army HQ to flee, disrupting communications, and capturing 500 supply wagons, and forcing 6th Army to evacuate the territory between the rivers Bug and Zbruch. This was taken as an example of cavalry's ability to influence the outcome of an Army operation.[106]

From 24 July to 21 August the same division carried out another such mission against 6th Army, against Stry, an important rail junction in the foothills of the Carpathians. The division operated for some time under the control of 1st Horse Army, and for some time independently. The orders were sophisticated and each of the division's three brigades took a different route, covering, on average, 40 kilometres a day with frequent battles. The raid destroyed numerous rail communications, all communications between Army HQ and subordinate formations, a large number of supplies, and killed about 1,500 troops. It also prevented the enemy forming a defensive line between the Zolotaya Lipa and Bug rivers.[107]

SIGNIFICANCE OF THE RUSSIAN CIVIL WAR FOR THE EVOLUTION OF OPERATIONAL MANOEUVRE FORCES

During the civil war 'raiding operations' by the Red Army got bigger and better organized. They were often specifically targeted on rail communications, but were usually linked with main forces' actions so that the raid formed the operational manoeuvre element. The Whites held the initiative in the development and employment of cavalry, but the Reds learned quickly. The White commander Wrangel, in particular, stiffened cavalry with machine guns and small-calibre artillery mounted on trucks, armoured car detachments, and squadrons of aircraft. Soviet commentators immediately after the war considered that a new arm of service was emerging: '*armoured cavalry*', which disposed of great firepower without losing its manoeuvrability, flexibility, speed of movement, and shock action. Wrangel was indeed considered to have

> turned the relationship of infantry and cavalry upside down. Armoured cavalry became his principal arm, and infantry acquired an auxiliary role. Infantry cleared the way for cavalry where cavalry was unable to proceed independently (through barbed wire entanglements, river crossings. . . .). Combat itself was a matter executed and decided by [armoured] cavalry.[108]

It was obvious that what were then described as 'technical forces' – armoured cars and armed lorries – were already beginning to replace cavalry. Other nations were drawing the same conclusions. Budënny's breakthrough, in particular, provided a startling insight into what operational manoeuvre forces might achieve. German analysis rightly concluded that horsed cavalry as such had little relevance for modern war, especially in the west. However,

> The question of how much swifter and more destructive Budënny's thrust would have been had he disposed of terrain crossing armoured vehicles and the possibility of co-operating with aircraft under the command of the Army Group may be resolved in the future . . . in the form of the fully armoured brigade as a substitute for horsed cavalry on the Budënny model.[109]

ENTER AIR, ENTER ARMOUR

The Russian Civil War provided ample encouragement for the advocates of 'deep raids' in future war. During the 1920s attention focused both on the assimilation of so called 'technical forces' – tanks, armoured cars and motorized troops – on the ground, and also of aircraft.[110] A notable work in the latter category appeared in 1928, K. Monigetti's *Combined Action of Cavalry and the Air Force*.[111] Monigetti examined many of the same issues that were to re-emerge when the OMG was identified in the early 1980s, for example, whether aircraft (in his case fixed-wing; in the 1980s, helicopters) should be based with the mobile force or fly missions in support of it from the main deployment. He also considered whether it would be possible to do away with the deep penetrating ground force altogether and achieve the effect with air power alone. A force on the ground would, he concluded, have a far more pronounced and enduring effect, whereas air strikes would only have a temporary and local influence. The political effect of a ground force in the 'enemy' depth in winning 'hearts and minds' was one important issue, which again has direct relevance to the OMG.[112]

Russian and Soviet experience with deep cavalry raids converged naturally with the need, highlighted by World War One, to convert the 'break in' to the 'break out', and keep the enemy off balance.[113] Deep operational penetration by armoured forces supported by air was advocated in the seminal works of Mikhail Tukhachevskiy, Vladimir Triandafillov, and G. S. Isserson.[114] Operational manoeuvre forces were central to the two principal possible variants of future large-scale war. If possible, Soviet planning sought to prevent the formation of solid fronts. In order to do this, by the early 1930s, they planned to organize 'highly mobile forces' (motor-mechanized and cavalry formations, supported by air), which would outflank the enemy, and be launched into the rear, ripping away part of his deployment, creating a breach with exposed flanks and inducing 'oscillation'.[115] This group, ahead of the main body, was called the 'advance-guard echelon' (in the modern context, the 'advance guard' of a formation or higher formation must not be confused with a forward detachment or OMG).[116]

If the advance-guard echelon failed to prevent the formation of a defended front, then the second variant would apply. It would be necessary, not to exploit a flank, but to create one, by breaking through (see also Chapters 1 and 3). In this case, the breakthrough exploitation echelon (ERP) would then re-establish manoeuvre behind the former enemy front.[117]

The pervasiveness of the cavalry raiding tradition is underscored by the personalities involved in this crucial and far sighted military thinking. The most prominent figures were Tukhachevskiy, Yegorov, I. P. Uborevich, I. E. Yakir and Ya I. Alksnis, now an air force officer. Yegorov, we saw, had commanded the south-west front in 1920, and worked out the plan for the Kiev operation, in which a deep-raiding operational-manoeuvre formation was so prominent. Uborevich had commended Primakov as a 'daring raider', and commanded the 14th Army in the 1920 operation. Yakir had commanded the Fastov grouping, again in the 1920 operation (immediately adjacent to Budënny's cavalry) and Alksnis, as deputy commander of Soviet forces in the Ukraine and Crimea, was not far away![118] Other officers involved in the evolution of deep-operation thinking were Uborevich's deputy commander of the Belorussian military District, Borbov, Yakir's Deputy in the Ukrainian MD, Kuchinskiy, and senior armour and chemical forces officers.[119] There is, therefore, a direct personal link between the Kiev operation in 1920 and the development of deep operation thinking in the 1930s. Although this link was abruptly severed after 1937, when many of the key figures (Tukhachevskiy, Yegorov and Yakir included) were shot, the idea lived on. Isserson continued to write in the restricted journal, *Military Thought*, during 1938, stressing the role of fast-moving armoured units in seizing rear defensive lines before they could be occupied, a prime function of modern OMGs, and in 'parallel pursuit'.[120]

THE GUTS OF THE ENEMY: THE GREAT PATRIOTIC WAR

The guts of the German Army have been largely torn out by Russian valour and generalship.

Winston S. Churchill, March, 1944[121]

The German offensive of June 1941, caught the Red Army unawares. It was reeling from the effect of the purges, endeavouring to adjust force structures to new technology and contradictory lessons from Spain, Finland and the campaigns in the west, a structural skeleton as yet not fleshed out by the necessary equipment norms, but configured primarily for the offensive. It was hardly surprising, therefore, that during the 'first period of the war' (up to November 1942), opportunities to revive the 'deep operation' concept of the 1930s were limited.[122] The employment of mobile formations in the operational-manoeuvre role only began in earnest again in the Stalingrad counter-offensive as the Russians gathered confidence and expertise, and new equipment, particularly the excellent T-34 and KV tanks came on line.[123] However, as early as the Moscow counter-offensive in the winter of 1941–2 used 'improvised mobile groups' comprising tank brigades and battalions and cavalry units and formations, as well as cavalry corps in their traditional role of exploiting success. The cavalry corps and improvised mobile groups helped increase Soviet forces' rate and range of advance, but there were no powerful tank formations (the giant July 1940 mechanized corps had been torn apart in the opening encounters of Barbarossa and disestablished). This limited the scale and success of the Soviet advance.[124]

The mobile group created in the 10th Army in the Moscow counter-offensive is of particular interest. Between 24 November and 5 December, the 10th Army regrouped in the area of R'yazan'. Its forces were split into two groups, one placed under the command of Lieutenant General Mishulin, the other, Lieutenant General Kalganov. The main army command post was at Shilovo with a temporary command post at Starozhilovo with the forward units between 20 and 75 kilometres ahead (see figure 5.10). Soviet forces pushed the Germans back until by 19 December they were on the approaches to Plavsk. By this time, the army commander, Lieutenant General Golikov had decided to form a mobile group from the 41st, 57th and 75th Cavalry Divisions in order to seize Plavsk – and important road and rail junction – more quickly. These three formations, dispersed over a total of 75 kilometres at the start of the operation, were committed on 17 December after the third line of enemy defences had been penetrated. Because of the extremely difficult terrain, the mobile group was, in fact, unable to overtake the infantry and one of the component cavalry divisions (41st) took part in the capture of Plavsk, along with two of the infantry divisions. However, this is a striking example of an OMG coalescing in the enemy rear from formations which were not only themselves separate but had been subordinated to separate higher commands (the two 'groups' within the 10th Army). As figure 5.10 also shows, 57th and 75th divisions moved across the left-hand boundary not only of their Army, but of the Western front. Army Headquarters could only communicate with Mishulin's group by radio.[125]

Nor was this an isolated case. In the same operation, the 30th Army deployed a Mobile Group comprising two tank brigades, a motor rifle regiment and an independent tank battalion. The 16th Army deployed two Mobile Groups, again, one comprising a tank brigade, cavalry division and a rifle brigade, the other tank and rifle brigades and an independent tank battalion.[126] There was therefore no reason why 1980s OMGs should not take a wide variety of forms.

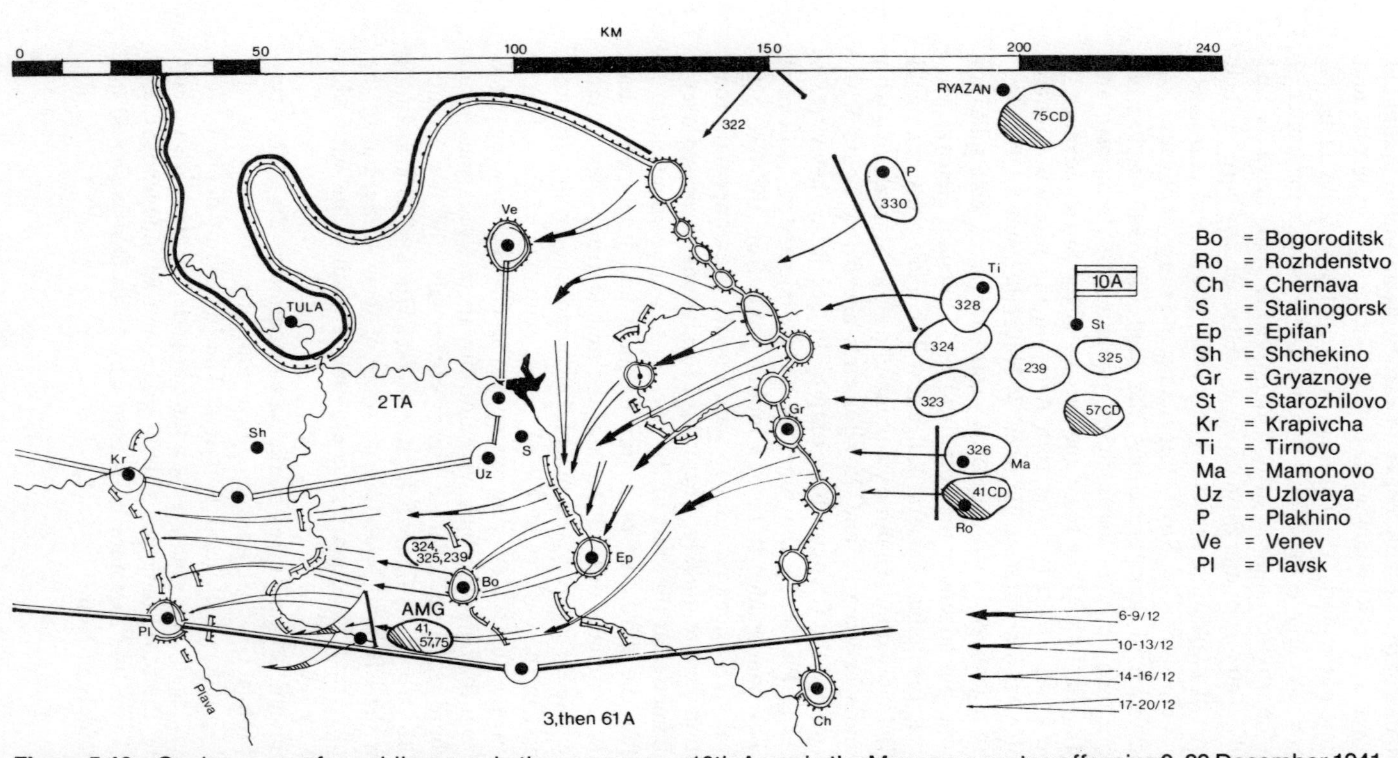

Figure 5.10 Coalescence of a mobile group in the enemy rear, 10th Army in the Moscow counter-offensive 6–20 December 1941

Note: Three cavalry formations (41st, 57th, and 75th cavalry divisions), widely separated at the start of the offensive, combine to form Army Mobile Group (AMG) over 100 kilometres beyond the start line. Other numbers indicate rifle divisions.

Different-sized arrow tips indicate movements between successive dates. Shaded symbols indicate mobile group and components.

PLANNING FOR DEEP PENETRATION

In the aftermath of the Battle of Moscow, the Soviets once again began to plan for great offensive operations along the entire Soviet German front.[127] Soviet industry's achievement in producing the newest medium (T-34) and heavy (KV) tanks was colossal: light tanks, which had featured in pre-war concepts of deep penetration were no longer manufactured from the end of 1943. In spring 1942, formation of the first tank corps began. A tank corps comprised 7,300 men and 168 tanks in 3 brigades each of 2 23-tank battalions. In April and May 1942 11 tank corps were formed within fronts and a further 14 in the reserve of the Supreme High Command. Twenty-fourth tank corps was formed in the Voroshilovgrad area in April. In May 1942 there were 25 Tank Corps and 114 independent Tank Brigades. Tank battalions and brigades were for direct support of infantry but the Tank Corps were explicitly for use as Mobile Groups.[128] The newly formed Tank Corps were first used in offensive operations conducted by the South-West Front in May 1942. Their employment was 'essentially in conformity with pre-war views', underlining the direct continuity between 1930s military thinking and Great Patriotic War practice.[129] The plan was to commit the Tank Corps into battle on the third day, but in fact they were not inserted until day six. The results were disappointing, due in part to the slow rate of advance of the main all-arms forces and the lack of artillery and air support for insertion.

Further experience of employing Tank Corps was obtained between July and September. Tank Corps were broken up and committed in dribs and drabs or, as in the case of 1st Guards Army in September, separately one after the other. These experiences vindicated pre-war views on the need to mass armour, and failures were attributed to all-arms commanders' lack of understanding of those principles. As a result, in October 1942 People's Defense Commissariat (NKO) order 325 was promulgated, directing that Tank Corps were Front and Army assets and forbidding their employment split up into brigades. This remained in force until the armoured and mechanized force regulations of 1944. Tank formations were only to be committed to battle when the main line of enemy defenses had been broken through, facilitating a 'clean break' and their action in the enemy depth. Meanwhile, in September 1942 the Red Army also began to form mechanized corps, 13,500 strong, each with 175 tanks and 96 guns.[130] These tank and mechanized corps were first committed to battle in accordance with the requirements of NKO order 325 in the counter-offensive at Stalingrad, when the South Western, Don and Stalingrad Fronts disposed of 15 such formations between them.[131]

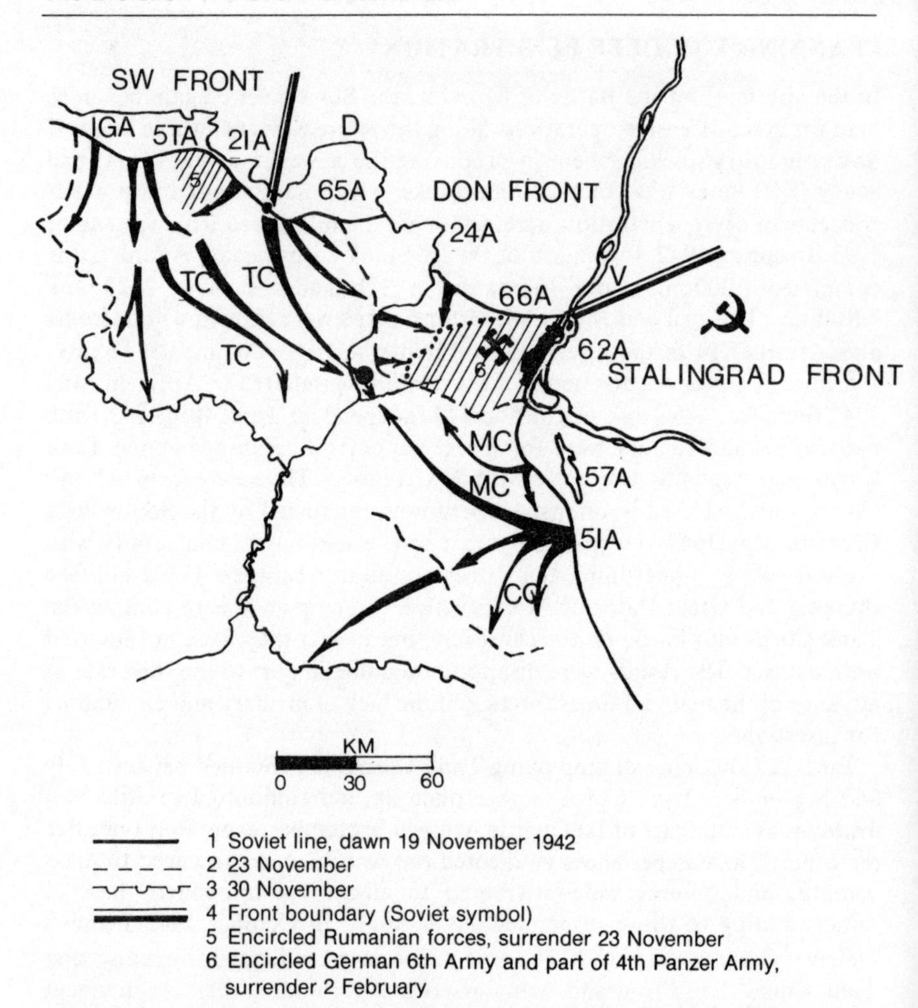

Figure 5.11 Battle of Stalingrad and the Middle Don operation, 1942

A = Army
CC = Cavalry Corps
GA = Guards Army
MC = Mechanized Corps
TA = Tank Army
TC = Tank Corps
D = Don River
V = Volga River

TWENTY-FOURTH TANK CORPS OF 1st GUARDS ARMY IN THE TATSINSKAYA RAID, DECEMBER 1942

On 19 November 1942 Soviet forces launched their carefully prepared counter-offensive against the over-extended German, Rumanian, and Italian forces which had pushed to the frontiers of Asia. By 24 November the Red Army had encircled the German 6th Army, elements of the 4th

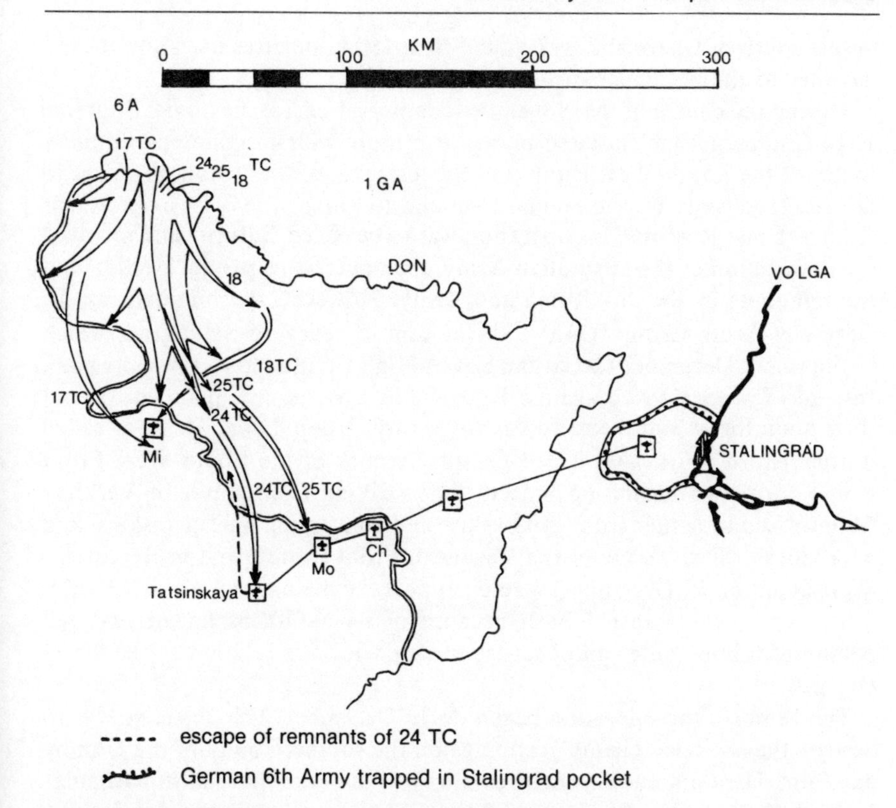

Figure 5.12 Cutting the air reinforcement and resupply route: detail of the Middle Don operation, 16–30 December 1942

A = Army
GA = Guards Army
TC = Tank Corps
Ch = Chernyshkovskiy
Mi = Millerovo
Mo = Morozovsk

Panzer Army in Stalingrad itself and to the west, destroyed the 4th Rumanian Army, pushed back the 3rd Rumanian Army to the River Chir and created a gap between the Germany Army Group B in the north and Army Group A in the Caucasus (Figure 5.11). On the same day Hitler declared that Stalingrad was to be a 'fortress', to maintain its position relying on air resupply which Goering had promised with colossal over-optimism.

This air resupply line presented a clear parallel with the railway lines of former wars (Figure 5.12). The German forces in the pocket required 600 tonnes of resupply for normal existence with 300 tonnes as the bare minimum. Bad weather, Soviet anti-aircraft guns, and fighters combined to prevent even this being attained, the average daily delivery being just over 100 tonnes. From the relative safety of the Germany Army Group, or so the Germans thought, 'Don' aircraft would leave Tatsinskaya airfield, with about 40 kilometres to the next field south of Morozovsk, then another 40 to one south of Chernishkovskiy, and then 150 kilometres over Soviet-held territory to the beleaguered pocket itself.[132]

During the course of the November counter-offensive the Soviet Supreme High Command had initiated plans for a more wide-ranging operation by forces of the South-West Front and the left wing of the Voronezh Front to destroy the enemy on the middle Don and to pursue the offensive towards Kamensk and Rostov. This operation was to be called 'Saturn', and involved the destruction of the 8th Italian Army, the operational group 'Hollidt' and the remnants of the 3rd Rumanian Army. However, the offensive by the Germany Army Group 'Don' with the aim of relieving Stalingrad, which began on 12 December forced the Soviet High Command to revise its plan. Instead of a deep strike against Rostov, the Soviets now planned to send their main forces south east to destroy Army Group 'Don'. This was called 'Little Saturn'. First and Third Guards Armies of the South-West Front would form two encircling pincers, one striking from south of Verkhny Mamon and the other from Bokovskiy, and converging on Tatsinskaya and Morozovsk. Sixth Army of the Voronezh Front (transferred to the South-West Front on 19 December) would support the main attack from the west. Tatsinskaya was a target both because of its nature as an air and rail communications centre and because of its position in relation to the forces engaged.[133]

The Middle Don operation began on 16 December. The objective was to destroy the encircled enemy groupings on the southern bank of the Don by day four. The tank and mechanized corps would take the lead in particular 1st Guards Army's 24th Tank Corps which was given Tatsinskaya as its objective to be taken by 24 December and 25th Tank Corps, targeted on Morozovsk, the other key airfield on the Stalingrad route. These objectives were clearly defined before the offensive, as was that of 6th Army's 17th Tank Corps, which was to cover the right flank and drive for Kantemirovka. Seventeenth Tank Corps would then continue to the airfield at Millerovo, which it would attack in concert with 18th Tank Corps by 24 December.[134]

Meanwhile, from 12 December, prior to the operation, Soviet Front aviation (17th Air Army) launched attacks on Millerovo and Tatsinskaya airfields and the railroad junction at Likhaya. Immediately before the operation they attacked Tatsinskaya, Morozovsk and the railroad between the latter and Likhaya. Night bombers attacked enemy headquarters and reserves.[135]

First Guards Army was deployed in two echelons for the attack, the three Tank corps (18th, 24th, 25th) forming the second 'breakthrough exploitation echelon'. The Tank Corps themselves were also deployed in two echelons before their insertion. Eighteenth and 25th Corps were committed to battle on 17 December and 24th Corps on 18 December. The latter coincided with the collapse of 8th Italian Army two days after the beginning of the Soviet offensive. Twenty-fourth Tank Corps under Major-General

Badanov tore into the gap created by the Italian collapse and on towards its distant objective. By 19 December, 17th, 18th, 24th and 25th Tank and 1st Mechanized Corps were cutting through German support elements and driving south-east in order to cut off the enemy's withdrawal routes to the south-west.[136]

German air made strenuous efforts to check the swift advance of 24th and 25th Tank Corps. On 24 December alone the Luftwaffe launched 500 sorties against 25th Tank Corps. By this time, the mobile groups were over 100 kilometres ahead of their supporting infantry, and had covered a total distance of up to 240 kilometres.[137]

Supplies for the Stalingrad pocket were brought into Tatsinskaya by both air and rail. They were stockpiled on the airfield and at the train station. Defending this key point were some 120 men of 62nd infantry division. The Germans had, apparently, realized that Soviet mobile forces might interrupt operations, but requests to move the airlift further west were refused. There were 180 Ju-52 transport planes on the field which, together with the He-111 bombers at the Morozovsk airfield, comprised the entire airflift capability for the Stalingrad pocket. At 0530 on Christmas Eve the tank corps' artillery opened up with a brief barrage, after which Soviet tanks rushed the airfield.

Twenty-fourth Tank Corps launched a concentric attack, – indeed, Marshal Rokossovskiy later commented on the corps' widespread use of enveloping movements. Fourth Guards and 130 Tank brigades attacked Tatsinskaya from the line of march simultaneously from the north-west, east and south (see Figure 5.13). A tank battalion from 130 brigade attacked the station and destroyed 50 German aircraft with all their fuel. Immediately afterwards, tanks overran the airfield proper from north and south, shooting up aircraft or driving over them. Fifty-fourth Brigade attacked the outskirts of Tatsinskaya town from the west and by the evening of 24 December the German forces surrounded in the area had been destroyed. However, some 124 aircraft managed to take off and a proportion of the Germans got away. Nevertheless, the effect on the already inadequate Stalingrad airlift was noticeable.[138]

The Germans reacted swiftly. On 24 December, even before the airfield and surrounding area was completely in Soviet hands, an advance detachment of 6th Panzer division recaptured the area north of Tatsinskaya. Sixth Panzer closed in, with 11th Panzer and 306 Infantry Division moving in from the east. By 27 December 24th Tank Corps had been encircled, and frantic radio messages in clear calling for 1st Guards Army to come to the rescue of the corps were to no avail. The corps had been refuelled from motor fuel and lubricants captured at the airfield but was dreadfully short of ammunition. An urgent radio message from Badanov on 27 December, and by 2300 hours on the same day Soviet aircraft had dropped 450 artillery shells, 4,500 rounds of rifle and 6,000 of submachine gun ammunition. The official restricted Soviet General Staff deductions from the operation

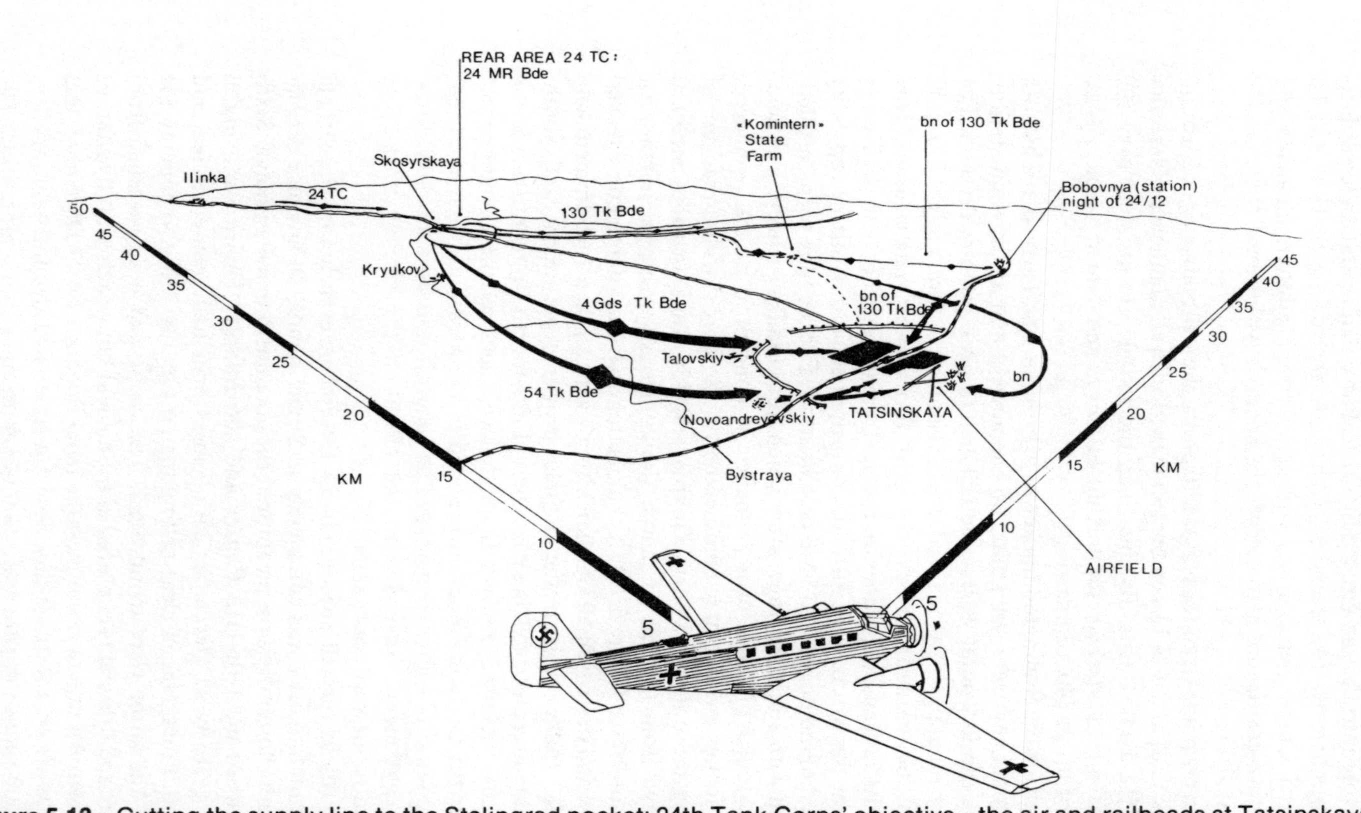

Figure 5.13 Cutting the supply line to the Stalingrad pocket: 24th Tank Corps' objective – the air and railheads at Tatsinskaya

24th Tank Corps moves to assembly area south of Skosyrskaya 23 December. Splits into component brigades which move to Tatsinskaya during 24 December. Assault on railway station and airfield, 1700 hrs, 24 December.

considered that it had only been possible to drop such limited quantities of ammunition because no provision had been made in advance for the resupply of a corps engaged in exploitation of a breakthrough, although 'the possibility of fighting in an encirclement had to be expected'.[139] The General Staff went so far as to assert that 'had it been possible for the Corps to receive larger quantities of ammunition it would have been quite able to bring into action its 39 T-34 and 19 T-70 tanks and hold out until the arrival of 25th Tank and 1st Guards Mechanized Corps which, by 29 December, had moved into the areas of Kachalin and Lesnoy.[140] By the night of 28 December 24th Tank Corps had no working tanks left and was running out of ammunition. The final hours were savage, the wounded on both sides freezing to death where they fell. Some of the Soviet troops, including general Badanov himself, managed to escape and rejoin their own forces; the rest perished.

Vatutin, commanding the South-West Front, ordered 25th Tank and 1st Guards Mechanized Corps to relieve Badanov, but it was too late. However, 24th Tank Corps' achievement was undeniable. According to Soviet sources, in the ten days (18–27 December) of the operation the Corps killed over 11,000 enemy troops, destroyed 84 tanks and 431 aircraft. It also took 4,800 prisoners although it is not known what the Russians did with them and how, or whether, they were evacuated from the battle zone. The corps was renamed 2nd Guards Tank Corps during the final desperate hours of the fight at Tatsinskaya and on 27 January 1943 received the honorific title 'Tatsinskiy'.[141] Its destruction was a tactical reverse for the Russians but the corps was not such a large element of its parent army that its loss was unbearable. 'The vacuum created by the loss of Italian 8th Army still existed and the destruction of 24th Tank Corps only eliminated the vanguard of one of the South-Western Front's advancing armies.'[142] The corps had been the spearhead of a thrust which, if successful, could have isolated Army Group A which was actually of greater military importance than the Stalingrad airlift. The Germans were also trying to relieve the Stalingrad pocket simultaneously and diverting 48th Panzer Corps against Tatsinskaya left only 57th Panzer to attempt to break the Stalingrad encirclement. The Russians were therefore able to use all their available reserves in the immediate Stalingrad area against the relief attempt. The synergic effect of a Mobile Group penetration and main forces operations is thus emphasized.

On the other hand, the way the Germans dealt with 24th Tank Corps' penetration is exemplary with a view to countering such deep attack formations in future. First, the Tank Corps (Mobile Group) was isolated from its parent forces (1st Guards Army) and, indeed, from any other OMG-type formations operating in the enemy depth (25th Tank Corps, for example). Next, it was fixed in place while information was obtained about its composition and nature, and simultaneously encircled. Finally, it was eliminated by a series of 'well planned, simultaneous and co-ordinated combined arms attacks'.[143] This denied the Soviet commander the opportunity to shift forces to deal with a succession of attacks. These attacks, by their violence and speed, also capitalized on the psychological vulnerability of a force surrounded and cut off in the enemy rear. On the other hand, the fact that they were 'deep in a hostile land' and had 'no alternative', as Sun Tzu realized, almost certainly made the Russians fight harder – until they ran out of ammunition, at any rate.[144]

Twenty-fourth Tank Corps was not strongly reinforced with other arms of service: only an extra Anti-aircraft Artillery Regiment and Rocket Launcher battalion were attached.[145] Had it been more strongly reinforced with infantry, artillery, engineer, or other specialist units, its ability to fight in the enemy depth would have been enhanced significantly. Substantial, dedicated air support would also have increased its resilience. The need to reinforce deep-penetration formations and increase self reliance was reorganized and acknowledged in the official General Staff deductions from the experience, published in autumn 1943. As well as acknowledging the potential value of air resupply organized in advance, the General Staff noted that most of the corps experienced shortages of motor fuel because the distance between supply bases and advanced mobile formations reached unexpected lengths although the most advanced of all, 24th was able to top up its reserves from the airfield it captured.[146] The General Staff also noted that the Corps' organic equipment was insufficient for salvaging broken-down transport or fighting vehicles. The experience suggested 'the necessity for reinforcing Corps operating far away from and without direct contact with the main forces with salvage companies equipped with powerful tractors'.[147]

The General Staff also drew lessons for the handling in battle and overall composition of mobile groups. In most cases, they had been inserted while the enemy was still holding out and this led to unacceptable casualties: 25th Tank Corps, for example, lost 27 tanks on unreconnoitred minefields. The corps had not been supported by aircraft in the breakthrough phrase: in future, such co-operation should be planned on an army or front scale. When mobile forces were acting in the operational depth, fighter and ground-attack aircraft should be controlled by the Tank or Mechanized Corps commander. The experience also taught that the Tank or Mechanized Corps' action was bound to achieve more success if their initial successes were exploited and consolidated by infantry. Motorized infantry or cavalry should therefore be organized for this purpose. In order to insure the continuous effectiveness of a thrust throughout the entire depth of the operation, Tank and Mechanized Corps should be merged into one mobile group comprising several corps (not less than two, at least one mechanized and the rest tank) and this group should be committed by echelons – two or even three. It would be more difficult to form and weld together an improvised Headquarters than in the case of infantry, and this suggested that Tank Corps Headquarters should be configured and receive their battle training as component parts of mobile groups. The train of thought leading to larger mobile groups and Tank Armies is clear:

> The operation carried out by the South West Front in the Middle Don area serves as an example of such employment of Mobile Groups. *The experience has shown that an operation of this kind can be accomplished only by a group of Corps placed under a unified command or merged into one Tank Army.*[148]

THE CRYSTALLIZATION OF THE OPERATIONAL MOBILE GROUP AND THE PROGRESS OF MECHANIZATION

According to the British observer, Martel, writing in May 1943, the operations south-west of Stalingrad in November 1942 had been the first opportunity for the Russians to put into practice their 'tentative' ideas regarding a breakthrough to allow a deep penetration by armoured forces, and the experience gained had since been embodied in their regulations. He felt, however, that 'Army Group' (front) staff had little experience of handling large armoured formations. The dispositions for breakthrough and penetration operations had been thought out 'with great care'. First would come motorcycle units to drive ahead fast and disrupt enemy communications, including airfields. Then came a reconnaissance group comprising light tanks, more motorcycles, anti-tank artillery (logically enough, because enemy armour was the first threat that might be encountered), and engineers (who might be needed to clear the way). Then came a group comprising motorized infantry, more anti-tank artillery and fast 76 mm guns to repel expected counter-attacks. Then came the main tank columns, moving 'in depth of about 25 kilometres', and usually in two main columns. Finally came cavalry.[149] Martel's description seems to be of a composite formation, a cavalry-mechanized group, but it makes sense and is not in principle, dissimilar to a modern Soviet formation, which would also have motorcyclists in front and probably move in two or three main columns, which, (Mishchenko's and Mamontov's raids), seems to offer maximum flexibility. Soviet Mobile Groups and Forward Detachments, the prototypes of modern Soviet OMGs, evolved along two parallel tracks. The first were the more predominantly mechanized formations, like the tank corps which performed so usefully at Tatsinskaya. In May and June 1942 the two first Tank Armies, numbers 3 and 5 were formed on the basis of the command structure of standard all-arms formations. In August 1942, the 3rd Tank Army in the Kozel'sk operation comprised two tank corps, an independent tank brigade and three rifle divisions. The 5th Tank Army in the Stalingrad counter-offensive had two tank corps, one cavalry corps, an independent tank brigade and six rifle divisions. This mix of tank and rifle formations was designed to enable these armies to 'operate independently to guarantee the breakthrough of the enemy defence and to develop tactical success into operational'.[150] It was a logical development of the pre-war breakthrough exploitation echelon (*ERP*), but with a large amount of unmotorized infantry and immobile artillery it could not penetrate as fast and far as its mission warranted. The Russians realized that the unmotorized infantry would have to be shed, the mobility of the artillery and the supply echelons increased and engineer assets substantially reinforced. Third Guards Tank Army was formed on 14 May 1943; 4th Tank Army in July and 6th Tank Army in January 1944. The composition of these armies was closer to a true front OMG.[151]

The second line of evolution was a development of the traditional use of cavalry for deep-raiding manoeuvre operations. The cavalry-raiding traditions, which we have seen can be traced directly to the American Civil War, was very much in the minds of Soviet Officers in this period and is directly linked to the modern OMG through the cavalry-mechanized group. Some of the most careful examination of the theory and purpose of OMG-style operations took place not in the context of tank armies, although it obviously applied to them also, but of cavalry mechanized groups. This emerges clearly from the testimony of Issa Aleksandrovich Pliev, an outstanding commander of such formations. Pliev was a distinguished officer, later rising to the rank of Army General (1962) and twice a hero of the Soviet Union.[152] In August–December, 1941, Pliev had used his cavalry to conduct raids on the rear of the advancing German Army Group Centre. Pliev played a central role in the evolution of the World War Two Mobile Group from earlier cavalry raids, a process which is precisely and consciously documented.

According to Pliev, there were widely varied views on the employment of mobile formations as late as the beginning of 1943. The lack of success of 4th Guards Kuban Cossack Corps in late summer, 1943 precipitated a discussion on the role of mobile formations. The corps commander, General Kirichenko, was of the opinion that 'the era of operational raids has passed', which hardly boded well for his corps' success.[153] This view was one extreme. The opposite was held by a group who enthusiastically supported the continued employment of cavalry formations with relatively light reinforcement. These cited the role of cavalry in the Moscow and Stalingrad operations, the North Caucasus and the Ukraine, in support of their view. The third group lay between the two extremes, supporting the formation of so-called Cavalry-Mechanized Groups, KMGs. These would combine the undoubted mobility of cavalry with the firepower of tank and mechanized corps, supported by aviation. They could also cite some experience in the employment of such formations, and it was this group which eventually won.[154] Pliev set out the argument clearly:

> The greater the density of fire, the more powerful the striking forces, the more mobile must forces become . . . With the breakthough of a Cavalry-Mechanized Group into the enemy rear the possibility of wide-ranging and decisive manoeuvre is created, as deployments there are not as dense as where defences are constructed in advance. In turn, the destruction of reserves, headquarters, communications centres, elements of the rear services, and so on, disorganizes command, increases apprehension in the enemy ranks, lowers their morale and will to resist.[156]

By 1943, cavalry alone could be of little use in modern war. Combined with other arms and reinforced with the most mobile firepower available it remained effective in this role to the end of the war. The logistic and administrative difficulties of handling forces comprising hooves and hay, as well as wheels, and tracks, were overcome. This must testify against those who now believe that the Russians would endeavour to avoid mixing wheels and tracks, or different types of tracked vehicle, in all cases. The mobile groups of the GPW were diverse in the extreme. There is also no doubt that the cavalry mechanized groups were to a very great extent the true prototypes of 1980s front OMGS.

Martel, the only western observer to spend any significant time with the Red Army clearly had the impression that mechanization had not permeated the gigantic Russian military machine to a great extent. 'Except for a partial success after Stalingrad', he wrote, 'the Russian armoured forces never had any success of any kind in mobile operations . . . the Russian armoured forces were a hopeless mix up of hastily improvised infantry in lorries and Tank battalions'.[156] Furthermore, he dismissed the idea of Russian 'mechanical spearheads. They certainly did not exist during the war. They are terribly short of MT (Motor Transport), which makes all these things very difficult'.[157]

This is an extreme view and quite wrong. All other reliable sources confirm that from the end of 1942 all Soviet counter-offensive and offensive operations were spearheaded by forward detachments and mobile groups, boldly and skilfully employed. Immediately after the war, German and US Intelligence experts agreed that: 'the mechanized corps invariably has been employed in every major offensive operation not only to exploit breakthroughs but to develop the breakthrough into pursuit'.[158]

The Russians were short of motor transport, it is true. The solution in these circumstances was to maintain the distinction between the army of quantity and the army of quality. Liddell Hart certainly did not accept Martel's observations, but formed the opinion that in the later stages of the Eastern Front, it was the 'fast-moving spearheads which carried out the decisive penetrations, while the massed army walked on behind'.[159] To the end of the war, the Russians still used cavalry, partly because of the shortage of motor vehicles.

The Germans bear witness to Soviet employment of 'mechanized spearheads'. Forward Detachments, Mobile Groups, Cavalry-Mechanized Groups, Tank Corps and Armies. However, some like von Manstein, actually considered that the Russians had learned the technique from them. Describing the Stalingrad counter-offensive of 19 November, von Manstein noted how 'at each of the two points of penetration strong Soviet tank forces had immediately pushed through in depth – just as we had taught them to do'.[160] Von Manstein cannot have been familiar with – or must have forgotten – the rich body of Soviet pre-war military literature which had advocated just this and, indeed, the Red Army's year of co-operation and experimentation with the Germans before the war.

By the summer–autumn campaigns of 1943 the function of developing the breakthrough at the operational level had become primarily a front responsibility, and would be carried out by 'at that time new, powerful mobile higher formations, like the *Unified Tank Army*'.[161] Relatively wide use was being made of independent tank and mechanized corps in cavalry-mechanized groups (KMGs). Tank corps formed the exploitation echelons of all-arms armies. The Soviets aimed to insert these only after the rifle formations accomplished the breakthrough of the main and, sometimes, second line of enemy defence but in fact they often had to complete the breakthrough of the tactical zone along with the main forces before proceeding.[162] However, the tank armies 'had a different purpose'. That was to act as a powerful asset under front control to 'develop tactical success into operational by dealing a strong determined tank strike' in the operational depth and to accomplish encirclement operations. Therefore, it was the front level tank armies and cavalry-mechanized groups which had a truly operational role. Soviet sources invariably class tank armies and cavalry-mechanized groups together when discussing their broad operational function.[163] Strictly speaking, the tank corps and brigades acting as part of the all-arms armies are not fully operational, although they have obvious relevance as OMG prototypes. The corps and brigades acting as forward detachments of mobile groups on the other hand, were clearly participating in an operational function. Twenty-four Tank corps, attacking Tatsinskaya in 1942 under effective front control, was also clearly an operational level mobile group. It was arguably too weak and this is one reason why the larger tank armies and KMGs assumed this role in 1943–5. Tank armies and tank corps both received their orders from front commanders and sometimes even from the Stavka of the Supreme High Command.[164] From now on, tank armies became the pre-eminent operational manoeuvre forces. Key characteristics of tank army operations are given in Table 5.1.

Table 5.1 Key indicators concerning tank army operations, 1943–5

Operation	Army	Composition	Number of Tanks and SPs	Width of pene-tration km	Depth of pene-tration km	Maximum rate of advance km/day	Maximum severance from main forces, km	Penetrate only: Width of initial corridor	Penetrate only: Number of routes	Penetrate only: Day on which inserted
Counter-offensive at Stalingrad November 1942	5 Tank	1 and 26 TC, 8 CC, 6 RDs, 1TBr	—	—	—	40–45	up to 40	—	—	—
Orlov (July-August 1943)	2 Tank	16 TC, 3TC, 11 GITBr	371	15–25	115	8	15	—	—	—
		12 TC, 15 TC, 2MC, 91 ITBr, 179 ITBr,	799	15–25	95	20	20	—	—	—
	4 Tank	5 TC, 30 TC, 11 TC, 25 TC	735	15–20	95	18	10	—	—	—
Belgorod-Kharkov (August 1943)	1 Tank	6 TC, 31 TC 3MC	562	10—20	150	35	40	4	2	1
	5 Gds Tank	29 TC, 18 TC, 5 GTC	350	10—25	120	28	25	5	2	1
Kiev (October-November 1943)	3 Gds Tank	6 GTC, 7 GTC, 9TC, 91 ITBr	621	up to 35	120	40	20	8–10	4	2
Zhitomir-Berdichev	1 Tank	11 GTC, 8GMC, 31 TC 64 GITBr	546	up to 40	180	35	80	—	—	—
	3 Guards Tank	6 GTC, 7 GTC, 9MC 91 ITBr	ca. 400	up to 40	180	35	20	—	—	—

Table 5.1 *cont.*

Operation	*Army*	*Composition*	*Number of Tanks and SPs*	*Width of pene-tration km*	*Depth of pene-tration km*	*Maximum rate of advance km/day*	*Maximum severance from main forces, km*	*Penetration only*		
								Width of initial corridor	*Number of routes*	*Day on which inserted*
Korsun' Shevchenkovskiy (January-February 1944)	5 Guards Tank	18 TC, 20 TC, 29 GTC	236	10–20	75	40	50	10	2	1
	6 Tank	5 GTC, 5 MC	204	10—15	80	40	none			
Proskurovsk-Chernovits (March-April 1944)	1 Tank	11 GTC 64 GITBr	8 GMC, 549	15–35	250	30	15	—	—	—
	3 Guards Tank	6 GTC, 7GTC, 9MC, 91 ITBr	—	up to 35	300	30	15	6	6	1
	4 Tank	10 GTC, 6 GMC, 93 ITBr	253	15–25	300	25	10–15	10	4	1
Umansk-Botoshansk (March-April 1944)	2 Tank	16 TC, 3 TC, 11 GITBr	231	20	250	65	30	12	4	1
	5 Guards Tank	18 TC, 20 TC, 29 TC	196	up to 40	300	40	—	10	4	1
	6 Tank	5 GTC, 5 MC	153	up to 40	300	40	15–20	11	4	2
Vitebsk-Orshansk	5 Guards Tank	29 TC, 3 GTC	562	up to 25	125	75	50	—	—	—
Minsk	5 Guards Tank	29 TC, 3 GTC	ca. 500	30–40	130	60	60	—	—	—

Belorussian (July-August 1944)	5 Guards Tank	29 TC, 3 GTC	—	—	—	55	40+	8	4	night of 4th
	3 Guards Tank	6 GTC, 7 GTC, 9 MC	—	—	—	60	40–45	—	—	
	4 Tank	10 GTC, 6 GMC	—	—	—	40–50	up to 55	—	—	—
Lwów-Sandomierz (July-August 1944)	1 Guards Tank	11 GTC, 8 GMC, 64 ITBR	419	20–35	400	60	40	10	4	5
	3 Guards Tank	6 GTC, 7 GTC, 9 MC	490	up to 100	330	60	40–45	6	1	4
	4 Tank	10 GTC, 6 GMC	464	20–40	350	55	55	6	1	6
Lublin-Brest	2 Tank	16 TC, 3 TC, 8 GTC	732	20	300	60	45	—	—	—
Yassy-Kishinëv (August-September 1944)	6 Tank	5 GTC, 5 MC	561	up to 35	300	65	70	15	4	1
Budapest (October 1944-February 1945)	6 Guards Tank	5 GTC, 9 GMC	229	15–40	200	30	3–8	—	—	—
Vistula Oder (January-February 1945)	1 Guards Tank	11 GTC, 8 GTC 64 GITBr	792	up to 60	610	75	80	10	4	2
	2 Guards Tank	9 GTC, 12 GTC, ITC	838	up to 40	705	90	90	20	4	3
	3 Guards Tank	6 GTC, 7 GTC, 9 MC, 91 ITBr	922	20–40	480	50	40–60	8	4	1
	4 Tank	10 GTC, 6GMC	750	20	400	60	60	6	4	1
Vienna (March-April 1945)	6 Guards Tank	5 GTC, 9 GMC	360	15–25	320	25	5–15	8	4	4
East Prussia (January-April 1945)	5 Guards Tank	10 TC, 29 TC, 47 IMBr	585	15–30	250	50	30	12–14	4	4

Table 5.1 *cont.*

Operation	*Army*	*Composition*	*Number of Tanks and SPs*	*Width of pene-tration km*	*Depth of pene-tration km*	*Maximum rate of advance km/day*	*Maximum severance from main forces, km*	*Penetration only*		
								Width of initial corridor	*Number of routes*	*Day on which inserted*
East Pomerania (February-May 1945)	1 Guards Tank	11 GTC, 8 GMC 11 TC, 64 GITBr	585	up to 35	140	30	20	8	4	1
	2 Guards Tank	9 GTC, 12 GTC, 1 MC	276	10–15	140	40	60	10	4	2
Berlin (April-May 1945)	1 Guards Tank	11 GTC, 8 GMC 11 TC, 64 GITBr	709	8–12	110	20	none	7	4	1
	2 Guards Tank	9 GTC, 12 GTC, 1 MC	672	25	130	50	none	7	4	1
	3 Guards Tank	6 GTC, 7 GTC, 9 MC, 91 ITBr	572	20	130	50	25–30	7	4	1
	4 Tank	10 GTC, 5 GTC, 6 GMC	395	15–30	170	50	25–30	7	4	1
Prague (May 1945)	3 Guards Tank	6 GTC, 7 GTC, 9MC, 91 IBTr	c. 400	10–20	180	90	100	10–11	4	1
	4 Guards Tank	10 TC, 6 GMC, 5 GMC	325	14–22	190	90	100	12–13	4	1
	6 Guards Tank	5 GTC, 9 GMC 2 GMC	ca. 400	up to 25	180	60	70	8	4	2
Manchuria (August 1945)	6 Guards Tank	5 GTC, 9 GMC, 7 MC	1019	200	820	150–180	100+	—	—	—

Source: Radzievskiy/Tankovy udar, Appendix 3 and Kurochkin, *Deystviya tankovykh armiy v operativnoy glubine,* VM, 11/1964, PP. 62–3; Losik, pp. 126, 149. In some cases the sources may conflict.

WAR IN THE NORTH: THE SVIR–PETROZAVODSK OPERATION, JUNE–AUGUST 1944

The final operation in the struggle to raise the siege of Leningrad and to secure the approaches to the city involved driving Finnish forces, which were still allied with the Germans at that stage, out of the isthmus between Lake Ladoga and Lake Onega, the centre of which lay some 200 kilometres east of Leningrad. This piece of land (see Figure 5.14) was allocated to the Soviet 7th Army. The Finns had cut this isthmus with a multi-layered, very deep defence comprising no less than six defensive lines. This would be the main axis of the attack into southern Karelia. Soviet 32nd Army would also attack through the Karelian isthmus to the west of Lake Ladoga. The Ladoga and Onega naval flotillas would also co-operate in the operation, making this an interesting example of the use of mobile groups, both in difficult conditions and, albeit indirectly, in concert with naval and amphibious forces. The Karelian Front in 1944 saw the employment of a wide variety of mobile groups, forward detachments and so on, of widely differing composition and under widely differing designations.

On 20 June Soviet forces overcame the last Finnish rearguard units holding on the south side of the Svir, and reached the river bank. The Finns then blew up the Svir hydro-electric power station dam, causing a sharp rise in the water level and width of the river, not the first difficulty the Russians were to encounter in this land of water, rocks and forest. On 21 June, Soviet forces pushed across the Svir, now 300–400 metres wide and 5 to 7 meters deep after a three-and-a-half hour air and artillery bombardment, using amphibious lightly armoured vehicles ('amphibians').[165] In the crossing, a prominent role was played by 92 Independent Regiment of *ampibious tanks* and 275 Independent Battalion of light amphibians. Two companies of 92nd Amphibious Tank Regiment with an infantry company of 300 Guards Rifle Regiment, and the other two companies with a company of 297 Rifle Regiment carried the far bank of the Svir to the West of Lodeynoye Pole in five to ten minutes (see Figure 5.14), while 275th Independent Amphibian Battalion carried 1,000 troops from 297th Rifle Regiment across the Svir in thirty minutes. We will return to the significance of this later. The Soviets energetically widened the bridgeheads on the north side of the Svir. Seventh Army as a whole forced the river crossing all along the front and by the end of 23 June the Russians had penetrated the first and, on the right flank, the second major lines of enemy defence.

After forcing the Svir, forward detachments of up to reinforced rifle battalion strength were pushed forward in pursuit of the withdrawing enemy, acting in concert with tank regiments which were part of the first echelon of rifle divisions. In addition, immediately after the crossing of the Svir an Army Mobile Group was formed, comprising 29th Red Banner Tank Brigade, 339th and 378th heavy self-propelled artillery regiments

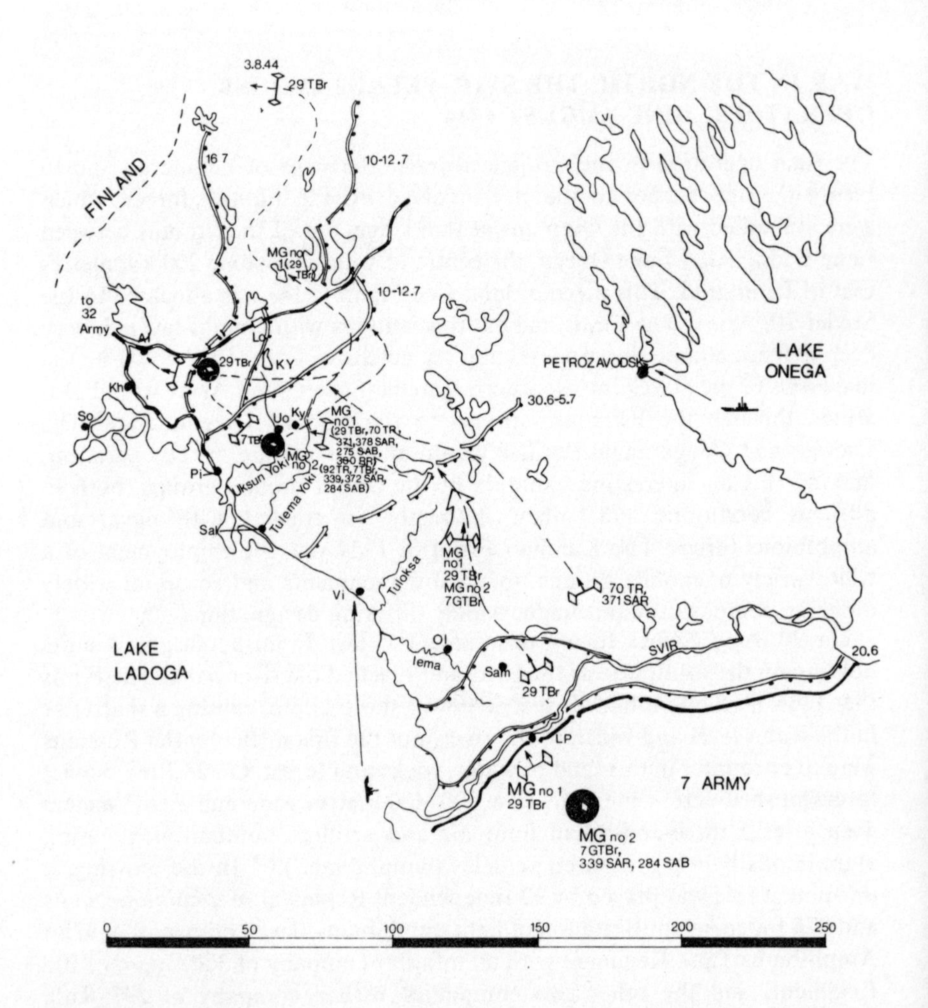

Figure 5.14 The Svir-Petrozavodsk Operation, 20 June–9 August 1944, 7th Army sector

Note: Figure simplified to show movements of Mobile Groups nos 1 and 2 only. Soviet forces were engaged across the whole front and depth of 7th Army's sector.

GTBr = Guards Tank Brigade (subsequently, just Tank Brigade (TBr))
MG = Mobile Group
TBr = Tank Brigade
TR = Tank Regiment
SAB = Special Amphibious Tank Battalion
SAR = Self Propelled Artillery Regiment
RR = Rifle Regiment
Al = Alattu
Kh = Kharlu
Ky = Kyasnasel'ska
KY = Korpi Yarvi (Lake)
Lo = Loymola
LP = Lodeynoye Pole
Ol = Olonets
Pi = Pitkyarant
Sam = Sambatuks
Sal = Sal'my
So = Sortavala
Uo = Uomas
Vi = Vidlitsa

(152 mm assault guns), and 275th Independent Amphibian Battalion plus some infantry. The mobile group was tasked to 'exploit success' (from tactical to operational), and ordered to advance in the general direction of the Olonets fortified region. Simultaneously, in order to cut off the withdrawal of forces from the Olonets fortified region towards Sal'my, two naval infantry brigades carried out a *sea assault landing* in the estuary of the Tuloksa river (Vidlitsa). They seized railroads and highways to prevent the Olonets group manoeuvring north-west.[166]

Meanwhile, 29th Tank Brigade was exploiting forward in co-operation with 300th and 298th Guards Rifle Regiments, with 378th Heavy Self-propelled Artillery Regiment and at 1500 hours on 23 June reached the river Iema. As we have just seen, 29th Tank Brigade had been designated as the core of the Army Mobile Group, along with 339th and 378th Assault Gun Regiments and 275th Amphibian Battalion. At this stage, 29th Brigade only had one of the latter supporting units attached to it, 378th Assault Gun Regiment. Instead of the others, it was teamed up with two rifle regiments. This is a clear indication from Soviet sources that although a number of formations and units may be designated as a 'mobile group', they may not coalesce until much later, and in the enemy depth.[167]

The Soviet reconnaissance elements found two fords over the Iema and the tank brigade pursued the enemy towards Sambatuks. There, the German-Finnish command had installed a powerful defence comprising permanent concrete pill boxes covered by minefields and stakes. 29th Tank Brigade, as the kernel of an army mobile group, was unable to overcome such a centre of resistance in a quick attack, and the brigade commander sent his reconnaissance detachments north and south west of Sambatuks. They discovered that the fortified positions north of this town were unmanned, and utilizing this fact the brigade executed an enveloping movement. The enemy defensive system was unhinged and Sambatuks was taken on 25 June. Units of 29th Tank Brigade then pushed forward to Megrozero and Kuitezhi, capturing the latter.

By 1900 hours on 25 June 29th Brigade had apparently joined up with 339th Assault Gun Regiment, and had reached the south-east edge of the town of Olonets. Another tank and assault gun regiment circled to the south west of the town and cut the road to Sal'my. The enemy began to withdraw in haste to the north west. By manoeuvring fast, the Russians had forced the enemy to stop defensive preparations in the Olonets region and withdraw. Another of the units designated to form part of the Army Mobile Group, 339th Assault Gun Regiment had joined 29th tank brigade, but the mobile group was not yet complete.[168]

In the next five days Soviet forces pushed forward to the line from which the second phase of the operation (1–20 July) was to start. At some stage, the mobile group forming around 29th Tank Brigade was designated 'Mobile Group No. 1', and 'Mobile Group No. 2', centred on 7th Guards

Tank Brigade began to be mentioned. This tank brigade had previously been held in the army reserve, but on 1 July, the first day of the second phase of the operation, it was committed to battle as both 7th Guards Tank Brigade and 'Mobile Group No. 2'. By the end of 3 July it had reached the river Tulema Yoki (see Figure 5.14). After crossing the river, 92nd Amphibious Tank Regiment was subordinated to 7th Guards Tank Brigade as part of Mobile Group No. 2, forming its forward detachment. Mobile Group No. 2 began pursuing the withdrawing enemy along the railroad towards Pitkyarant. At the same time, the Army command brought 29th Tank Brigade into the area in order to take enemy forces in the flank and rear towards Kyasnyasel'ka. This was Mobile Group No. 1, now comprising in addition to 29th Brigade, 378th and 371st Assault Gun Regiments, an assault engineer battalion, 275th Amphibian Battalion, and 300th Rifle Regiment of 99th Guards Rifle Division. Mobile Group No 1, now a very strongly reinforced brigade, crossed the Tulema Yoki under fire, captured Kyasnyasel'ka and by the end of 6 July had reached the defile of Uomas, between two lakes. The enemy had fortified this narrow isthmus very strongly with mines and granite 'dragon's teeth' (anti-tank obstacles), and the mobile group's first attack on it failed. Under cover of fire from the assault guns, a company of the engineer battalion supported by infantry worked their way up to the obstacles. At the same time, 275th Amphibian Battalion, again supported by assault guns, crossed one of the lakes and attacked the defenders from the north. The main forces of the mobile group, utilizing the lanes created by the engineers, attacked from the south, penetrating the defence. However, the Finns launched a spirited counter-attack with units from an officer training school, and the Russians had to pull back. These incidents underline the principle that mobile groups, forward detachments or OMGs should not get bogged down reducing strongly fortified positions, but should avoid them. However, in country like this, punctuated by lakes which create narrow defiles, this may not always be possible.

While this was going on, Army Mobile Group No. 2 had sent a forward detachment (three companies of tanks, a battery of ISU-152s, a sapper and a sub-machine gun company) towards Pitkyarant. They flanked the railroad bridge over the Uksun-Yoki, causing the bridge guard to withdraw. Then the main forces of Army Mobile Group No. 2 crossed the river and by the end of 9 July had reached the outskirts of the Pitkyarantsk fortified region. The Army Commander, (Lieutenant General A. N. Krutikov) ordered Mobile Group No. 2 and a unit from 4th Rifle Corps to envelop the region from the north. During the night of 10 July they carried this out and at 2100 hours the next evening stormed Pitkyarant. It should be remembered that in those latitudes there is virtually no darkness at that time of year.

On 13–14 July, the mobile groups were re-grouped. Twenty-ninth Tank Brigade, 70th Guards Tank Regiment. 378th and 371th Assault Gun Regiments,and 275th Amphibian Battalion were designated Army Mobile Group No. 1 and ordered to act in co-operation with 99th Rifle Corps in the advance on Loymola. Seventh Guards Tank Brigade, 92nd Amphibian Tank, 339th and 372nd Assault Gun Regiments and 284th Amphibian Battalion were designated Army Mobile Group No. 2 and ordered to work with 37th Guards Rifle Corps in the advance or Kharla. Mobile Group No. 2 participated in the capture of Sortavala, and on 19 July 29th Tank Brigade with 70th Guards Tank and 339th Assault Gun Regiments (part of Mobile Group No. 1) pushed forward into the area north west of lake Korpi-Yarvi. Their orders were to 'roll-up' the enemy in the direction of lake Ranki-Yarvi. They met ferocious resistance, the tanks and assault guns fighting without infantry support through the night of 19 July and the morning of the 20th. The Army Commander ordered them to return to the start point and that the eight damaged tanks and two damaged assault guns be evacuated. The army engineer recovery company did this, under cover of a smokescreen. They were all repaired by 27 July, an indicator of the capabilities of the Army's vehicle repair 'brigades'.

On 28 June the Russians renewed the attack on Alatt. By this stage 7th Army had effectively attained its final objective in the operation. However, this was not the end for 29th Tank Brigade. The right flank of 32nd Army, pushing up through the Karelian isthmus to the west, continued their operations until 9 August. In the area around Ilomantsi there were nine enemy infantry and Jäger battalions and a cavalry brigade. The Soviet front Commander, Army General Kirill Meretskov, ordered 29th Tank Brigade which as we have seen, was the core of one of 7th Army's mobile groups, to move across into 32nd Army's area. This necessitated a move of 245 kilometres lasting 45 hours, from 1 to 3 August. The high speed of advance through 'marshy-forested terrain pitted with lakes' was accomplished thanks to the presence of a sapper brigade, three battalions strong, and two battalions of infantry. The sappers and infantry, riding on the outside of the tanks and in armoured personnel carriers, assisted the passage of the tanks by restoring the route. In narrow defiles and in marshy areas they constructed planking using poles, branches, and special wooden screens. At 1800 hours on 3 August the brigade, plus an extra tank company, deployed from the march in the Liusvara area to attack enemy forces, driving them back to Kuolisma. The tank brigade linked up with other Soviet forces in the area and on 4 August renewed its advance towards Longonvar. Shortly after this, the right flank of 32nd Army stabilized, as its operation, too, came to an end.[169]

It is necessary to examine the operations of 7th Army and, indeed, of the Karelian Front in detail to appreciate the complexity and variety of the actions of Great Patriotic War Mobile Groups, OMG prototypes. As in the Vinnitsa operation, we see an army pushing forward not one but *two* brigade-strength mobile groups. Furthermore, although their composition was envisaged in some way before the operation commenced, they coalesced in the enemy rear, after one major line of defence had been penetrated, and their composition constantly changed as more defence lines were overcome or infiltrated. The fact that mobile groups could and did operate between defence lines of great complexity and depth, and in forested marshy terrain, consisting of alternate lakes and defiles, is revealing, as is the Soviets' ingenuity in overcoming terrain problems. Both mobile groups were on occasions used against strongly fortified positions, which is not their ideal or optimal role. This may be a reflection of a general shortage of tank forces in the area. Both mobile groups comprised specialized assets to help them fulfil their role: engineers and, in particular, the amphibian and amphibious tank units. These further underline the need perceived by the Russians, to use specialized technology to deal with particular circumstances. Lastly, 29th Tank Brigade was sent, on front commander's orders, far away across a wildnerness to join up with forces of another army. All this indicates that future OMGs would not necessarily be homogeneous, would not be allocated to armies and fronts in any uniform way, might appear and disappear in response to circumstances, and could be employed in a highly flexible and unpredictable manner.

KEY ASPECTS OF THE VISTULA-ODER OFFENSIVE OPERATION, JANUARY-FEBRUARY 1945

Soviet authorities consider that in its scale and military-political results the Vistula-Oder Operation of 12 January to 3 February 1945 was one of the most important of the Great Patriotic War. It was executed by 1st Belorussian and 1st Ukrainian Fronts, with the left wing of 2nd Belorussian and the right of 4th Ukrainian. The aim, broadly, was the conquest – or liberation – of the former territory of Poland, in three weeks. It succeeded. The fronts disposed of numerous mobile groups, both tank and cavalry formations.[170] The effect of these mobile formations ranging fast and far into the enemy depth was not only apparent to observers expert in military strategy, as Figure 5.15 shows. First Belorussian Front's first echelon comprised seven armies. Third Shock Army formed the second echelon, while the front mobile group comprised 1st and 2nd Guards Tank Armies and 2nd Guards Cavalry Corps. First Ukrainian Front had six armies in the first echelon, two in the second and 3rd Guards Tank and 4th Tank Armies as the front mobile group. The mobile group was therefore quite distinct from the second echelon.[171]

The tank armies had forward corps and these corps, in turn, all deployed forward detachments. First Ukrainian Front's Commander, Marshal Konev, took a special interest in briefing and targeting these forward detachments, an example of the way in which senior Soviet commanders would bypass the chain of command where mobile and specialized formations were concerned.[172]

"PSST!"

Figure 5.15 The OMG is not a new phenomenon. Cartoon inspired by the Vistula-Oder operation, 18 January 1945

Note: This cartoon by the popular British illustrator Carl Giles was published in the *Daily Express*, 18 January 1945, as Soviet mobile forces drove fast and far into German-held territory as part of the Vistula-Oder Operation (12 January to 3 February). Awareness of the scope and sweep of Russian operations was not confined to the senior professional military; nor was historical perspective (note that the Cossack is portrayed carrying a Tartar reflex bow!).

One of the most problematical questions about OMGs is resupply. With its penetrations of up to 500 kilometres in three weeks, the Vistula-Oder operation presented an enormous logistic challenge. The commanders of tank armies and their deputies for rear services had studied the experience of their own and other fronts' mobile groups in previous operations, but this one was particularly difficult, not least because the mobile formations were separated from the main forces by almost unprecedented distances. In the L'vov Sandomir Operation, 3rd Guards Tank Army had operated 30 to 60 kilometres ahead, but in the Yassy Kishinev Operation (August 1944), with 6th Tank Army operating 60 to 70 kilometres ahead was the only comparable example. Now, four armies were operating at such distances ahead. Logistics was a problem for all mobile forces, but the logistics of the tank armies were clearly the most complicated. A tank army's daily requirement of fuel alone was 600 to 750 tonnes, which required 270 to 300 trucks to carry it.[173]

One way of overcoming logistic problems is the use of captured supplies and munitions. The Russians captured many fuel dumps during the Vistula-Oder Operation. However, this fuel could not be used until it had been tested to ensure compatibility with Soviet vehicles and that it had not been tampered with. Testing was carried out in Army Field Laboratories (PL or PSL) at army field POL (Petrol-Oil-Lubricants) depots. This detached metallic particles and water in diesel fuel, established the viscosity of oils and established octane numbers for fuel. It is all very well to say that Soviet engineers have equipment to draw fuel from civilian and captured military stocks, but the problems do not stop there. In 1945, the Russians only permitted the use of captured fuel 'after all kinds of laboratory testing, and this took some time'.[174]

Detailed analysis of the operation shows that logistics were a real problem. Special motor transport parks of about 600 vehicles were created in 1st, 2nd, and 3rd Tank Armies, but 2nd Guards Tank Army, for example, could only carry half of the standard 'unit of fire' of ammunition and one-third of a 'refill' of fuel in these vehicles on each trip. This meant repeated trips for the transport fleet and, as tank armies pushed on, covering 500 kilometres in three weeks and moving between 60 and 90 kilometres ahead of the main forces, front-line vehicles had to be enlisted to carry supplied over stretched and (had there been serious enemy opposition in the air), increasingly vulnerable lines of communication.[175]

FULL CIRCLE: PLIEV'S CAVALRY-MECHANIZED GROUP IN THE MANCHURIAN OPERATION, AUGUST 1945

This is a fine example of a mobile group being selected for a special mission in especially difficult terrain. Pliev's cavalry mechanized group was also an international (Soviet-Mongolian) formation, without precedent in Soviet

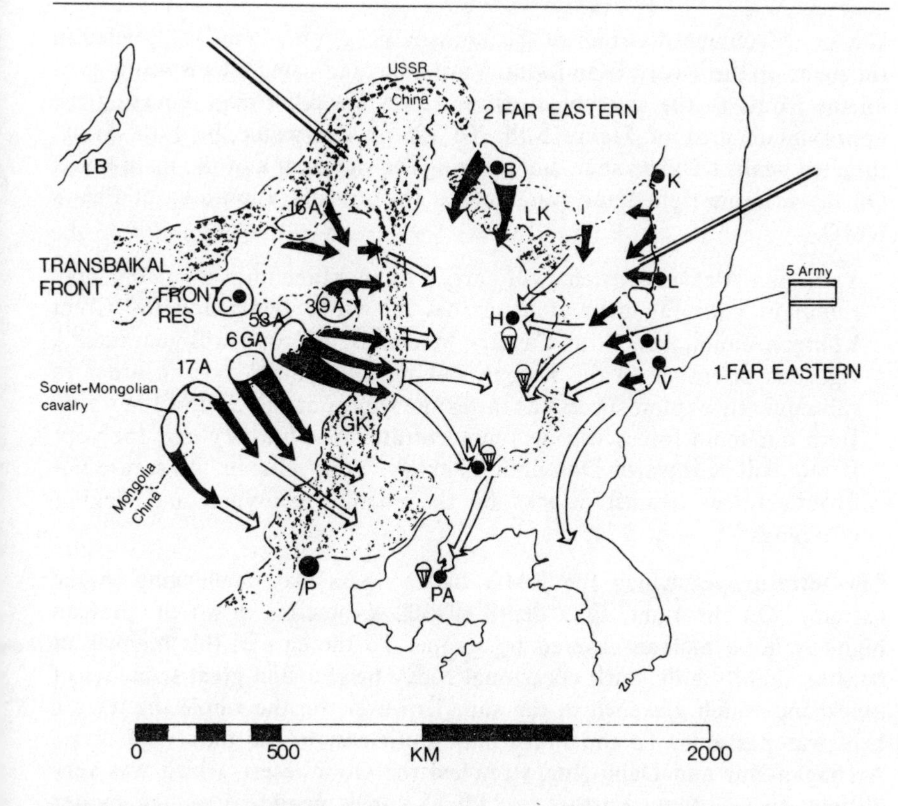

Figure 5.16 The Soviet strategic offensive in Manchuria, August 1945

Note: This was a prototype theatre strategic operation. For simplicity, only component armies of Transbaikal front are in full, plus 5th Army mentioned in text.

Dark arrows indicate advance by Soviet forces to 14 August.

Light arrows subsequent advances. Parachutes indicate main airborne landings.

B	= Blagoveshchensk	P	= Peking (Beijing)
LB	= Lake Baikal	H	= Harbin
C	= Choybalsan (Mongolia)	PA	= Port Arthur
LK	= Lesser Khingan range	K	= Khabarovsk
GA	= Guards Army	U	= Ussuriysk
M	= Mukden	L	= Lesozavodsk
GK	= Grand Khingan range	V	= Vladivostok

military history. Furthermore, its equipment was diverse in the extreme. Commanders and formations were selected with a view to their experience in Europe. One of them was Issa Pliev, who had played a cardinal role in the evolution of the cavalry-mechanized group concept and the theory and practice of operational-level deep penetration. Now, Pliev was ordered to take command of an international formation, the Soviet-Mongolian

Cavalry Mechanized Group of the Transbaikal Front. Pliev was briefed in the plane on the way to Ulan Bator. Vasilevskiy indicated the dispositions of all the fronts to the assembled officers on a big relief map, covering the approximate area of Figure 5.16. To Pliev's left would be 17th Army, targeted against Daban'shan and forming the right flank of the main body. On the extreme right flank was marked the concentration area of Pliev's KMG.

> You, Issa Aleksandrovich, will carry out in Manchuria a *raiding operation* in your favourite style, across the Gobi desert and the Great Khingan mountains. Your Cavalry Mechanized Group will guarantee a vigorous offensive on the Kalgan–Peking operational axis, in order to subsequently exploit success as far as the shores of the Gulf of Liao-dun. Here our main forces will also concentrate. A *subsidiary axis* for your Group will be towards Dolonnor-Zhekhey. Your mission – to secure the Front's forces against attacks by the enemy's Sui-yuan and Peking groupings.[176]

The terrain over which the KMG had to pass was challenging in the extreme. On the right, to a depth of 300 kilometres down the Kalgan highway lay a plateau covered by steppe. To the east of this highway it became slightly hilly, with occasional rocky heights and great stretches of quicksand which glistened in the sun. However, on the whole the terrain here was passable. To the south and south east of the bitter salt lakes, Archagan-Nur and Dalai-Nur, stretched the Gobi desert, which was very difficult to negotiate. Further on, Pliev's forces would encounter greater difficulties. On the line Chzhanbey-Dolonnor lay the foothills of the Great Khingan then the mountain chain itself, and the passes through it. In the mountains, movement off the road was absolutely impossible, and the tracks which existed were difficult going for motor vehicles, especially tanks, and often crossed fast-flowing mountain rivers. The end of the Kalgan road, extended by the Japanese on the eve of the war, passed along the Great Wall of China.[177]

Pliev's group advanced in two columns, 200 kilometres apart with 25th Mechanized Brigade and 43rd Independent Tank Brigade as their two forward detachments. By nightfall on 9 August they had covered 90 kilometres. On 15 August they ran into heavy opposition from Inner Mongolian units fighting for the Japanese. The forward detachment of the southern column, now 27th Motor-Rifle Brigade, engaged the enemy and two days later the main forces of the southern column converged to defeat the Inner Mongolian forces.[178] In all cases Pliev's KMG deployed its forward detachment to rush enemy opposition, making maximum use of surprise and maintaining momentum. In two cases, the engagements at Dolonnor and Chzhanbey, the forward detachment endeavoured to envelop the objective, a town, simultaneously deflecting reinforcements.[179] On

18 August the KMG reached the edge of the Kalgan fortified region. Although the Japanese High Command announced the capitulation of the Kwantung Army on 18 August, the units in Kalgan did not cease resistance until 21 August.[180]

Within the KMG, forward detachments were used to maintain momentum. The maintenance of a high rate of advance was especially important in the Manchurian operation, not only because the internal dynamics of military action mean that the faster you go, the fewer casualties you sustain, and the more likely you are to win. Speed was also vital for political reasons, to secure the capitulation of the main Japanese land force quickly enough to offset and complement the effect of the dropping of the first atomic bombs and to secure the USSR a role in determining the future of the region. 'While we have heard of blundering swiftness in war, we have not yet seen a clever operation that was prolonged'.[181] Pliev's group was also the southernmost of the Soviet formations, passing closest to Peking. One of the most tempting possible circumstances for the exercise of the OMG concept would be in a Soviet invasion of China in a future USSR/China war, a lightning seizure of the Chinese industrial north east. There is obviously much to be learnt from Pliev's KMG and, indeed, from the Manchurian operation as a whole in that regard also.[182] Pliev's KMG comprised a heterogeneous mixture of tanks, armoured cars (the Mongolian 7th Motor-mechanized Brigade, for example), Armoured Personnel Carriers, trucks and horses.[183] Nowadays, many believe that the Russians would go to great lengths to avoid mixing different families of tracked vehicle (BMP and BMD), or tracks and wheels (BTR) in a composite unit or formation, in order to keep the logistics simple. While that might be an ideal, Pliev's KMG presents a successful example of a formation with a requirement for hay and horseshoes as well as petrol, diesel, lubricating oil and spare tracks. Pliev's KMG also showed Soviet troops could work well and effectively with allied armies, and indicated, perhaps, that a joint Soviet–East German OMG, for example, while unusual, would not have been an impossibility, especially if the units had similar equipment. Finally, Pliev's KMG drew together many fine military traditions from widely separated places and times. Russian Tanks and Mongolian cavalry acted together, deliberately and effectively, as a component of a prototype Theatre Strategic Operation in August 1945, merging the techniques of Genghis Khan into the nuclear age. There could be no clearer evocation of the historical continuity and unbroken chain of development which this study has revealed.

CONCLUSION: MILITARY HISTORY AS A MYTH-BUSTER

The Soviet OMG was therefore a manifestation of classic principles of war, understood by competent soldiers and strategists in all countries and all times. In its 1980s form, if it could be disentangled from the imperatives of troop movement and mobile battle at all, it could trace its ancestry back to the raids of the American Civil War.

From the historical examples cited, lessons could be drawn about how a modern OMG might work, and, if necessary, how to counter it. The actions of the German command in countering the Tatsinskaya raid are exemplary at the operational level: those of Pilsudski to counter 1st Horse Army's penetration in 1920 at the strategic. Above all, as Pilsudski coolly demonstrated, do not panic. The OMG should be cut off from its parent formation and attacked energetically. OMGs would be extremely vulnerable in their isolation, and everything possible should be done to exploit this.

At the same time, this study has cast new light and sometimes major doubts on prevailing assumptions about OMGs. In particular, the author would dispute the astonishing assertion that 'the OMG concept is untested and would run counter to characteristic Soviet command and control practices'.[184] There are countless examples of OMG-type formations throughout Russian and Soviet military history. Even the most autocratic, authoritarian, and undeniably Russian commands, whether the Imperial Russian Army or Stalin, were prepared to set broad objectives for mobile formations and leave the precise method of execution to the mobile group commander. The command and control procedures for OMG-type formations have been exhaustively analysed in Soviet military publications, and their relevance to current and future operations was stressed in the most unsubtle terms.

Second, when the OMG concept re-emerged in the early 1980s, it was believed that Soviet Armies would deploy division-sized OMGs, and fronts, army-sized OMGs. The restructuring now believed to be in progress has made possible a more flexible approach: historically, 'front' OMGs have tended to be of 'corps' or equivalent strength, whilst 'army' subordinated OMGs have equated more with brigades. A World War Two tank army was more like a modern corps in strength. We have also seen examples of armies throwing forward not one but two 'forward detachments', of brigade strength (Vinnitsa, Svir-Petrozavodsk), and of two forward detachments leap-frogging ahead of one another (Svir-Petrozavodsk). A large mobile group (OMG) must itself have a forward detachment, which in a rapid advance will exploit every opportunity to unbalance the enemy, and thus itself acquires the nature of an OMG. One must agree with Richard Simpkin that 'to anyone with a feel for the linear imperative of troop movement, the OMG is essentially evolutionary in nature'.[185] However, one must take issue with Simpkin placing a Soviet tank army on a single route with its head between Brussels and Ostend and its tail just clearling Berlin.[185] The modern corps, roughly equating to a World War Two tank army, is a more manageable beast, and if we examine 2nd Guards Tank Army, in that war, we see that it was inserted, two divisions up, on seven routes, a far more feasible proposition.

More theoretical questions have been posed. A Polish officer suggested, correctly, that World War Two mobile groups were to a large extent a function of the 'technical differentiation in the ability of troops to manoeuvre'.[187] Mechanization and motorization had permeated the Red Army to a limited extent and it made sense to distinguish certain formations which could move faster or more easily. Nowadays, however, the officer considered that 'the capacity of all formations to manoeuvre is similar and the potential of such action is in doubt'.[188] The author would dispute this: there will always be some types of military vehicle which are more manoeuvrable than others, and it is likely that the most manoeuvrable will also be the most expensive and scarce. In the modern context, the helicopter has perhaps usurped the position of the tank in World War Two. A modern OMG might include a substantial air component and other especially mobile vehicles. Another more philosophical question (but one with immense practical implications) that has been asked is 'whether gaps [for OMGs] should be opened by force or sought as coincidentally occurring opportunities'.[189] Anyone who has read this book will, it is hoped, have come to appreciate that war is a most complex business, involving chance, luck, quick thinking, and exploiting any opportunity that arises. The distinction drawn by this worthy commentator is therefore artificial. If an opportunity arises coincidentally, a clever combatant will exploit it. It is sometimes also alleged that Soviet officers may be temperamentally unsuited to exploit opportunities that do occur. In the author's view, this is rubbish. One only has to recall another example, from 1st Tank Army's 'deep raid' on Berdichev during the Vinnitsa operation in December 1943. Quick-witted Lieutenant Petrovskiy tagged on to a line of German tanks to slip inside the German minefields.[190] Military history provides an infinite number of examples.

The question of exploiting opportunities leads to another topical issue. When experimental 'corps-type' formations were identified at certain locations within the USSR, some believed that they were intended for employment as OMGs.[191] In the author's view, this runs counter to the whole nature of military operations, and to Russian and Soviet experience. It would be unwise, surely, to rely on easily identifiable discrete masses like these experimental corps for OMG missions, and how could any commander be sure that such a specialized formation would be available exactly where and when it was needed? These 'corps' were more likely part of a move to a corps–brigade structure generally, although having said that, World War Two experience (and particularly the combination of tank corps in a tank army) suggests that a corps would probably be the most suitable formation for employment in a front OMG role. But it would need to be indistinguishable from other corps, and perhaps to be designated, or tasked, at the last minute.

The more one examined the OMG in historical perspective the less clearly identifiable, the less distinctive it became. Some formations will always possess greater mobility and/or greater resilience than others. There will always be a requirement for forces to manoeuvre swiftly to prevent enemy withdrawal and thus pin him for potential destruction. OMGs were but a manifestation of classical principles of war. Take, for example, the artificial distinction between coincidentally occurring opportunities and opening gaps by force. This mirrors the distinction in classical military theory between the direct and the indirect approach, a distinction emphasized by Liddell-Hart. In its aim of unbalancing the enemy, crumbling him from within, paralysing his operational and, perhaps, strategic command centres and lines of communication rather than attacking forces *per se*, the OMG was the 'indirect approach' incarnate. However, in order to get forces into a position to do that, it might be necessary to blast a hole for them. This was certainly the lesson of the Great Patriotic War where gaps had to be created or, at any rate, enlarged and secured by force. That is a straightforward application of the 'direct approach'. Being prepared to use a measure of direct approach in order to create conditions for applying the indirect approach does not render the concept, as a whole, any less 'indirect', as Fuller understood.[192] For, as Sun Tzu said, 'he who knows the art of the direct and the indirect approach will be victorious. Such is the art of manoeuvring'.[193]

Chapter six

Case study two: Don't get involved in a land war in Asia

ENVIRONMENTAL FACTORS

Asia is the highest and largest of the continents, being nearly five times as large as Europe, half as large again as Africa, and larger than North and South America put together. Its main geographical feature is the large amount of table land, centred on the Pamirs, which forms a letter 'y' laid on its back: to the west, the bottom of the 'y' extends to Turkey. To the east, the mountain systems diverge: the Himalayas to the south and the Tian Shan, Altai, and Yablonoi mountains to the north. The centre of the continent is very dry, causing extremes of climate, and many of the rivers do not find the sea, but empty into lakes. Asia's coastline forms a number of peninsulas which bear an odd similarity with the corresponding smaller peninsulas of Europe (the theory of chaos may have some explanation): Arabia, like Spain, is high and relatively dry; India, like Italy, comprises a low rectangular plain surrounded by mountains in the north and a triangular plateau surrounded by sea in the south; the Himalayas correspond to the Alps; while the peninsulas and mountains of South-East Asia breaking up into an archipelago resemble the Balkans, Greece, and the Ionian islands. In South-East Asia the populace has depended on the rivers and adjacent swamps for wealth and subsistence, while Vietnam (Annam) is largely mountainous. This suggests some interesting comparisons: Iran (Persia) occupies much the same place as France; Mongolia and Central Asia take the place of Germany; Afghanistan that of Switzerland; while Japan is a 180-degree reflection of Britain. The last four analogies certainly bear superficial examination where their military past is concerned. The geographical factors have all influenced military operations profoundly and must continue to do so.

Since 1945, large-scale warfare and substantial guerrilla conflicts liable to burst into such warfare, have largely been confined to Asia. Most prominent are Korea; the three Middle East Wars (1956, 1967, 1973); the two Vietnam wars involving the French and the Americans; the Indian–Chinese and Indian–Pakistan wars; the 1979 Chinese invasion of Vietnam; the 1979–89 Iran–Iraq war; and the overt Soviet involvement in Afghanistan also 1979–89.

The author believes that the Middle East, South and South-East Asia and the Far East are the most likely theatres for major conventional war and the higher stages of guerrilla warfare over the next quarter to half century.

The military history of these areas is of obvious importance as future warfare must take place on the same terrain and under the same climatic conditions. Furthermore, Asia was for thousands of years the setting for warfare which surpassed that of contemporary Europe in scale, sophistication, and subtlety. The great Asian empires of the Arabs, Persians, Indians and Chinese had for millennia surpassed Europe in culture, science and prosperity, and this inevitably affected civilization's dark side (see Chapter 1), warfare. The first great work of military thought known to us, its concise aphorisms of enduring and timeless relevance and infinite application, is from China, the work of Sun Tzu, dating from between 400 and 320 BC.[1] Unfamiliarity with Asia's military past, bred by Eurocentrism and lack of linguistic competence, may not only lead to unnecessary repetition of previous mistakes, but also to neglect of some of history's most successful military commanders and a rich repository of principles of warfare. Western military institutions evince a surprising and rather worrying ignorance of the military past of the vast bulk of the Eurasian land-mass. It is true that in the last century European military technics have permeated Asian armies, but this has been a slow process and has certainly not led to European attitudes (why should it?). There are grounds for arguing that in Asia the delicate balances between the simple principles and concepts introduced in Chapter 1 are altered. In this respect, the art of war resembles cooking.

This study traces the origins and development of distinctive Asiatic styles, flavours, and combinations of flavours in warfare and highlights Asian examples of key military principles, some of which have occasionally been recognized as such by scholarly western soldiers. It traces the emergence of the Asian style of guerrilla warfare (with which western soldiers and analysts have been more familiar), the impact of western technology and its assimilation, the influence of scale, teeming populations and terrain. The conclusion is, perhaps, summarized in this chapter's title.

THE INFLUENCE OF THE STEPPE

The soldier fears the arrow and swift flight of the Parthian[2]

Since at least the time when heavily armoured, and frustrated, Roman infantry failed to get to grips with the elusive Parthians, western soldiers have noted the swift movement of Asiatic adversaries. Frequently, as Horace observed, this was combined with devastating fire- (or arrow-) power.[3]

The wide belt of grassland known as the steppe runs east–west from Poland to Central Asia, and to the south, from the Caspian to the Chinese plain runs a strip of desert and semi-desert. It is hardly surprising that solutions to military problems developed here transcended historical and political divisions, or that they have something in common, with those applied in not dissimilar conditions in North Africa, the Middle East and Persia. For thousands of years, since man first began to ride horses somewhere in Central Asia sometime after 3000 BC, waves of nomadic horsemen swept across this vast land. A style of warfare developed which was characterized by the horse–bow (mobility–firepower) combination, in contrast to the slower-moving and heavier forms of the west. This combination probably originated in the need to protect the grazing herds of the various nomadic peoples who roamed across the heartland of Eurasia: Huns, Avars, Petchenegs, Khazars, Magyars, Eastern Slavic peoples, and Mongols. The 'oriental manner of combat', concisely expressed by the Hungarian term *keleti harcmodor* was successfully applied in the fifth century AD by the Huns and from the eighth to the tenth centuries by the Magyars.[4]

The environment had other crucial effects. Whereas in the west a feudal and urban society grew up which was wedded to certain areas of ground, here ground was just an element across which one moved. It was not sufficient to drive the enemy from the field: he must be pursued, fixed, held in place if necessary, and eliminated. This may be one reason for the tremendous emphasis on pursuit in the Asiatic method of waging war. It is still particularly pronounced in the Soviet style of war today.[5]

GENGHIS KHAN 1155(1162)–1227 AND THE EARLY MONGOLS

When Temujin, meaning 'man of iron', was born in 1155 or 1162, the child was seen to clutch a large ruby red clot of blood. He grew up to be probably the greatest warrior this planet has ever known. Under his command and that of his wily generals Chebe and Subedei, the peoples whom he united, known collectively as the Mongols or Tartars, conquered not only the largest continuous land area ever subjugated by a combination of diplomacy and military force, but also the most advanced civilizations in the world at that time, defeating armies almost always more technologically advanced and numerous than their own. Not for nothing have Genghis and his proteges been called 'masters of war upon the stage of the world' (Figure 6.1). Had the Mongols' political and cultural development been as advanced as their military organization, they might easily have conquered all Europe as well.[6]

Genghis's pre-eminence as a leader is due as much to his sagacity as a diplomat and administrator as to his genius for war. Genghis only survived his fraught childhood with considerable help from his friends, and he always preferred making allies to making enemies. He was as cruel as any other leader of his time to those who threatened him, but was almost always lenient to the defeated rank and file and so they joined him in droves. Mongol operations and tactics were overwhelmingly offensive in character but were frequently carried out with aims that were strategically defensive.

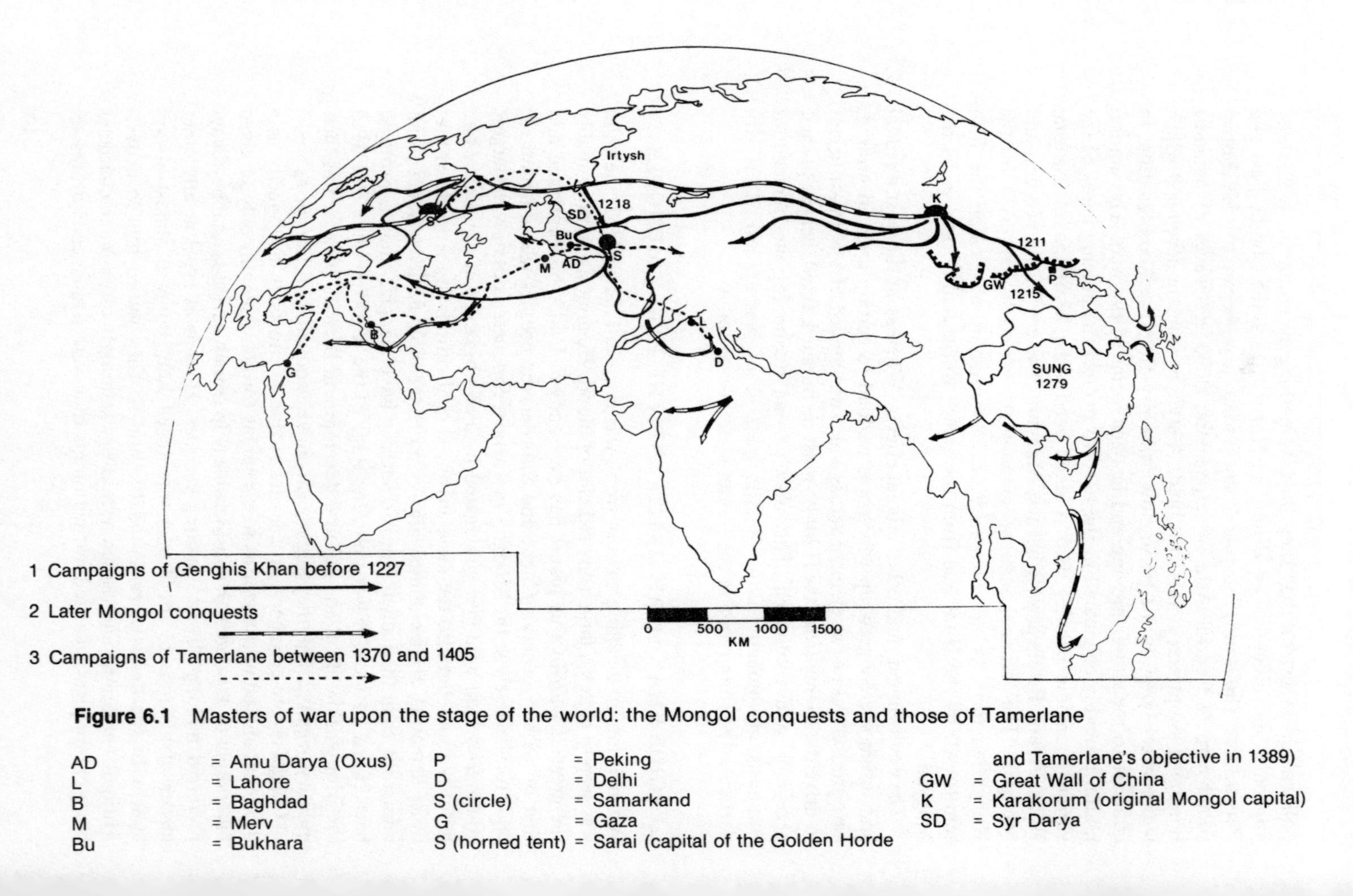

Figure 6.1 Masters of war upon the stage of the world: the Mongol conquests and those of Tamerlane

AD = Amu Darya (Oxus)
L = Lahore
B = Baghdad
M = Merv
Bu = Bukhara
P = Peking
D = Delhi
S (circle) = Samarkand
G = Gaza
S (horned tent) = Sarai (capital of the Golden Horde and Tamerlane's objective in 1389)
GW = Great Wall of China
K = Karakorum (original Mongol capital)
SD = Syr Darya

This policy extended to what is now called the pre-emptive strike. Genghis's attack on the Chinese Chin Empire in 1211, for example, was precipitated by the knowledge that the Chins were seeking to divide the tribes which he had united with so much effort.[7] Forseeing danger, Genghis always took the initiative, and he understood the value of this as well or better than any commander in history. Expertise in the interconnected and indistinguishable activities of war and politics has been a hallmark of many of Asia's great captains: Mao Tse-tung and Vo Nguyen Giap, two of the greatest of recent times, are remembered as much as political and guerrilla leaders as commanders in more conventional war, although both commanded operations of the latter kind.[8] One characteristic of the Asian style is perhaps a more all-embracing view of war and peace and the many shades between, less of a propensity to create clear-cut divisions.

Genghis's upbringing as a nomad in the broad sweep of the steppe and desert bred in him an instinctive concept of scale and contempt for distance. In the first operation he undertook against a highly sophisticated adversary, the Chins, he moved his armies on an enormously wide front, dictated largely by the need to live off the country. The main army was directed at Peking while three others crossed the Great Wall a hundred miles further west. Like Napoleon, he realized that to concentrate his armies on the field of battle itself was the acme of operational skill (using the term *operational* in the modern sense of the operational level of war), but he moved his forces on a far wider canvas than, for example, Napoleon's Ulm-march-manoeuvre in 1805.[9] Furthermore, Genghis controlled his operations from a supreme headquarters well back, communicating with his widely dispersed armies by his famous system of 'arrow' messengers, but generally allowing the army commanders great freedom. In addition, strategic command and control was achieved by strict, indeed, draconian reliance on timing. Thus, there is good evidence that in the invasion of Europe (1241) (Figure 6.2) the converging forces probably arranged to meet in the centre of the country near Buda at some predetermined time.[10] Such devices lent themselves to Genghis's style. He had complete confidence in his commanders, relied on their initiative, and did not interfere. Genghis was, therefore, the classic exponent of *Auftragstaktik*, the mission-oriented command system, and a study of his methods and a comparison with Tamerlane (see below) provide cardinal and indispensable lessons for command and control today and in the future.

The need to move large forces over great distances like this and to allow commanders great *operational* freedom overrode considerations of low-level *tactical* initiative. Instinctively, Genghis espoused the principle enunciated by Sun Tzu 'the clever combatant looks to the effect of combined energy and does not require too much from individuals'.[11] The Mongols may have encountered intellectualized theories of war among captured Chinese but the scope and prescience of their operational and

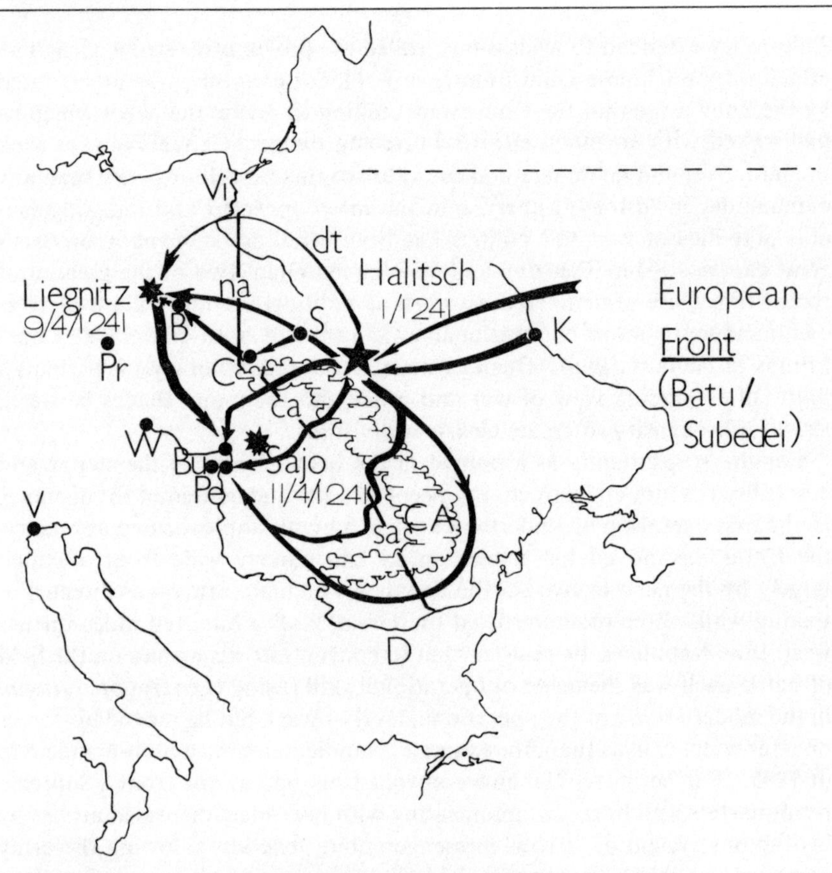

Figure 6.2 The Mongol invasion of Europe, 1241

5-pointed star = concentration of Mongol forces January 1241

B = Breslau
Bu = Buda
C = Cracow
CA = Carpathian Mountains
D = Danube
K = Kiev
Pe = Pest
Pr = Prague
V = Venice
Vi = Vistula
W = Vienna
ca = centre army (Batu/Subedei)
na = northern army (Kaidu)
sa = southern army (Kadan)
dt = detached *fumen*. Kaidu sent this to protect his Baltic flank. It was recalled after the Battle of Liegnitz, so at that engagement Kaidu had only two *fumens* or 20,000 men.

strategic vision must be ascribed primarily to instinct. This led them to create a military system unsurpassed at the higher levels of command until World War Two. Genghis's early campaigns were undoubtedly opportunist and motivated by considerations of survival, but later Mongol leaders conceived a grand strategy embracing the entire known world and actually put it into practice. After Genghis's death in 1227, the 1235 *kuriltai* or council of war planned simultaneous attacks on Poland and Korea. Obviously, planning on this scale could only consist of a broad intention to conquer a given country in a given time, and here Genghis's policy of allowing operational commanders great initiative reaped rewards.

The Mongols saw no contradiction between mass and mobility. They realized instinctively that 'force is the product of mass and the square of velocity'.[12] Unlike most armies, the Mongol forces consisted entirely of cavalry and to assure mobility at all times each trooper had one or more spare horses. Thus provided, Genghis's army covered 130 miles in two days in 1221, without a break. In 1241, Subedei's army travelled 180 miles in three days through deep snow. The Mongols' ability to traverse ground with what the ancient Chinese philosopher Sun Tzu called 'supernatural speed' and their mobility generally was studied by the Russians when contemplating operations in the same regions. Thus, the Russian general Ivanin (1801–1873) noted that in 1218 Genghis had carried out a winter expedition from the river Irtysh to the Syr Daria (see Figure 6.1) with an army of 250,000 men (this figure is probably a gross exaggeration), and that if Mongols could conduct such winter expeditions in Central Asia, then so could the Russians. Ivanin noted that Genghis had moved 50,000 men along the road from the latter river to Khiva, and this was a guide to its capacity in his day. It is fascinating to note that analysis of rates of advance done by the Israel Defence Forces on the basis of comparable frontages, breakthrough battles, and so on, has revealed that the Mongols, with an average rate of advance of 27 kilometres per day in battle conditions were the fastest-moving army in history, outstripping Rommel and the German armies in France, Poland, and Russia during World War Two.[13] We can learn a great deal about speed and mobility from the Mongols. This extraordinary fluidity gave rise to the stories of the Mongols using vast numbers of troops. In fact, the so-called 'hordes' were usually numerically weaker than their adversaries, but infinitely more efficient. The Mongol army which invaded Europe in 1241 (Figure 6.2) was only three divisions (*tumens*) strong, or 30,000 men, and only two of these were present at the battle of Liegnitz when the Mongols decisively defeated a German–Hungarian army about twice its own size.[14]

This enormous mobility also enabled the Mongols to concentrate firepower (or, rather, arrow power), at the decisive point. The Mongols' ability to move in widely separated columns and then concentrate quickly, like the fingers of a hand, has particular lessons for operations in a nuclear-scared environment, and the Soviets certainly emphasize the same quality. In some respects it is similar to the 'fan' analogy made by the great American naval strategist, Alfred Thayer Mahan (1840–1914). Indeed, the movements of these bodies of horsemen without the defensive ability to hold ground, passing swiftly over the open expanses of land mass and the enormous turning movements which they executed can be likened to a kind of naval war on land. Nowadays, war on land is once again acquiring certain characteristics associated with war at sea: the movements of discrete, armoured vehicles, ranging and firing at other, discrete moving targets over great distances, the increasing expense and complexity of land platforms, perhaps necessitating active defence measures such as – anti-missile missiles, like those already installed on warships – all these developments make the Mongols' experience of continuing, maybe even increasing relevance.[15]

At the tactical level, it was not to manpower or ingenuity of manoeuvre that the Mongols owed their success, but to firepower and battle drills. Each Mongol trooper carried 60–90 arrows with armour piercing heads shot from composite bows whose range and power equalled or exceeded those of any others in the world. After initial contact with the Persians and Chinese, the Mongols also rapidly adapted and used vast quantities of mechanical artillery. They were, indeed, the first army to use heavy-weapons fire systematically to pave the way for an assault. Tactical manoeuvre was limited to the stereotyped battle drill called the *tulughma* or 'standard sweep'. The first Mongols to make contact with the enemy would be the scouts, operating up to 110 kilometres ahead of the main body. Similar screens operated on the flanks and rear. Once contact was made the main body would extend its front over a wide area so as to envelop the enemy. The lightly armed scouts would probe the enemy force. If it were small, the Mongol vanguard might destroy it, but otherwise the scouts would fall back, luring the enemy with them. The Mongols were always most dangerous when withdrawing, recalling the 'swift flight' of Horace's Parthians.

When the main body of the Mongol force joined battle the *tulughma* was employed. This force was divided into light and heavy cavalry. Some light cavalry would move forward and pour a devastating fire into the enemy, like Napoleonic skirmishers, while that on the wings carried out a double envelopment. Then the heavy cavalry engaged the enemy centre while the light cavalry pincer closed on flank and rear. The Mongol tactical system was therefore an almost perfect combination of firepower, shock action, and mobility, based on thorough training and rigid discipline. When the enemy broke, the Mongols conducted a ruthless and aggressive pursuit, not stopping until the enemy was hunted down and destroyed. After the defeat of the Mamelukes at the battle of Salamiyet, for example, Mongol troops were seen as far south as Gaza, some 500 kilometres from the battle.[16]

Speed was crucial to Mongol success, and they avoided being tied down by the need to besiege fortified positions. If an inital attempt to rush a city failed, they would bypass it, leaving it to be taken later. They could afford to do this, untrammelled by European concepts of lines of communications. The Mongols travelled very light, carrying nothing with them except weapons and emergency rations of dried milk curd, millet, and a little dried meat. The horses were sustained by the very land over which they passed although, as noted, this did itself impose dispersion on Mongol armies. This policy of travelling light was imitated by the young Soviet general, Tukhachevskiy, in his lightning thrust towards Warsaw in 1920, which almost succeeded. Although detailed comparisons between the conditions of the Mongol conquests and those of the twentieth century can hardly be valid, given the totally different ammunition and fuel requirements of post-industrial armies, the general policy of travelling light and being as self-contained as possible is one to which all great commanders have adhered.

Lastly, the Mongols instinctively followed Sun Tzu's maxim: 'know your enemy, and know yourself'. They made extensive use of psychological warfare, exaggerating their own strength and sowing dissension in enemy countries. They had a highly developed network of reconnaissance patrols and spies. They were realistic about their capabilities, and declined to fight when they did not expect to win. In Genghis's first invasion of China, the Chins tried to lure him to attack Peking, hoping that his horsemen would be shot to pieces as they battered in vain against its stone walls. Genghis would not be drawn. When the Mongols prepared to invade Europe, they came expecting to fight for eighteen years: a fairly conservative estimate of the time it would take to subjugate the continent.[17]

The relevance of Genghis's career in the twentieth century was admirably summarized by another great Captain, the American general Douglas MacArthur, in his Annual Report of the Chief of Staff for fiscal 1935:

> Were the accounts of all battles, save those of Genghis Khan, effaced from the pages of history . . . the soldier would still possess a mine of untold wealth from which to extract nuggets of knowledge useful in molding an army for future use . . . He devised an organization appropriate to conditions then existing; he raised the discipline and the morale of his troops to a level never known in any other army, unless possibly that of Cromwell; he spent every available period of peace to develop subordinate leaders and to produce perfection of training throughout the army, and, finally, he insisted upon speed in action, a speed which by comparison with other forces of his day, was almost unbelievable [and with forces ever since]. Though he armed his men with the best equipment of offense and defense that the skill of Asia could produce, he refused to encumber them with loads that would immobilize his army. Over great distances his legions moved so rapidly and secretly as to astound his enemies and practically to paralyse their powers of resistance. . . . So winnowed from the chaff of medieval custom and of all other inconsequentials, they [the unvarying necessities of war] stand revealed as kernels of eternal truth, as applicable today in our efforts to produce an efficient army as they were when, seven centuries ago, the great Mongol applied them to the discomfiture and amazement of a terrified world.[18]

TAMERLANE, 1336–1405

Timur Beg, popularly known as Timur the Lame or Tamerlane (1336–1405) was a distant relation of Genghis Khan through the female line. He was not a Mongol himself, being of Turkic extraction: his name, predictably perhaps, means 'iron' in Turkish.[19] Like Genghis, his early life was a struggle for survival, and his military organization was of the classic Mongol type. It is alleged that Tamerlane personally drew up the scales of equipment for each soldier and detailed regulations for such things as the siting of camps.[20]

Like Genghis, his forces covered vast distances in extremes of climate. The most spectacular example is Tamerlane's daring penetration from his capital at Samarkand into the territory of the Golden Horde as far as their capital on the river Volga in 1389. His opponent, the Khan Tokhtamysh, was the direct heir of Genghis Khan and enjoyed vastly superior resources. Rather than wait to be attacked, Tamerlane decided to press Tokhtamysh back on his capital in the heart of Russia. This decision was based on knowledge of Tokhtamysh's character; the Khan of the Golden Horde had refused battle twice before. Rather than launch a counter-attack against Samarkand, Tokhtamysh did indeed fall back, and although his lines of communication were getting shorter and Tamerlane's men more tired, he was forced to conform to Tamerlane's plan. Tamerlane's army reached the Volga after a march over 2500 kilometres which had lasted for four-and-a-half months. When he finally came up against Tokhtamysh, the latter launched a reckless attack, which was trapped in a carefully laid scheme of mobile defence. Thus, Tamerlane had combined the strategic offensive with the tactical strength of the defensive, exactly what the great German strategist von Moltke was to do in the Franco-Prussian war.[21] The campaign was also a classic example of the efficacy of the relentless pursuit.

In 1400 Tamerlane met the Arab historian Ibn Khaldun outside Damascus, in what must be one of the most interesting interviews ever recorded. Ibn Khaldun quickly realized that Tamerlane's oblique questioning was intended to divine as much strategic intelligence as possible about Egypt and the dominions further west. Intelligence had always been a strong preoccupation of the Mongols: the early Mongols had used spies in the guise of merchants and diplomats (*plus ça change*) and long-range reconnaissance patrols to capture prisoners in order to complete their intelligence picture. Tamerlane then asked Ibn Khaldun to write a detailed description of the west, 'in such a manner that when the conqueror read it, it would be as if he were seeing the region'. There can be few more graphic descriptions of the effect of good and complete intelligence, compiled by an expert and presented to a military genius. Ibn Khaldun completed the assignment in a few days.[22]

There were however crucial differences between Genghis Khan and Tamerlane. Whereas Genghis had stood back and allowed his trusted generals to execute the movements which he had sketched on the broadest of canvases, Tamerlane exercised much more rigid control, concentrating his armies and leading them in person. This may have been a difference in policy or style, or it may simply be that Genghis was better served. None of Timur's generals were first-rate commanders in their own right, like Subedei and Chebe. Genghis was extremely cautious, avoiding battle if unnecessary and moving directly to the enemy's centre of resistance. Timur was as meticulous in planning and preparation as Genghis, but once a campaign had started he was impetuous. He once rode into the suburbs of Baghdad with a small band of followers and then pursued the Sultan Ahmed out of his kingdom. Genghis would have told someone else to do it. General Sir Richard Gale, the distinguished soldier and historian, noted that the differences between Genghis and Tamerlane were in some ways like those between Eisenhower and Montgomery.[23]

In his insistence on doing everything himself, Tamerlane emerges as a classic exponent of *Befehlstaktik*, a dictatorial imposer of his authority down to the lowest level, as opposed to Genghis's *Auftragstaktik*, the delegator with his mission-oriented command system. A comparison of the two grand masters' methods is of particular relevance today as improved command and control technology, and command, control and information systems (CCIS), give higher commanders the opportunity to interfere in their subordinates' affairs so that initiative and response to the pulse of the battle are stifled. Genghis Khan's methods of communications would not have permitted him to interfere in the detail of his generals' conduct of operations, even if he had wanted to, but his style is to be emulated, especially as modern battle over vast frontages and depths is increasingly less susceptible to *Befehlstaktik*. *Befehlstaktik* worked for Napoleon, but in modern conditions Genghis's methods are more appropriate than ever. Genghis and Tamerlane, the two 'world conquerors', are the most perfect exponents of antithetical principles of command and control, the two diametrically opposed principles which like Yin and Yang pervade the essence and application of military skill. It may fairly be said, however, that of the two, the subtler, the delegator, the avoider of combat where possible, was the greater. In terms of the scale and scope of his victories, Tamerlane was almost comparable with Genghis.

> Much as we admire the campigns of a Caesar, the exploits of a Hannibal, or the inspired strategy of a Napoleon, upon reflection it is becoming clear that these two conquerors from Asia are, with Alexander, masters of war upon the stage of the world. Their feats of arms may have been duplicated by others in miniature, but never upon the earth as a whole.[24]

VERDANT FASTNESS:[25] VIETNAM (ANNAM) AND THE ORIGINS OF MODERN GUERRILLA WARFARE IN ASIA

Throughout their history, the Vietnamese, or Annamites as they were once known, have continually had to fend off attacks by larger and more powerful neighbouring states, particularly China. In that sense, Douglas Pike's analogy, 'the Prussians of Asia'[26] is appropriate. Unlike Prussia, however, their country was extensive and mountainous. Whereas the open steppe and dependence on a moveable resource – flocks and herds – had made speed and sudden concentration for pitched battle the only possible method of conducting operations, a predominantly agricultural population in a comparatively rich environment was almost bound to evolve different styles of fighting. According to Vietnamese sources, what we would recognize as guerrilla warfare was of 'strategic importance' in the struggle against the forces of the Chinese Chin Empire in the third century AD.[27] The guerrilla warfare conducted by the Viet Minh and the Viet Cong against the French and Americans in the third quarter of the twentieth century was arguably not so much a product of Communist ideology as part of an Asiatic (and particularly Vietnamese) tradition going back for centuries. As Mao Tse-tung also stressed, guerrilla warfare was not an end in itself, and was not decisive in itself. It was forced on the weaker side because they lacked the conventional strength to oppose the stronger. Mao argued that guerilla warfare was a possible, natural and necessary development in an agrarian-based revolutionary war. Furthermore:

> It is a weapon that a nation inferior in arms and military equipment may employ against a more powerful aggressor nation. When the invader pierces deep into the heart of the weaker country and occupies her territory in a cruel and oppressive manner, there is no doubt that conditions of terrain, climate and society in general offer obstacles to his progress and may be used to advantage by those who oppose him. . . . During the progress of hostilities, guerrillas gradually develop into orthodox forces that operate in conjunction with other units of the regular army.[28]

In the case of the Vietnamese, the war was not so much 'revolutionary' as a war against foreign invaders. Similarly, the Spanish guerrillas of 1808 to 1813 were not 'revolutionary' they simply wanted to help Wellington throw the French out of Spain. They fought as guerrillas because their conventional army had been beaten.[29] In the same way, the Vietnamese adopted a defensive, 'guerrilla' posture because of an initially unfavourable balance of forces. 'The measure was only temporary and designed to turn the tables on the enemy and set up an advantageous position to regain the strategic offensive'.[30] The artful combination of 'guerrilla' and 'orthodox' forces has been a distinctive feature of Vietnamese military art:

> That strategy took various forms, with two main modes of operation: small-scale battles with scattered forces and large-scale battles with massive troop concentration.
>
> These two modes of operation have always been well combined on the strategic as well as on the tactical plane, which has led to resounding victories, since they have permitted the mobilization of the entire people in the fight against an enemy superior in numbers and material.[31]

Simultaneous small-scale attacks on many points would wear out enemy forces, causing constant tension, interfering with their supplies and forcing them to spread their forces thinly. Thus, neither their technical proficiency and strength nor numerical superiority could be exploited and their weaknesses and shortcomings could be shown up. However, 'if the offensive is confined to these small-scale combats with scattered forces we cannot win a decisive victory'.[32] As another of Vietnam's great generals, Giap, said in 1973, the fundamental problem of all wars lies in the annihilation of the enemy's armed forces. It was therefore necessary 'to engage in large-scale battles with concentration of mobile forces in order to achieve decisive strategic annihilation'.[33] Vietnamese military history anticipated and underpinned the classic theory of guerrilla warfare, which must be anathema to certain groups of anarchists, that guerrilla warfare is a second best and that the aim is to build up to regular military organization and decisive large-scale battles of annihilation, to 'start with small-scale operations, ending up with medium size or large-scale operations'.[34]

The military history of Vietnam in the pre-industrial era provides many examples of the skilled blending of small- and large-scale actions. The battles tend to be more dispersed and cover a larger area than those of contemporary Europe. In 1077 the Sung Chinese troops reportedly occupied a position stretching 30 kilometres along the southern bank of the Nhu Nguyet river: 'the correlation of forces and the enemy's disposition of his defences were such that we could not encircle the Sung troops and destroy them at one blow'. This was a foretaste of the dilemma faced in World War One[35]. They are also instructive in the ways of the principal adversary, the Chinese, who knew the terrain well, fielded competent and experienced generals, and, with some exceptions, respected Vietnamese expertise in war. The *Rules of the Foot Soldiers of Annam* were frequently mentioned by a historian of the Chinese Sung dynasty, before the Mongol invasion.[36] The first Asian military academy was established in the thirteenth century, in what is now Hanoi, and shortly afterwards the Vietnamese produced the first military handbook in the South-East Asia region. This apparently contained an innovative strategy which was instrumental in enabling the defeat of the formidable armies of Kublai Khan, with their combination of Mongol and Chinese expertise.[37] Vietnamese military history is equally illuminating about the opposition. As early as 1077, we read of a Chinese attempt to conduct a secret, lightning campaign against Vietnam before turning north to face threats from the Liu and Hsia, the classic grand strategic dilemma of the 'two-front war'. 'Therefore, we should settle the Annam affair quickly'.[38]

The warm, moist climate, and dependence on the numerous rivers and surrounding marshes and paddy fields for food and communication meant that riverine (fluvial) operations acquired a particular prominence, and fleets at sea often operated in close co-operation with land forces of both sides. Both these features were also prominent during the 1965–75 American involvement in the Vietnam War. The engagements at Nhu Nguyet (1077, against the forces of the Chinese Sung dynasty) and Bach Dang (1288, against those of the Mongol-based Yüan dynasty) involved extensive riverine and coast operations.[39] During the Tay Son rebellion (1771–89) the rebels scored one of the greatest naval victories in Vietnamese history, in the form of a riverine ambush between the Rach Gam and Xoai Mut canals, in 1785. The rebel vessels emerged from the canals to trap the enemy fleet (Vietnamese 'reactionaries' and their Siamese allies) sailing up the My Tho river to destroy the Tay Son army. The enemy force was then destroyed by point-blank[40] gunfire from the river bank and islands in the river, as well as by gunfire and boarding parties from the ambushing vessels.[41]

THE LAM SON UPRISING, 1418–1428, AND THE BATTLE OF TOT DONG-CHUC DONG, 1426

The Lam Son insurrection is an important event in Vietnamese history, and was chosen as the code name for one of the first great heliborne airmobile operations, in 1971 (Chapter 4). It provides many key precedents for the Vietnamese style of offensive-defensive war, and is therefore worth closer examination.

Taking advantage of a revolution and establishment of a new dynasty in Vietnam in 1400, the Chinese Ming Empire carried out a massive invasion in 1406. There was a widespread popular resistance, and in 1418 opposition crystallized in a mountain area of Thanh Hoa province (Figure 6.3). The first Lam Son insurgents were few in number, and suffered heavy losses at the hands of Ming armies which allegedly numbered up to 100,000. As with contemporary European history, such numbers should be treated with caution: in Chinese, 'many tens of thousands' can mean 'a lot'. At one point the insurgents were reportedly down to 100 men. By early 1426, however, after eight years of fighting the insurgents dominated a large area from Thanh Hoa down to Thuan Hoa (roughly, the twentieth to the sixteenth parallel). Ming troops only held five isolated bases, 'like islands in the sea', a popular analogy in Asian warfare.[42] North of Than Hoah, the Ming were in control, centred on the capital Dong Quan (Hanoi).

In April 1426, the Ming sent a further 60,000 troops to launch a strategic offensive against the rebel-controlled area, Lam Son. However, this took six months to complete, and meanwhile, in September, the rebels launched three columns northwards, one (3,000 men) west of Hanoi, one (4,000 men) east, into the Red River delta, and one (2,000 men) to make a show of strength south of Hanoi. The columns received substantial assistance from the populace, and this relatively small force was thus able to operate freely over a large area of the delta and highlands of northern Vietnam. The left (western) column pushed as far north as Nhan Muc bridge, intercepting a Ming column and effecting encirclement of the capital (Figure 6.3). They were insufficiently strong, however, to prevent 50,000 Ming troops under General Vuong Thong moving down what is now Highway 1 from Kwangsi through Lang Son and into Hanoi by 31 October. Twenty thousand of Vuong Thong's men were seasoned soldiers, the rest porters. As in contemporary Europe, both sides employed cannon and crossbows, but also rockets and rudimentary flame-throwers. As now, three roads ran into Hanoi from west and south west, and with a naval force on the Red River he could control all road and river communications. Vuong Thong knew that the Lam Son forces in the north were few in number, and that the main force still lay to the south, around Thanh Hoa. He resolved to attack the latter, down two roads, the 'Mail Road' (modern Highway 1) and the 'Mountain Road' (modern Highway 6 and 21, via Chuc Dong and Nho Quan). Vuong Thong decided to take the latter route. The western and centre insurgent columns, now totalling 10,000 men with local reinforcements, had to stop him.

The Ming forces were divided into three columns, 10 to 15 kilometres apart, south west of Hanoi. Splitting one's force so that it cannot concentrate again is a classic error, although it is sometimes unavoidable. The insurgents attacked and routed two of the isolated columns, the remnants of which seem to have withdrawn toward Hanoi.[43] This left the main column astride what is now Highway 6, the 'Mountain Road' to the capital, and the Day River. Vuong Thong resolved to attack Ninh Kieu, an important position south west of Hanoi where the Mountain Road and the river converged (Figure 6.3, main map and insert). The rebels launched a probing attack from Ninh Kieu against the advancing Ming, but the Ming were now well prepared. The Lam Son forces were repelled, but Vuong Thong did not pursue for fear of falling into an ambush. The insurgents then withdrew to Cao Bo.

Vuong Thong planned to envelop Cao Bo, the main force conducting a frontal attack along the Mountain Road, while the subsidiary force, an elite corps with cavalry and artillery, would take a short-cut to attack the Lam Son force from behind. When in place, it would use its artillery to give the signal for the main force to attack (a fitting comment on the utility of the artillery of the time in open combat). The insurgents, outnumbered almost ten to one, knew the ground well and that the surrounding paddy fields were flooded, so that the Ming force would be limited to the road. Like the Finns in 1939–40, they planned to cut the much more numerous road-bound Ming troops into what the Finns, five hundred years later, called *mottis*.[44] When the vanguard of the main force reached Tot Dong, the rearguard was approaching Chuc Dong. Both would be amputated, the force ambushing the rearguard also acting as a 'cut-off' to prevent the Ming force withdrawing to Hanoi. A small force was left in Cao Bo as a deception, and the Lam Son command had learned of the plan to use gunfire as a signal. Therefore, the small force in Cao Bo would fire shots at a set time to confuse the enemy and lure his main force into the ambush. The plan worked. On 7 November, the main force, toiling through rain and mud, heard shots from Cao Bo and

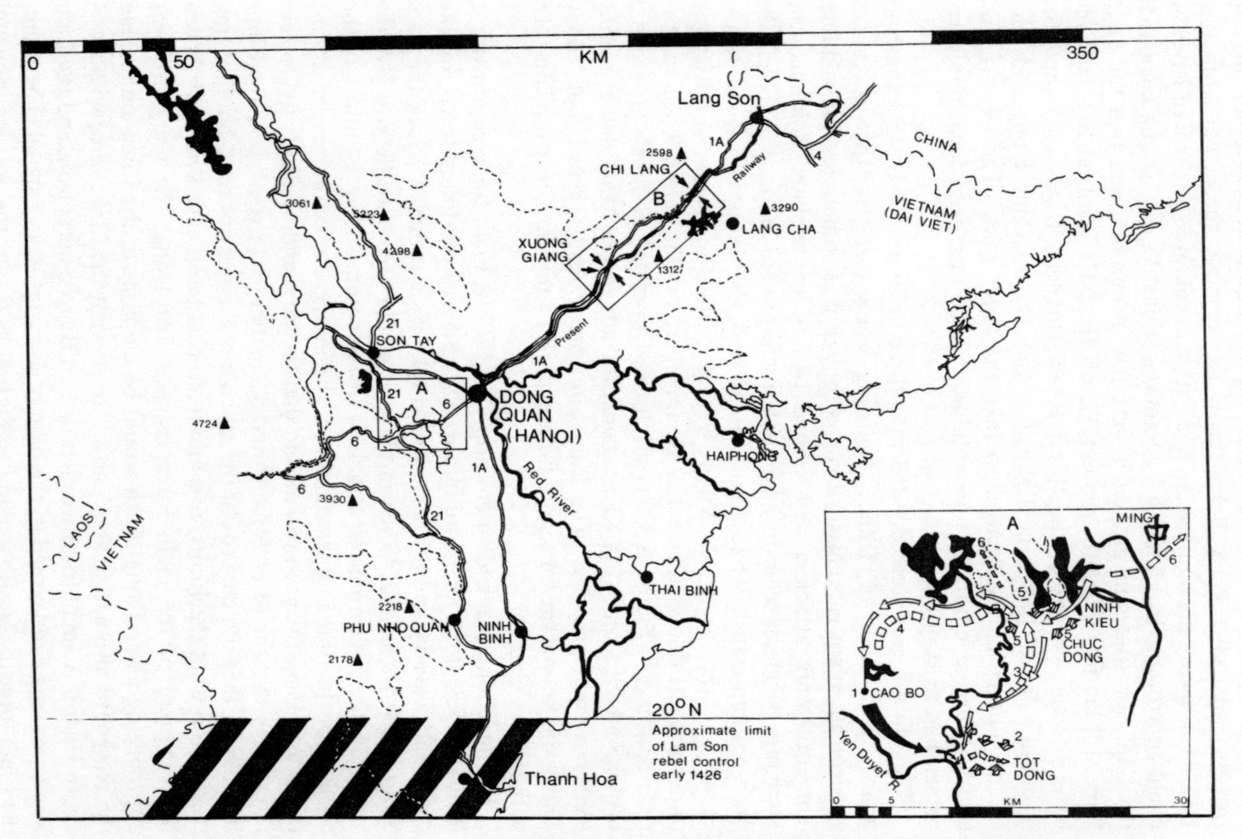

Figure 6.3 The Lam Son Rebellion and the Tot Dong-Chuc Dong Battle, 1426

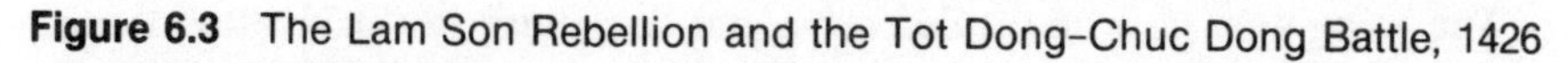

Figure 6.3 The Lam Son Rebellion and the Tot Dong–Chuc Dong Battle, 1426

present roads, with route numbers

present railway

lakes

edge of mountainous area (over 500 metres above sea level, approx.)

present national boundaries (medieval boundaries corresponded roughly)

selected high peaks, height in metres

Note: Inset: Tot Dong-Chuc Dong, 5–7 November 1426 (approximate position indicated on main map)

1 Lam Son HQ fires false signal shots
2 Chinese vanguard (main column) advances into ambush at Tot Dong
3 Remainder of main column withdraws
4 Flanking column hears of fate of main column and withdraws
5 Remnants of both columns fall into second ambush at Chuc Dong
6 Surviving remnants escape north and towards Dong Quan (Hanoi)

ambushes by Lam Son rebel forces

Lam Son movements in black, Chinese in white

B = Approximate area of Chi Land – Xuong Gian battle, 8 October–3 November 1427

Lam Son ambushes

thought that the flanking column had arrived. They therefore advanced into the ambush at Tot Dong and were cut to pieces. Vuong Thong was wounded but survived to retreat to Ninh Kieu.

Meanwhile, the flanking column advanced towards Cao Bo as ostentatiously as possible, to attract the defenders' attention, but were surprised to hear gunfire. Then they heard that the main column had been ambushed, and immediately turned to withdraw to Chuc Dong. Here they, and the withdrawing main force, fell into the second ambush. In the interim, the Lam Son insurgents had destroyed the Ninh Kieu bridge. Many drowned trying to escape across the river: others fled north along the 'right'[45] bank. Total Ming casualties are estimated at 50,000 dead and 10,000 prisoners. The Tot Dong-Chuc Dong victory completely changed the face of the war, forcing the Ming onto the defensive. According to modern Vietnamese analysis, the battles fought between 5 and 7 November running up to the Tot Dong-Chuc Dong battle constituted:

> a real military campaign of a decisive character. It illustrated one of the characteristics of our national military art which consisted in 'defeating large armies with only small numbers and successfully opposing powerful forces with weaker ones'.[46]

The Ming were now on the defensive, but the insurgents lacked the strength to take Hanoi, 'well supplied and solidly entrenched'. Since 1425, the insurgents had been attempting to reduce encircled citadels, always combining 'political work' – persuading the enemy to surrender – with military operations. As a result, several citadels surrendered without a fight. The Ming had a system of fortified positions along the route linking Lang Son with Hanoi and other citadels at key points to block land and water communications. They also sent a force of 150,000 men in two columns, the smaller (60,000) under Moc Thanh, an old general of great experience. The Vietnamese knew that he was cunning and cautious, and would not be drawn into a trap easily. It was the game of 'know your enemy'. Instead, the Vietnamese ambushed a force under the arrogant Lieu Thang at Chi Lang, (straddling the present Hanoi-Lang Son railway), on 10 October 1427. Ming forces pushed on south west, and were attacked again on 18 October in a hilly region between Can Tram and Xuong Giang, and the Vietnamese finally surrounded a large force in Xuong Giang. Moc Thanh, alarmed, retreated towards Yunnan in China. In flight, his forces were more vulnerable (the 'golden bridge') and the Vietnamese harried him all the way to the border. The encircled force was destroyed on 3 November 1427. On 3 January 1428 the last Ming troops left Vietnam.[47] The powerful, well organized and disciplined Ming force had been split up in the constricting terrain. Then, bit by bit, the weaker Vietnamese forces had chewed it away. The Vietnamese campaigns of 1426 and 1427 are replete with lessons for the art of war in general, and for the interaction of large-scale conventional and smaller-scale 'guerrilla' actions in particular. The Vietnamese consider these historical examples highly relevant. They have pointed out that even in the Middle Ages, large enemy formations took several days to wipe out (Tot Dong-Chuc Dong) or even a month (Chi Lang-Xuong Giang). Breaking up enemy forces on the march was made easier by the terrain, 'it is easy to attack one several thousand league long expedition in exposed areas'.[48] Like Sun Tzu, the Vietnamese realized that although the enemy might be persuaded to surrender, he would most likely do so 'only if they created a position of strength so overwhelming that if fighting occurs it will certainly result in the enemy's defeat'.[49] They paid attention to the different capabilities of different types of troops. The 'court' or guards units were regular troops and were used as operational manoeuvre forces, committed at the decisive place and time, where their mobility, discipline, superior arms and reliability were used to optimum effect. Provincial troops and feudal levies were only used in their own areas. The mutually supportive value of conventional and special forces is stressed:

> The conventional forces operate along classical lines: on well defined fronts according to the principles of positional and mobile warfare. The special forces act more freely, launching surprise or rear attacks without

> regard for classical principles. Our ancestors combined these two categories of troops laying stress on the special forces. For we must often oppose a numerically inferior force to a more numerous enemy. These special forces are sometimes used on a large scale, and sometimes even the conventional forces are used as if they were special forces.[50]

This was exactly the technique the North Vietnamese used in 1965–75, and, perhaps even more importantly, it is the way that armed forces are likely to develop in the twenty-first century: 'conventional' and 'special' forces will become indistinguishable.

THE INDIAN SUB-CONTINENT

The great historian Arnold Toynbee described the area between the rivers Oxus and Indus as the eastern crossroad of history. Alexander the Great realized that after his lengthy campaign to pacify the region. No one can stay in that inhospitable area, and all those who occupy it are inevitably propelled south east, south, or south west. To the south east lie the alluvial Indo-Gangetic plains; to the south, the coast of the Arabian sea, and to the south west, the Iranian plateau leading towards the Euphrates valley and the Persian Gulf. The majority of invaders from Central Asia chose the plains of the South Asian sub-continent – India – as their objective. Between 1221 and 1398, twenty-three Mongol invasions of India took place (mostly failures). The Afghan crossroad funnelled and accelerated both political and military ambitions, and also ideas. It was through this region that Buddhism passed into Central Asia and on to China, and, moving in the opposite direction, Islam entered India. The first Muslim invader to reach India through what is now Afghanistan was a Turk, Sultan Sabuktugin of Ghazni. With his son, Mahmud, the Sultan destroyed the Rajput kingdoms of northern India, although the Rajputs continued a guerrilla war from mountain strongholds into the early sixteenth century.[51] It is against this background that two of the sub-continent's great military leaders appeared.

Babur 1483–1530

A descendant of Tamerlane, Babur succeeded to the throne of Fergana, with its capital at Herat, in about 1495. Like so many others, he undertook military expeditions to enhance the security of his vulnerable little state, and began to accrete territory. He took Samarkand and Kabul. In 1506 he set out northwards to check Uzbek aggression and it is during the march back over the mountains, in deep snow, that his qualities as a first-rate general become apparent. The advance slowed to two or three miles a day, with snow up to the men's chests. Like all great commanders, he insisted on exactly the same conditions for himself as his men, refusing to take shelter when there was none for them. Babur was the founder of the Moghul dynasty which subsequently ruled India. Having made himself master of the Punjab, Babur set out to conquer Hindustan. He met the ruler, Ibrahim, in battle at Panipat, in 1526. Babur knew Ibrahim to be impetuous and negligent, and his battle plan played on this. Babur adopted an order of battle devised by the Osmanli Turks, which has much in common with the mobile forts of the Hussites in fifteenth-century Europe and the Russian *gulai gorod* (mobile town). This consisted of guns tied together with ropes and chains, and musketeers behind field defences and obstacles. Gaps were left between the defences to enable company strength groups to counter-attack. This was quite out of character for a descendant of the Mongols (albeit a very distant one), and Babur's advisers counselled against it. Babur stuck to his guns, and sent out scouts to draw Ibrahim onto the defended position, after which the plan was for cavalry to issue from behind it, encircle and destroy them. At first, Ibrahim would not be drawn, and Babur noted that his men were getting jumpy. Like all great commanders he was acutely sensitive to his men's morale. Then Ibrahim attacked, but his forces wavered as they saw the strength of the position. At this juncture, the flanks wheeled to encircle Ibrahim's forces and then the infantry in the centre advanced under the covering fire of its guns. The same technique was used against the Rajputs in the following year. The appearance of a European-style infantry–artillery defensive position in Asia at this time is interesting; it reflects an awareness of the value of the tactical defensive in concert with the strategic offensive, which Tamerlane grasped, and also the high state of technology and the industrial base which the settled civilizations of that region possessed. As far as we can tell, Tamerlane had made no use of firearms, even though they were known in Europe throughout his lifetime. However, Babur's use of this static fire-base was a response to a particular opponent and particular circumstances.[52] The light cavalry-bow combination remained very significant in India, and dominant throughout the rest of Asia.

Akbar 1542–1605

Akbar was Babur's grandson. Like Babur, he was a wise ruler, but if anything he was greater than his grandfather as a warrior. Akbar's military successes were astounding. In 1567 he took the enormous Rajput stronghold of Chitor, standing on a huge rock eight miles in circumference, by a combination of skilled siegecraft and luck. An excellent example of the methods of Genghis and Tamerlane is provided by Akbar's lightning campaign against the rebellious state of Gujarat, in 1572. On hearing of the rebellion, Akbar immediately started out in the fierce August heat with 3,000 cavalry. He raced across Rajputana, covering 80 kilometres a day, or 900 in eleven days. The rebels, numbering about 20,000 were astonished by the appearance of Akbar's force, which they had been informed was still a fortnight away. Attacking at once, the momentum of Akbar's force carried the enemy's main body before them. A second wave of rebel forces coming up saw the main body retreating in disarray, and in turn panicked. Another outstanding example of generalship is Akbar's advance on Kabul in 1581. Meticulous about detail, Akbar ensured that all ranks received an eight months' advance of pay (it will be remembered that this was one reason why the Ottoman Turks wielded a more cohesive and efficient military force than contemporary Europe, at about the same time: see Chapter two). The advance with 50,000 cavalry, a large body of infantry and many elephants, proceeded methodically. Akbar knew the time and place for supernatural speed, and where it was neither necessary nor possible. The mere approach of Akbar's army was sufficient to induce the rebel leader to flee. In 1587, Akbar took Kashmir, in 1592 Sind, and, in 1594, Kandahar.

The year 1599 saw the siege of Asirgarh, the strongest fortress in all India and probably in the whole world at that time. Asirgarh stood on a rock rising 300 metres above the plain. There was a constant supply of water from wells and provisions to support the garrison, estimated at 60,000, for ten years. The dimensions and scale of warfare in Asia in this period dwarf anything known in Europe. There is something fantastic about both the scale and the name of Asirgarh, which sounds more like a creation of Tolkien than a fact of history. The fortress mounted 1,300 guns, whereas the artillery of the Moghuls was very weak. There is one other interesting fact. Many of the defending gunners were Portuguese. Although Europe could still not match the immane scale and sweep of Asiatic operations, it was clearly drawing ahead in the crucial field of military technology and associated technique. Akbar investigated the possibility of getting siege artillery from the Portuguese, but they refused to help him, perhaps realizing that an all-powerful Akbar would be a threat to their settlements in Goa. In the end, Akbar succeeded in bribing the garrison of this impregnable fortress. Akbar never fully subdued the Deccan, but by the time he died the Moghul Empire stretched from the Bay of Bengal to Afghanistan and from the edge of the Deccan to Kashmir. Akbar evinced all the attributes of a great captain: an understanding of the value of speed and surprise, meticulous planning, concern for his troops' morale, and economy of effort. His army was typically asiatic in style, predominantly cavalry, and weak in artillery.[53] The same combination characterized that of the last great military leader of Asia before the advent of global civilization.

NADIR SHAH (C. 1688–1747)

Babur and Akbar were both highly cultured men, and this aspect of their characters and their upbringing in settled if not entirely secure communities affected their approach to war and diplomacy. Nadir Shah began as a shepherd boy and worked his way through domestic service to high office

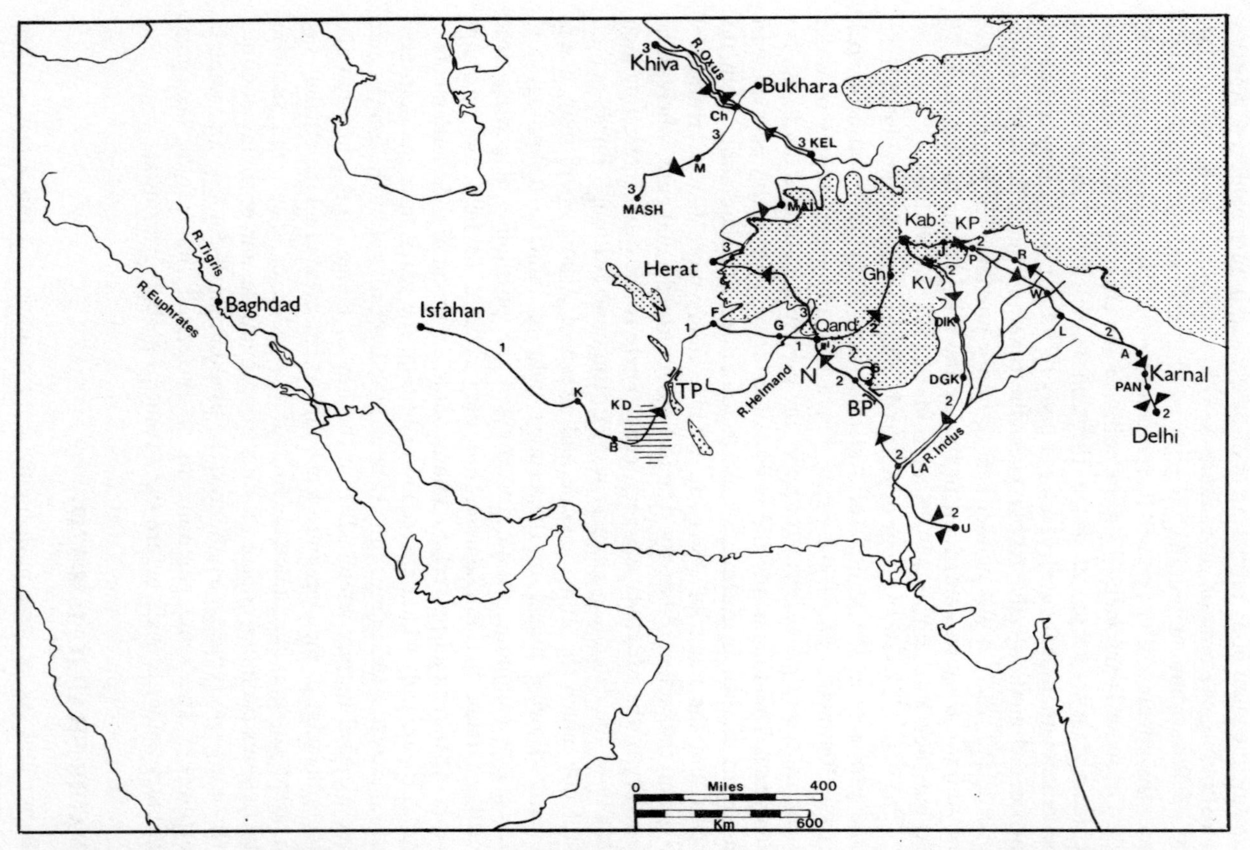

Figure 6.4 The conquests of Nadir Shah. (Nadir Shah's Campaigns in Afghanistan, India and Turkestan)

Figure 6.4 Conquests of Nadir Shah. (Nadir Shah's Campaigns in Afghanistan, India and Turkestan)

1 Reconquest of Kandahar, 1736–8
2 Indian campaign, 1738–40
3 Turkestan campaign, 1740–41

A	= Ambala	G	= Girishk
KV	= Kurram Valley	Pan	= Panipat
B	= Bam	Gh	= Ghazni
L	= Lahore	Q	= Quetta
BP	= Bolan Pass	J	= Jalalbad
La	= Larkana	Qand	= Kandahar
Ch	= Chardchou	Kab	= Kabul
M	= Merv	R	= Rawalpindi
DIK	= Dera Ismail Khan	Kel	= Kelit
Mash	= Mashhad	TP	= Tabarkand Pass
DGK	= Dera Ghazi Khan	KD	= Kelit Desert
Mai	= Maimana	U	= Umarkot
F	= Farah	KP	= Khyber Pass
N	= Nadirabad	W	= Wazirabad

Dotted area: high mountains (over 2,000 metres) posing significant obstacles during these campaigns.

and then leadership. His career and personality are more akin to those of Genghis Khan and Tamerlane than the two Moghuls. The parallels between the careers of Tamerlane and Nadir are indeed so numerous that there can be no doubt that Nadir deliberately modelled his career on that of the former. Nadir has been called the last great Asiatic conqueror and if it was a great achievement to rise from shepherd boy to Shah, it was a greater one still to lift Persia from the grip of powerful enemies to the foremost power in Asia in her own right. Nadir Shah's principal campaigns are shown in Figure 6.4. Like Genghis Khan, Nadir was quite illiterate and his understanding of war was innate. Lord Curzon, the famous viceroy of India, likened him to Napoleon, and if the parallels with Tamerlane are obvious, Nadir also resembled Frederick the Great in his position of master strategist of his state, his extensive recruiting beyond the frontiers of his country, his careful training of his men and belief in the importance of mobility. It is known that he usually thought out his campaigns beforehand although sometimes he was impetuous and in a disastrous battle with the Turks on the banks of the Tigris on 19 July 1733 he attacked a strongly entrenched Turkish position, losing 30,000 dead and all his artillery and baggage. He proved himself greater in defeat than previously in victory; he set about the stupendous task of rebuilding his army from nothing, an enterprise which was completed in two months. The moral courage evinced by this episode is another quality of all great captains, and the need to test future senior commanders in dealing with the unexpected, with foreseen disasters, has been stressed by many perceptive commentators on command and war.[54]

Like all great captains, Nadir showed that he could cover great distances with 'supernatural speed'. He excelled at the swift cavalry attack, delivered from an unexpected quarter. His defeat of the Indian forces guarding the Khyber pass by his forced march through the Tsatsobi pass and Bazar valley (Figure 6.5) using his favourite device of making a detour via an unexpected and difficult route was a masterpiece in the history of war. It is also a variant of the classic principle that the best way to take a bridge, or a mountain pass, is from both ends at once, a device also employed in exemplary fashion in the capture of the Shipka pass in the Balkans in 1877 (see Chapter 5). Nadir's capture of the Khyber pass was exhaustively analysed by the Russians at the end of the nineteenth century. Lieutenant General Kishmishev's analysis of Nadir Shah's campaigns in Herat, Kandahar, and India was published in 1889.[55] Plans for a possible invasion of British India were a preoccupation of the Russian General Staff at this time and there can be little doubt that this was the reason for Kishmishev's study. Although written as a straightforward military-historical narrative, it was accompanied, not as one might expect by historical maps, but by the latest Russian General Staff maps of the area, complete with railways. Kishmishev's description of Nadir Shah's encircling manoeuvre round the Khyber pass not only evokes classic and timeless principles of war, but also detailed and specific lessons for future operations on the same piece of ground:

> Some plan had to be decided upon and Nadir the genius gambled on one of the most audacious military operations: to go around the Khyber pass, towards Sari-Chub, over unknown country, inhabited by wild tribes who regarded any foreign irruption as an encroachment on their independence, which they valued so highly.
>
> Nadir kept the plan for an encircling movement a tight secret and in order to complete the deception, on 17 November [1738] he sent engineer detachments in the direction of the Khyber pass to improve the condition of the approach. From dawn to dusk under his personal direction the soldiers toiled to widen the roadway, clear it of rocks and to construct gentler gradients. Nadir, not dismounting from his horse all day, spurred them on with money and presents. His personal involvement in this enterprise convinced everybody that he intended to advance through the Khyber pass and Nasir Khan [Nadir's opponent] evidently believed it too as he began to reinforce his forward positions, blocking the way to the pass. The work continued unabated until on the 18th, before sunset, when everyone in the camp was convinced that there would be no action that day, Nadir summoned Adji-khan Bek and ordered him to move out immediately with a forward detachment [*peredovoy otryad* – see Chapter 5].
>
> Behind him, Nadir sent all the cavalry while the remaining forces stayed at Daka, until further notice. The cavalry set out in the dusk, not knowing, naturally, the final object of the movement. With the indefatig-

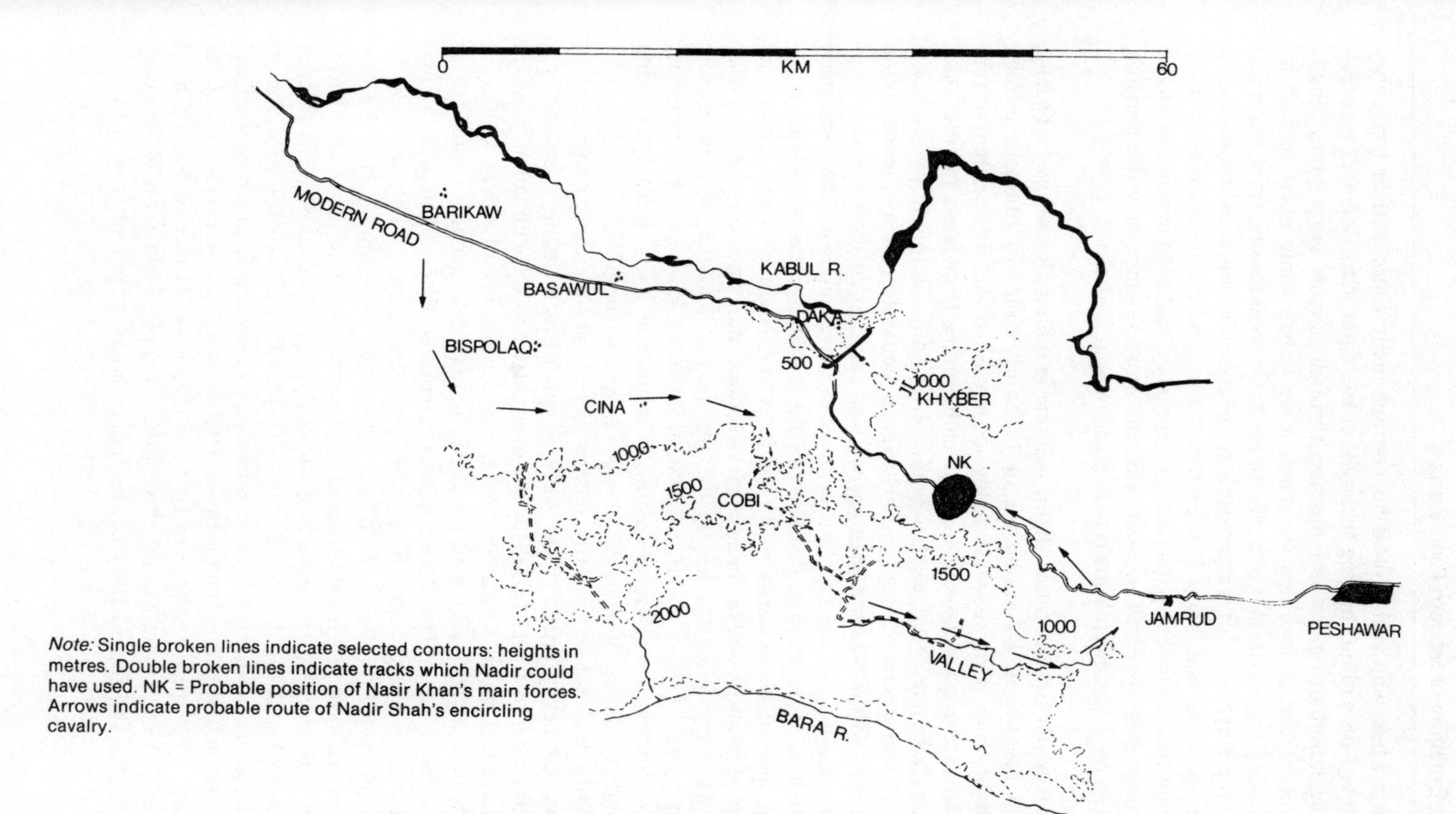

Note: Single broken lines indicate selected contours: heights in metres. Double broken lines indicate tracks which Nadir could have used. NK = Probable position of Nasir Khan's main forces. Arrows indicate probable route of Nadir Shah's encircling cavalry.

Figure 6.5 Nadir Shah's encirclement of the Khyber Pass and attack from the rear, 1738

able Khan who had proved his courage many times at their head, they moved on without halting throughout the night. Crossing over inaccessible mountain spurs and pushing through narrow gorges they finally crossed the main chain by dawn after unbelievable exertion and by midday appeared unexpectedly on the main road leading from Peshawar to the Khyber . . . it is impossible to describe the confusion in the Indian camp. Cries that Nadir had encircled them and cut them off from their main base produced panic. At first nobody could believe that the Shah's troops had appeared in their rear, the more so since only the previous evening it had been known that Nadir was at Daka.'[56]

The Russian General noted that it was rare in military history to find such a rapid movement: 80 kilometres in 18 hours, over mountainous terrain, without roads. This was presumably accomplished with non-existent or very inadequate maps and, being mid-November in the Himalayas, in bitter cold. Unless helicopters were used, it would be very difficult to improve on Nadir Shah's performance today, and helicopters would depend very much on the clemency of the weather. Kishmishev also emphasized the rigorous deception plan and the use of engineer detachments to improve the road, even though, as it turned out, this was to be a subsidiary axis (see Figure 6.5).

In spite of his enthusiasm for light cavalry, Nadir also appreciated the value of infantry, and he had a corps of trained marksmen called *jazayirchis*. He was very quick to size up the situation in a battle, that invaluable military gift of '*coup d'oeil*', what Suvorov called *glazomer*. A contemporary noted that it was 'scarce credible how quick he is in discerning the odds on either side and how active in succouring his troops'.[57]

Nadir was not particularly successful in sieges for three main reasons. First, his heavy artillery was weak in quality and quantity; second, the requirement to travel vast distances over extremely difficult country, which conditioned much warfare in Asia, made transport of heavy artillery a difficult matter. Nadir was obviously reluctant to sacrifice mobility for heavy equipment. The Russians later discovered that rockets were a good way of deploying heavy firepower in similar terrain, as these could be carried where artillery could not penetrate. In addition, the Persian engineers were not highly trained, and were much inferior to the Turks. Lastly, fortress warfare gave Nadir little opportunity to exploit his greatest gifts: his knack for shock and surprise, especially with cavalry. Also, the art of fortification, in Asia as in contemporary Europe, was on a high level and in a battle of position the means of defence were much superior to those of attack. The same was true of field fortifications. At one of Nadir's most crushing victories (at Karnal, during the invasion of India, on 24 February 1739), the centre and right of the Indian forces advanced from a heavily fortified camp, lured on by cavalry patrols feigning retreat. The left remained inactive because of political divisions within the Indian army.

Nadir ambushed and encircled those Indians who had advanced, achieving a crushing victory in the field, but the other faction withdrew to the fortified camp. Nadir declined to attack, but carried on negotiations with the Indian emperor until lack of food forced the beleaguered camp to capitulate.

Although Nadir's heavy artillery was weak, his light artillery was apparently very good and certainly the best in Asia. This owed a good deal to French and Russian experts in his employ, another indicator that Asia was now falling further behind in the field of pure military technology.

Nadir's achievements, like those of most great military leaders, were due in large measure to his ability to train, organize and inspire his men. Previously, the Persians had been thought of as a most un-martial race, yet Nadir made them fight and beat the ferocious Afghans and Turks – and win. He turned the Persians into excellent fighting material, restoring the morale which they had lost under a series of incompetent commanders. Like Frederick, he also infused large numbers of foreigners into the ranks, particularly Afghans and Uzbecks. By rigid enforcement of discipline and insistence on drill (like Frederick again), he welded the Persian army into a formidable fighting machine. His charisma as a leader was enhanced by his ability 'to recall all the principal officers in his numerous Army by their Names'.[58] He could also remember most of the private soldiers who had served under him. Nadir's astonishing ability to inspire the Persians and turn them into splendid soldiers recalls the recent example of the Egyptian army in 1973. Conditioned to despise the fighting ability of the Egyptians after inflicting a series of decisive defeats on them, the Israelis were taken by surprise by an Egyptian army which fought with aggressiveness, enthusiasm, and skill.

Although brought up far from the sea, Nadir Shah was quick to see the value of sea power. Control of the Persian Gulf was obviously crucial as it flanked any operations against the Turks in the west, and his forces also acted on the other side of the gulf, in Oman. Nadir Shah attempted to build up a navy in the Gulf, again with considerable European help, and he also had a small fleet in the Caspian Sea.[59] At its simplest, the sea (including major inland seas) provides a means of outflanking the enemy even in a primarily land operation. The last two Shahs of Iran also endeavoured to build up a strong naval force.

There can be little doubt that Nadir Shah was the greatest general in the world at the time, his career spanning the period between those of Marlborough and Frederick, and contemporary with de Saxe. He was the last prominent commander with the distinctive Central Asiatic style, before global civilization and the increasing dominance of European technics (on which Nadir himself relied to some extent) made the distinction inappropriate. He was the last of the great Asiatic conquerors in the steppe tradition of Genghis and Tamerlane. However, the more complex but equally extensive terrain and teeming masses of China and South-East Asia were breeding a kind of warfare with which the twentieth century has become horribly familiar.

ELEVEN MILLION DEAD. THE TAIPING REBELLION, 1851–64, PROTOTYPE OF MODERN WAR IN ASIA

The events of summer 1989 in Tiananmen Square in Peking came as a rude shock to western observers. Whilst the West had believed China to be on a liberalizing, democratic course, the Chinese leadership directed its army to annihilate the student opposition movement with a clinical ruthlessness that would have done credit to the Mandarins of Imperial China, though, mercifully, without the latter's ingenuity for prolonging human suffering. Then, it was reported that another Chinese Army was marching on the capital and that civil war might break out. The picture was a familiar one to students of Chinese military history. The Chinese Army is the world's largest, and its evolution in modern times is of obvious importance. Modern times arguably begin in 1851. The significance of the Taiping Rebellion which lasted for thirteen years, from 1851 to 1864, was soon recognized by contemporaries. In August 1853, *The Times* described it in thoroughly modern terms as 'one of the most important and remarkable movements of mass protest in modern history . . . the greatest revolution the world has yet seen'.[60] With hindsight, the Taiping Rebellion 'initiated a century of rebellions and revolutions and marked the beginning of the modern era of Chinese history . . . a sound knowledge of the rebellion is indispensable for an understanding of twentieth century China'.[61] The rebellion had 'definite influence on Dr Sun Yat-sen and other nationalist leaders in the 1911 revolution'.[62]

A common-sense indicator of the relative importance of conflicts in military history is the number of people killed. When many dead are combined with warfare of a new type, replete with military lessons, on the terrain of a major land power, the significance of the conflict can be in little doubt. According to statistical studies of armed conflict, the Taiping rebellion cost 11 million lives, and is therefore a 'class 1' war, of the same order of magnitude as World War One.[63] These would appear to be people actually killed (including massacres of civilians, which far exceeded battle casualties), as opposed to population loss through famine, displacement of the population, children who were not conceived and so on. Demographers use non-conception as a measure of population loss attributable to a war. Western observers have estimated the *total* population loss in the Taiping rebellion as 20 to 30 million, although Chinese researchers have estimated that loss in the four provinces of Kiangsu, Chekiang, Anhwei and Kiangsi alone far exeeded this.[64] The Taiping Rebellion is an example of 'internal conflict with external intervention' (British, French, and unofficial American), which tend to be relatively the most costly in human life.[65]

The Taiping Rebellion evinces many similarities to the American Civil War, with which it overlapped. It began with an insurrection in the south; the armies on either side were not large enough to command the full area available, but were forced to move about constantly in a country 'too big for the armies'; riverine communications were of cardinal importance; and much of the war revolved around a struggle for fortified positions, with massive sieges and elaborate entrenchments. The physical devastation was immense. 'The [Yangtze] valleys, notwithstanding the fertility of their soil, are a complete wilderness . . . [and in] ruins.'[66] The rebellion was 'probably the most destructive civil war of modern times before the twentieth century'.[67] Although it lacks the interest of the American Civil War from the technological viewpoint, it is arguably just as important as the beginning and prototype of modern land warfare. It led directly to the strategy adopted by Mao in the twentieth century, for example the phenomenon of the 'long march', of which the Taiping Rebellion saw several. Mao considered that of all the wars preceding his own struggle with the Japanese, the Taiping war was 'particularly' relevant, 'when guerrilla operations were most extensive'.[68]

The origins of the rebellion were, in part, religious, presenting an immediate parallel with the role of Islamic fundamentalism today. Resenting the repressive and hierarchical Chinese system of government, a proletarian revolutionary movement crystallized around a monotheistic religion, a form of pseudo-Christianity. The Taipings (from 'Great Peace') believed in complete equality of men and women and an egalitarian social structure. This was reflected in their more co-operative attitude to foreigners, in their abolition of the dreadful custom of binding and deforming women's feet, and in their employment of women as combat troops. Women from Kwangsi and Kwantung provinces had never had bound feet, which enabled them to fight on the battlefield.[69] The movement was led by Hung Hsiu-chan, an unsucessful academic and bureaucrat. After failing the examinations for the last time in 1843, he raised a group of revolutionary fighters in Kwangsi province in spring 1844.

The Taipings' early success was largely due to a strict disciplinary code, first published in January 1851. The five points of the code bear a striking resemblance to the code practised by the 8th Route Army under Mao.[70] Such puritanical codes have been concomitants of military effectiveness, from Genghis Khan through Cromwell to the Waffen SS. Wine, gambling, swearing and, above all, 'indulgence in the fumes of opium' were all forbidden.[71] In 1937, Mao Tse-tung, engaged in a guerilla war against the Japanese, concurred. 'The opium habit must be forbidden, and a soldier who cannot break himself of the habit should be dismissed. Victory in guerrilla war is conditioned upon keeping the membership pure and clean.'[72]

By 1851 the Taipings were strong enough to seize Chiangkouhsu, a major town on the Tahuang river, and on 25 September their vanguard captured Yung-an. By this time the ruling Ching government was alarmed and

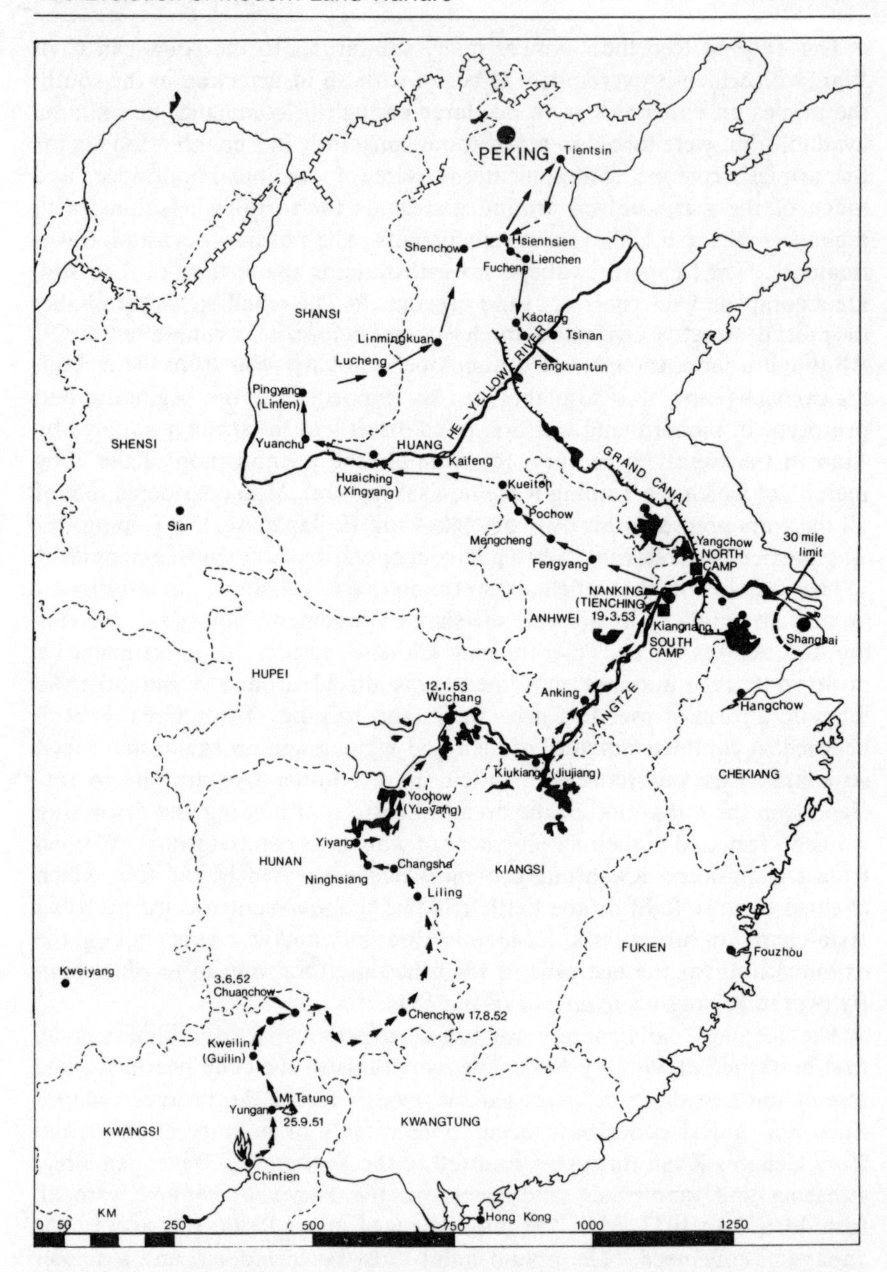

Figure 6.6 Taiping Rebellion, 1851–64, showing long marches of the Taipings

- - - - - Provincial boundary (only provinces mentioned in the text are shown)
Great Wall of China
Rivers and lakes (the Yangtze River's system is shown in simplified form)
Grand Canal
■ Great North and South Camps

besieged the city, but the Taipings broke out and headed east to Kusuchung (Figure 6.6). They concentrated their forces on Tatung mountain, some 7 kilometres outside the city, and ambushed the Chings, then headed to Kweilin, tried to take it, failed, and turned north.[73] By this time, they were able to engage in large conventional battles.

The Taiping forces were organized into armies (*chun*), each with a strength, on paper, of 13,156. Armies comprised 5 brigades, brigades 5 companies and companies, 4 platoons each divided into 5 squads of 5.[74] The effective 'span of control' in military organizations is usually reckoned to be 3 to 5: automated C^3I had increased the popularity of 5s (the pentomic division).[75] A recent Chinese history described this (perhaps predictably) as a 'completely new system of military organization', although the Manchu governor of Kwangsi province, having obtained a 'rebel book', told the British consul that it was the Sze-Mar system of the Chow dynasty. The Taipings 'cannot by any means be likened to a disorderly crowd, their regulations and laws being rigorous and clear'.[76] However, as with all tables of military organization throughout military history, armies in the field almost always fell short of their paper strength and attrition reduced them further. This explains why the numbers of Taiping forces are frequently grossly exaggerated: the paper strength of a given number of armies would be much in excess, perhaps many times, the actual number of troops. The Taipings also had four auxiliary armies: 'earth battalions' (miners, very important in the numerous sieges) 'water battalions' (boatmen, very important in the numerous riverine operations), the 'Young Boys' Army' and the 'Women's Army'. The latter reportedly reached 100,000 strong, and women fought in close combat at the battles of Yung-an and Yangchow, another precedent for twentieth-century developments (certainly in the Vietnamese and Soviet armies).[77] Within the Taiping community, trades and crafts were organized so as to serve the needs of the army.[78]

After capturing Chuanchow on 3 June 1852, the Taiping forces moved north, taking their first provincial capital, Wuchang, on 12 January 1853 and advancing eastwards by land and water, capturing Nanking on 19–20 March. This was renamed 'Tienching' ('Heavenly Capital'), and would remain the centre of the Taiping effort until 1864. Within twelve days they took and garrisoned the adjacent cities of Chinkiang, Yangchow, and Kwachow, a move of great military importance as these gave them control

(*key to figure 6.6 continued*)

- Towns and cities mentioned (as then named; where name has changed completely, new name in brackets).
- 30-mile limit initially imposed on military operations by western garrison of Shanghai
- Beginning of the Taiping Rebellion, Chintien, Kwangsi Province
- March of the Taipings to the 'Heavenly Capital' 1851–3, with dates of seizure of towns
- Northern expedition to Peking, May 1853–May 1855 (failed)

of the Grand Canal, the main line of communication between the southern provinces and Nanking.[79]

The weak points of the Taiping strategy were already becoming evident. They generally raced ahead, bypassing strongly garrisoned cities (compare the Mongols) and abandoning areas previously conquered without maintaining lines of communication and defences. Thus, after taking Nanking they had to make a western expedition to reconquer the Yangtze valley through which they had just passed. They also had to launch a northern expedition to overthrow the Manchu government in Peking, and to defend their headquarters at Nanking, to which the government forces immediately headed. During these campaigns, 'they were easily victorious at the beginning, encountered setbacks in the middle period, and reached the apex of their military power in 1856'.[80]

Meanwhile, Imperial forces set up two 'Great Camps', one at Kiangnan, (the 'South Camp') directly outside the city walls of the Heavenly Capital (former Nanking) and one (the 'North Camp') (Kiangpe) at Yangchow, 96 kilometres east of Nanking. These remained a constant threat to the capital. In 1855 Nanking was encircled and began to run short of food, but in April 1856 the North Camp collapsed and in June the South Camp outside Nanking was destroyed – for the moment.[81]

The Northern Expedition to Peking lasted from May 1853 to May 1855. Had the Taipings succeeded in taking Peking their chances of establishing firm control of all China would have been enhanced, although, like the Imperial troops besieging their capital they seem to have lacked the strength to take a most extensive city like Peking, which was the size of contemporary London. Although the expedition failed, it tied down Peking government forces and shielded the Taipings' own capital (the North Camp collapsed, in part, because forces from there had gone chasing after the northern expedition).[82]

The western expedition began in May 1853, with the immediate objective of attacking Changsha, the Hunan capital, in a pincer movement. Here the Taipings came up against a formidable adversary, a competent and ambitious intellectual, Tseng Kuo-fan. The Red Chinese histories are understandably derisive about 'Tseng the head-chopper', but he was a brilliant exponent of counter-revolutionary and counter-insurgency warfare. As twentieth-century theorists on counter-insurgency operations have advocated, Tseng realized that he must win the battle for 'hearts and minds'. A good essayist and propagandist, he inspired general resistance to the Taipings and attacked their pseudo-Christianity. Both sides, and their European allies, indulged in remorseless propaganda, which may explain some of the stories of incredible and perverted atrocities, but Tseng was unusually successful. He was a 'far sighted commander, who planned everything carefully before taking action'. His officers, most of whom were his students, showed him more loyalty than they did the Emperor. Tseng began a punitive expedition in February 1854, and in less than two years had eradicated the Taipings from Hunan, Hupeh, and part of Kiangsi.[83]

The next phase, after the northern, western, and greater Nanking campaigns was marked by internal dissension within the Taiping capital. The western 'king', Wei Ch'ang hui, was regarded by the Taipings as a rebel, with the result that the rebels lost his forces, including very substantial naval forces, and lost the means to control the Yangtze. During this period one Taiping general, Shih ta-Kai, left Nanking (June 1857) for Anking and Kiangsi, but in the course of wandering several thousand miles his force's strength was dissipated and he failed to score any signal victory.[84]

Nanking was besieged again in July 1857, with the re-establishment of the Great South Camp. A prolonged struggle for possession of various fortified camps raged until May 1860, with Taiping forces attempting to break out of the capital and others trying to break the siege from the outside, much as the Leningrad and Volkhov Fronts were to attack both sides of the German ring round Leningrad in 1941–4. The siege was raised for the second time when over 50 Ching camps and the Great South Camp were destroyed.[85]

There followed the last great Taiping expedition, this time eastwards, conquering a large area of the lower Yangtze and threatening the great foreign trade emporium at Shanghai. The British and French resolved to defend Shanghai, and foreign armed intervention against the Taipings began in the Shanghai area in July to August 1860. The Taipings captured Hangchow late in 1861, and began a five-pronged advanced on Shanghai, but were not strong enough to take it. Although the European forces were not sufficiently numerous to have a decisive effect in defeating the Taipings, they made mincemeat of the poorly armed Taipings in a series of small tactical battles, and western leadership (General 'Chinese' Gordon) gave Imperial troops an advantage from 1861.[86]

Tseng Kuo Fan took command of all Imperial forces in South-East China. Implementing his counter-revolutionary strategy step by step, Tseng began to close in on Nanking again. The scholarly soldier recalled that 'since ancient times the strategy for pacifying the lower Yangtze has been to establish a strong position in the upper region and then to press downstream'. In other words, he regarded Anking as a vantage point from which to seize Nanking. The Taipings were thus forced into a life or death struggle at Anking. Taiping forces from the east headed west towards Anking in two columns, north and south of the Yangtze. The siege of Anking lasted a year, the city ringed by trenches and forts in lines of circumvallation and contravallation. The Taiping relief force arrived late in April 1861, the two armies becoming 'locked together in jagged rings of encirclement and counter-encirclement'.[87] There were ten sorties from within the city, then it ran out of grain and on 5 September 1861 it fell. At Luchow, a similar 'encirclement and counter-encirclement campaign lasting three months', the Taipings did better, breaking out in February 1862 and sending 4,000

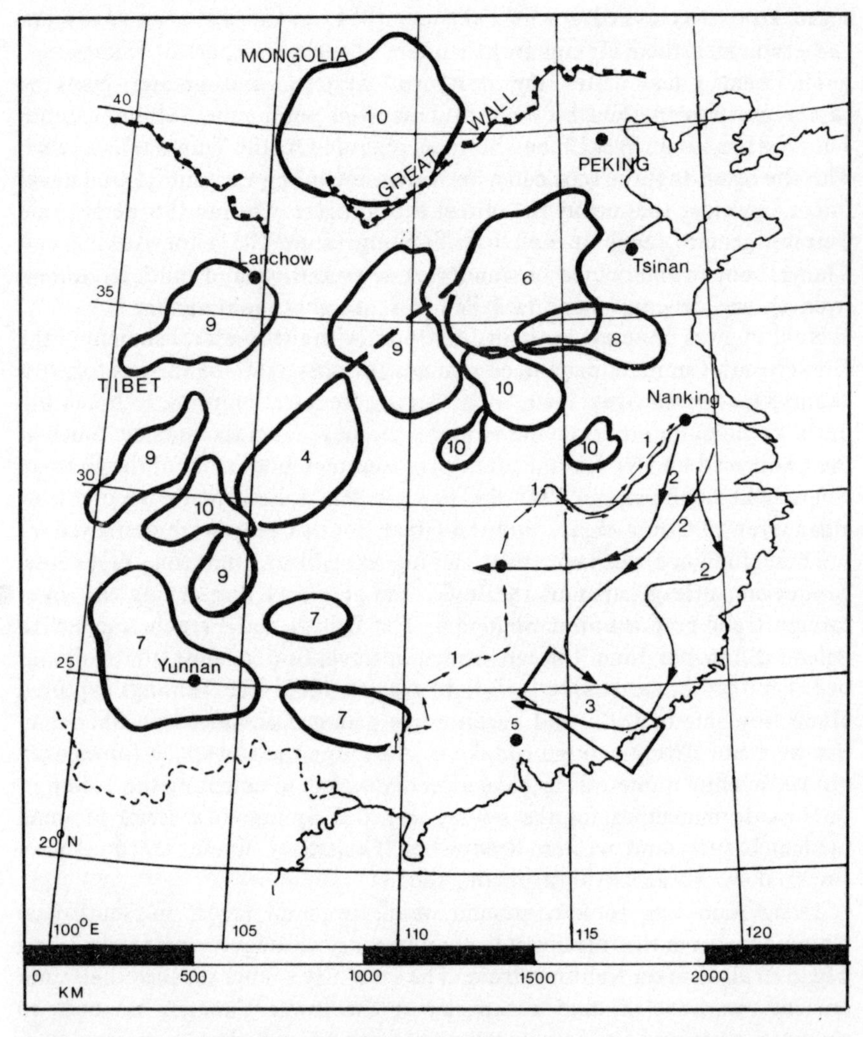

Figure 6.7 Positions of rebel groups at the end of the Taiping Rebellion, 1866

Note: same year as the Battle of Königgratz

1 Original march of the Taipings, 1851–3 (see Figure 6.6)
2 Retreat of Taipings after fall of Nanking (Tiencheng), 1864
3 Position of Taipings under Kan Wang, Shi Wang, 1866
4 Taiping under I Wang
5 Birthplace of Hung Hsiu-chan, the Taiping king (d. 1864)
6 Amalgamation of Taipings from I Wang's Command, Yellow River rebels (Nien) and Muslim rebels about July 1865 threaten Peking at same time
7 Miautze and independent mountaineers
8 Yellow River rebels (Nien: see text)
9 Muslim rebels
10 Local bandits, 'Honan brigands', etc.

seasoned troops speeding north.

By March 1864 the Taipings had lost all the cities and towns of Kiangsu and Chekiang, leaving no room for manoeuvre round the perimeter of the capital and no possible base from which a relief force could be launched. On 1 June, Hung Hsiu chuan, the Taiping leader, died, but the rebels fought on. The strongholds forming a defensive perimeter round the capital fell one by one, a reminder that a fortress is only defensible if besieging forces are kept outside artillery range of the objective itself. In July, the third peak of the Chungshan mountains fell, giving the besiegers control of the high ground overlooking the city, inaugurating the last stage of the battle (compare the assault of Port Stanley, in 1982). On 4 July the Imperial General, Tseng Kuo-chuan, commanding the central front under the overall command of Tseng Kuo-fan 'savagely ordered his battalions to attack the city in rotation day and night' – continuous combat operations, *à la russe*. The city wall was breached on 19 July, and the city burned for seven days.[88]

The fall of Nanking was the end of the Taiping Rebellion proper, although the North-West Taiping Army and other groups of rebels fought on. The position of the surviving rebel groups in 1866 is shown in Figure 6.7. The Nien were particularly elusive, and Tseng Kuo-fan, the counter-insurgency expert, was sent to deal with them. Tseng tried to do what the British did to pin down Boer guerrillas in the later phases of the Boer war: box them in, not with barbed wire and block-houses, but between the Grand Canal, the Shaho, Huai, and Yellow Rivers. He would then advance slowly to crush them. However, the Nien forces had already escaped. Fighting went on until August 1868, four years after the fall of the 'Heavenly Capital'.[89]

The overwhelming impression of this war is one of exhaustion and inadequate strength of the forces, brought about by the sheer size of the terrain. This is evident from S.Y. Teng's superb analysis and from contemporary reports. The ignorance and naivety of the Taiping leadership was one factor in the rebellion's failure, but the other was military weakness. The overall Taiping fighting strength in early 1853 was between 80,000 and 140,000: it then fell, and rose again.[90] Lindley, for example, made an accurate estimate of a Taiping army based on the number of river vessels carrying them, and their size, and reckoned it numbered about 30,000. 'It has been the invariable habit to immensely exaggerate the strength of the Ti-ping armies, and this force . . . was supposed by Europeans to number several hundred thousand.'[91] Lack of manpower would explain in part why the Taipings took the unusual step of using women combat troops. It would also explain why they left no garrisons in the cities between Kwangsi and Nanking, a distance of 1,000 miles that would have required a minimum of 100,000 men to secure the lines of communication. It explains why they were unable to wipe out the Imperial forces stationed outside Nanking for several years, and why they did not make use of or unite with other revolutionary forces.[92] On the northern expedition to Peking there were only 20,000 men (although the paper strength based on the number of divisions was 80,000). Half a million would have been a minimum realistic number to take Peking and occupy other parts of China. The total number of combatants on both sides at any one time is estimated at perhaps 200,000 a considerable number continuously engaged over 13 years.

The daunting scale of the country was fully apparent to contemporaries and the Taipings' opponents. Writing in 1857, Engels saw how the type of force which the British might commit to this theatre would just get lost:

> Canton may be totally destroyed and the coasts nibbled in at all possible points, but all the forces the British could bring together would not suffice to conquer and hold the two provinces of Kwantung and Kwang-si . . . the only point of attack which might lead to a decisive result is Peking. But where is the army to form a fortified and garrisoned base of operations on the shore, to overcome every obstacle on the road, to leave detachments to secure the communications with the shore, and to appear in anything like formidable strength before a town the size of London, a hundred miles from its landing place?[93]

The British Consul, writing to the Foreign Secretary in 1861, expressed the same view. To suppress this particular Chinese rising in the provinces where it was most active:

> would require a large fleet of steamers, operating throughout some 1,500 to 2,000 miles of the Great River [Yangtze] and its larger branches, and some 20,000 troops, operating in three or four complete small armies in different parts of the tract . . . more or less in the occupation of the Tae-Ping forces, and which extends about 800 to 900 miles from north to south and 1,000 to 1,100 from east to west. It would prove one of the most troublesome and costly wars that England ever engaged in.[94]

Anybody contemplating getting involved in a land war in Asia should be aware of such historical precedents. The spatial scale affected the timescale. The campaigns in the Taiping rebellion lasted two to three years. After the initial Taiping victories, there were no quick results in this war.

The Taiping rebellion took place at a time when European and North American armies were undergoing fairly rapid and accelerating technical transformation. In the small engagements with European forces, the technical backwardness of the insurgents told, although commentators at the time were never as dogmatic about technological superiority as we might be. Gordon, for example, agreeded that ' "given the same advantages in arms, ammunition, etc. etc., as the disciplined force had", the tables would have been turned, or at all events the fighting would not have been so one sided'.[95] The most spectacular western advantages lay in artillery (the British had the Armstrong gun, the French the new Napoleon gun), and in naval forces. The war was largely a struggle for domination of the Yangtze river.

Between 1851 and 1857, the Taipings had a very considerable fleet, and this coincided with the height of their power. From 1857 to 1864 the government navy (with western technics) controlled the Yangtze and the war was mainly fought on land. Deprived of river communications, the Taipings were deprived of food, ammunition, and above all, mobility. The Taipings were aware of their shortfalls and tried to obtain western guns, rifles, pistols and even steamships, but many of these were cheap, fraudulent imitations (a point which developing countries purchasing western equipment today should bear in mind).

The Taiping armies, and their Imperial adversaries presented an exotic and, to us, grossly outmoded appearance, but we must not dismiss the experience as irrelevant. According to Lindley:

> their firearms were light matchlocks, and *European muskets or pistols when they could obtain them.* The musketeers carried matchlocks, useless in wet weather, and European made *double barrelled guns, muskets and pistols*, generally of very inferior quality . . . many men from the northern provinces were armed with the tartar bow, which was a much more accurate shooting weapon than either musket or gingall [a very large musket carried by up to four men and fired from a tripod].[96]

Once again, Mao Tse-tung, fighting the Japanese in sophisticated, industrialized 1937, would have recognized the Taipings' problems as his own:

> Equipment cannot be furnished immediately, but must be acquired gradually. . . . Guerrilla bands that originate in the people are furnished with revolvers, pistols, *bird guns, spears, big swords*, and land mines and mortars of local manufacture. Other elementary weapons are added and *as many new-type rifles as are available are distributed.* After a period of resistance, it is possible to increase the supply of equipment by capturing it from the enemy.'[97]

The Taiping Rebellion is therefore replete with lessons about the conduct of war in general, and the conduct of war in China and East Asia in particular. Pure historians are told that they must address what was important *then*, not what seems important *now*. As the largest civil war before the present century, involving vast operations over a colossal area leading directly to the death of some eleven million people and dwarfing the contemporary American Civil War, the Taiping rebellion fulfils those criteria. It is also important now because it foreshadowed the struggles in China of the 1930s and 40s, the Vietnam Wars and other conflicts veering from 'low' to 'mid' intensity and back again. It contains many specific lessons: assimilating new, western technology; a coherent counter-revolutionary strategy; the employment of women in combat; pentomic force structure; riverine operations; encirclement and counter-encirclement, the latter presaging Stalingrad and Leningrad in the Soviet-German war. At the Taipings' capital, visitors were rigorously interrogated and searched before being allowed to enter, authorized persons having to wear 'a little wooden ticket at the waist, which had to be exhibited to the guard',[98] an early version of the laminated passes worn on one's clothes in high-security installations today. As with many aspects of our daily lives, the Chinese did it first.

THE LONG MARCH AND PEOPLE'S WAR

In 1893, a generation after the defeat of the Taiping rebellion, Mao Tse-tung was born in Hunan Province. By 1920, he was a confirmed communist and, inspired by the recent events in Russia, planned to create a new China governed according to Marxist and Leninist principles. In 1926 General Chiang Kai-shek took command of the 'National Revolutionary Army', but Chiang was eager to retain support of the gentry and began to purge the more left-wing communist elements. A number fled south to avoid Chiang's secret police, establishing a haven in the rugged terrain of the Fukien-Kiangsi borderlands. As with the Lam Son and Taiping rebellions, control was established over a wider area and in early summer 1930 the Communists began military operations against cities held by the Nationalists. However, they suffered a serious defeat at Changsha in September. In October 1930 Chiang announced a campaign to crush the Communists. This was unsuccessful, but in 1933 Chiang adopted a policy, on German advice, of forcible evacuation and using blockhouses and wire to control the countryside and isolating the Communists from their base, the peasantry whom they had laboriously converted.

For the first time, the Communists found themselves deprived of food and other essential supplies, and of information, the most pronounced advantage of a guerrilla movement based among the populace. In the face of the Nationalists' methodical and co-ordinated advance, the decision was taken to move the base of operations to Shensi province, right on the other side of China (see Figure 6.6). Thus the famous 'Long March' of almost 6,000 miles (9,600 kilometres) began.

The purpose of the Long March was to preserve the military power of the Communist Party by ensuring that problems of terrain and supply were solved on the Communists' own terms. The 'Long March' was thus a paradox: continuing the offensive by 'running away', or, rather, by conducting that most difficult of military moves, a *withdrawal in contact*. The number of skirmishes and larger pitched battles fought along the 6,000-mile route cannot be established for sure. The columns were under air attack for days on end, crossed rivers and mountain passes, and veered between near-tropical and subarctic climates. As they approached the borders of Tibet and swung north, caches of ammunition and supplies were carefully concealed for the return. The Long March is exemplary with regard to the importance of terrain, and its exploitation, and of morale and determination. The exhausted remnants of the Chinese 'Red Army' eventually reached shelter and safety in the caves of Pao An. Later, the base was shifted to Yenan, where Mao reflected on his experiences and developed the theory of guerrilla war which he embodied in *Yu Chi Chan* (*Guerrilla Warfare*). In this, he drew on sources as diverse as Sun Tzu, the Taiping Rebellion which ended thirty years before he was born, Clausewitz, Lenin, and Gusev's *Lessons of the* [Russian] *Civil War*.[99] He rather arbitrarily dismissed the idea that anybody other than Communist revolutionaries could conduct successful guerrilla warfare: in fact, military history has many examples of straightforward patriots and even counter-revolutionary forces conducting 'guerrilla' warfare with considerable skill and success. However, Mao was undoubtedly right that one cannot win without offering a political solution that is at least marginally preferable to the alternative.[100]

Meanwhile, in September 1931, anti-Japanese guerrilla campaigns were launched in the three north-eastern provinces of China. Mao divided this activity into two periods: sporadic attacks, up to January 1933 which were relatively unsuccessful because of lack of co-ordination and political aims, and the period from January 1933 to 1937, when seven or eight guerrilla regiments were formed, which necessitated disproportionate effort on the part of the Japanese. Writing in 1937, Mao predicted that 'the guerrilla campaigns being waged in China today are a page in history that has no precedent. Their influence will not be confined solely to China in her present anti-Japanese struggle, but will be world wide'.[101] The hallmarks of this influence include the premium placed on intelligence, readily provided by what the guerrillas hope will be a sympathetic peasant population; the use of deception, mobility, and surprise (all classic aspects of 'conventional' operations) in different and unorthodox ways; and the emphasis on educating the inhabitants in the ideological basis of the struggle in which they are to be enlisted.

The last element – an organized and prolonged campaign to establish a broad base of popular support – was the most prominent technique adopted by other revolutionary leaders throughout the world up to the 1960s. However, once obviously repressive, unjust and predatory regimes are overthrown, that is as far as the guerrilla can go on the basis of popular support. The 1970s and 1980s have arguably seen the tide turn, and 'guerrillas' seeking to change the political structure in liberal and democratic countries have two, interrelated options. One is to force revolution on a population most of whom do not want it, by the armed action of a dedicated minority, and then persuade them that they did want it (as arguably happened in Russia in and after 1917 and in Iran in and after 1979); the other (the classic theory of the urban guerrilla) is for the dedicated minority to use terrorism to create a situation where normal democratic processes and freedoms become inoperable, so that repression is created, against which the population can then be induced to rebel.

In Asia, however, where the general standard of living and expectations remained low, Mao's theories continued to be applicable throughout the 1960s and into the 1970s. Mao placed great emphasis on co-operation between guerrillas and more 'orthodox' groups. This, according to Mao, fell into three categories; strategical, tactical, and 'battle' co-operation, which, in accordance with the levels of warfare used in this book, can be renamed strategic, operational, and tactical.[102]

Strategical co-operation involves harassing the enemy's rear installations and hindering transport, forcing him to divert precious resources away from the main theatre of war. As an example of what he called 'tactical' co-operation, what we shall call 'operational', Mao cited the operations at Hsing K'ou, where guerrillas north and south of Yeh Men destroyed the T'ung and P'u railway and motor roads, thus complementing the action of regular forces in Shansi and in the defence of Honan. To be effective in this role, guerrilla forces acting in co-operation with regular forces need good, preferably radio, communications. In the immediate battle area, guerrillas hinder enemy transport, gather information and act as outposts and sentinels. These missions should be assumed 'even without precise instructions from the commander of the regular forces'.[103] All regular forces have relied on local intelligence and assistance of this sort to some extent. Mao's views on the co-operation and interplay between guerrilla and more orthodox forces is particularly relevant for the subsequent development of the art of war on land in Asia.

GENERAL GIAP: A GREAT CAPTAIN OF MODERN TIMES?

Although World War Two produced some charismatic commanders evincing a good deal of individual flair (Chapter 4), warfare among the highly industrialized powers had clearly become an affair of governments and their impersonal agencies, rather than of great commander-strategists. The situation was, and perhaps still is, less clear in Asia. Mao was undoubtedly a master of the interlinked arts of war and politics, but is remembered primarily as a politician and guerrilla leader, rather than as a great captain in conventional warfare. The Vietnamese General Vo Nguyen Giap, born in 1912, belongs to the latter category, as well as the former two. The second Vietnam War of modern times, from 1964 to 1975, was concluded when the North Vietnamese Army (NVA) reached a decision in conventional large-scale land battles.[104] Giap's writings contain little on orthodox land operations, but he clearly understood their principles and essence. He once boasted that the only military academy he had attended was that of the bush. That had not bothered the likes of Genghis, either. Giap cannot perhaps be placed in the first rank of military commanders, and in the early years his lack of a formal military education showed. For example, in 1950 he transgressed one of the classic principles of military art by attacking the

French in a strong position of their own choosing in the Red River delta (see Figure 6.3). Had Giap studied the precepts of Sun Tzu, he would almost certainly not have done this but 'offered the enemy a bait to lure him'.[105] There are many variants of this principle which Giap could have used to lure the French out of this position. Giap did, however, show considerable tactical skill at the battles of Vinh Yen and Mao Khe in 1951, but on both occasions was defeated by enemy air and sea power. He therefore resolved to act as far as possible out of range of naval guns and in conditions precluding the application of air power. This is a further example of how 'land' warfare had become critically dependent not only on air power but also on the ability of naval forces to project power inland. Since then, the advent of long-range ballistic and cruise missiles has, as the Soviet Admiral of the Fleet Gorshkov has observed, enabled naval forces to dominate not only a narrow littoral or riverine stip but the hinterland as well.[106] For these reasons, Giap shifted his main efforts to Tonkin in 1952–3. Giap's first world famous success was the capture of the strategic valley of Dien Bien Phu in 1954. Dien Bien Phu controlled communications between Tonkin and Laos, which were essential for Giap to conduct a prolonged offensive against French forces further south. The French therefore decided to seize it in operation Castor, commencing in November 1953. Giap, mindful of the need to score a victory over the French before a peace treaty was drawn up, and knowing that Dien Bien Phu was isolated and dependent on air resupply, decided to wipe it out. The French underestimated their opponent and had not correctly appraised the terrain, believing that the flat valley floor would be a lethal killing ground for anyone attempting to cross it. In fact, it was covered with thick bush giving excellent cover for attackers. Similarly, the forward slopes of the surrounding hills appeared to be the only suitable positions for besieging artillery, which would therefore be exposed to the defenders. Once again, these were covered in dense vegetation which in fact concealed the guns. Finally, the French greatly overestimated the ability of their air force to resupply the defenders, as the Germans had at Stalingrad.[107] The weather was often so bad that no aircraft could land and parachutes could not be dropped with any accuracy. All the classic military principles of knowing one's enemy, terrain and weather are applicable in the case of Dien Bien Phu (see Figure 6.8). Giap himself had learned caution; the rash attacker of Vinh Yen, Mao Khe, the Day River Line and Na San was now determined to advance slowly and surely. Knowing that the French had created a skilful defence with interlocking fields of fire he determined to overcome them using a method of 'progressive attack'. He constructed 'a whole network of trenches that encircled and strangled the entrenched camp, thus creating conditions for our men to deploy and move under enemy fire.'[108]

Giap's forces sapped closer and closer to the French defences, a reminder that the traditional techniques of siege warfare may well be applicable in the

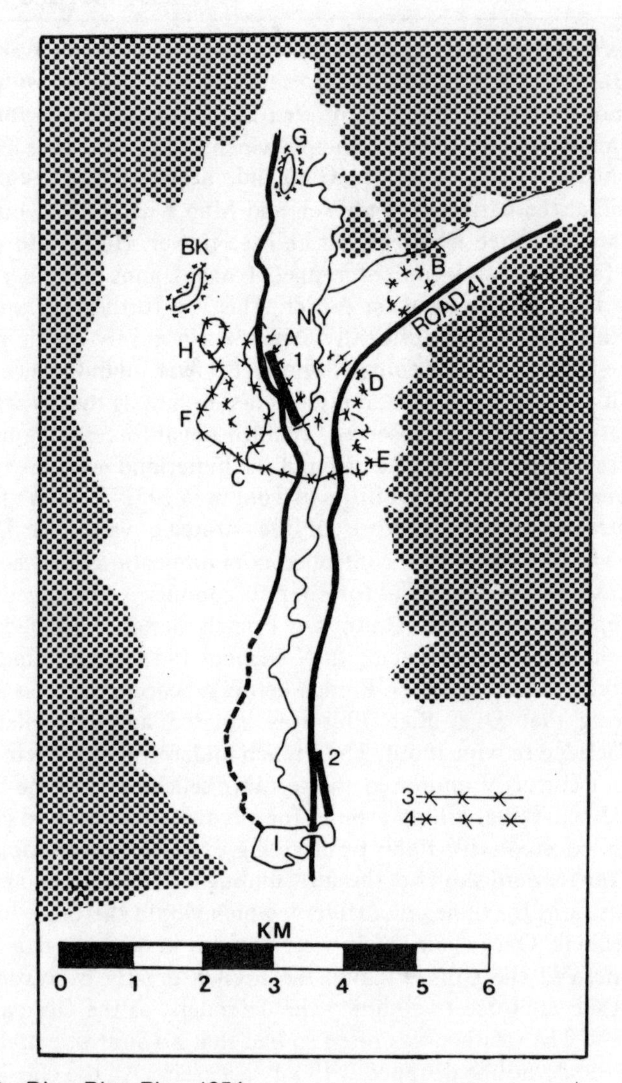

Figure 6.8 Dien Bien Phu, 1954

1 = main air strip
2 = subsidiary air strip
3 = main barbed wire defences
4 = barbed wire defences of individual positions

The individual strong points have girls' names:

A = Anne Marie
B = Beatrice
C = Claudine
D = Dominique
E = Elaine
F = Francoise
G = Gabrielle
H = Huguette
I = Isabelle

BK = Ban Keo (local name)
NY = Nam Yum River

Surrounding shading indicates forested hills. Heavy black line indicates roads, dotted heavy line indicates impassable roads.

age of the aircraft. Anti-aircraft guns forced French aeroplanes to fly so high that parachute resupply drops were inaccurate. With the aid of a terrifying Soviet-style artillery bombardment Giap overwhelmed the commanding positions of Béatrice and Isabelle on 13 and 14 March 1954, although the dénouement did not occur until 7 May.[109] Interestingly, Giap had to work very hard to maintain the morale of his troops during this slugging match. It is quite clear from Giap's own account that only strenuous 'party political work' and the constant influence of the party machinery at all levels kept the Vietnamese army fighting.[110]

Although guerrilla operations continued to divert and harass the enemy. North Vietnam committed its regular army to combat in 1964. Giap intended to slice South Vietnam in half. He reckoned that three divisions launched from secure bases in Cambodia and supported by Viet Cong guerrilla units would be superior to any countering force which the Americans and South Vietnamese could field. In fact, this classic military exercise in mass, manoeuvre, and concentration of force showed that the Americans' competence in such conventional operations far exceeded that of the North Vietnamese at this stage. The offensive ended in November 1965 when it was defeated by the First Cavalry Division in the Ia Drang Valley.[111]

Giap was military commander in chief throughout most of the ensuing Vietnam War. He was also a member of the ruling Politburo and deputy prime minister. Political and military leadership were thus almost perfectly combined, a key factor in the North Vietnamese victory. The Soviet Union enjoyed the same unity of political-military-psychological leadership in the Great Patriotic War, and might do so again. This becomes clear when we examine the Tet (Lunar New Year) offensive of February 1968. Although an operational-tactical failure, it was a strategic success for Giap. As yet still unable to match the Americans and South Vietnamese operationally, tactically, or materially, Giap went for a strategy which would have the maximum effect on American and world public opinion: a strategy directed at overrunning towns and thus achieving a high political and media profile. It had the desired effect, revealing Giap's awareness of the higher and more diffuse elements of strategy at which, for example, the Mongols were so expert.[112]

The 1968 Tet offensive is generally acknowledged to have been the turning point of the war. It was a series of hammer blows on a strategic scale, not unlike the Soviet offensives in the later stages of World War Two in configuration, though on a smaller scale. In the first week, after 30–1 January, the communist forces enveloped 34 provincial capitals, all 7 autonomous cities, and 64 district towns. Casualties were about 60,000 Communist and 10,000 US and South Vietnamese. The Tet offensive was intended as a blend of major conventional and less conventional operations. First, a shock assault by local (predominantly southern) communist forces would bring the war to the cities. This would lead to the second stage: uprisings by the people and defections by South Vietnamese (ARVN) forces. Finally, regular North Vietnamese units would enter the cities to mop up remaining ARVN forces and force the Americans to withdraw. In fact, the offensive did not pass stage one. For the North Vietnamese, Tet was, however, the end of the strategy of 'people's war' and a committal to major conventional operations. For the Americans, who refused to sanction an increase in troop numbers by 200,000, it was the end of the possiblity of conventional military victory, a realization reinforced by operation Lam Son 719 in February 1971, when South Vietnamese forces failed to achieve decisive success in spite of generous American air support (see Chapter 4).[113]

Giap continued to learn. In the Easter invasion of 30 March 1972, he committed four fronts at staggered intervals. Second, after overrunning Quang-Tri, he ordered a three-week respite. The effect of these two strategic mistakes: lack of concentration and delay, was that the ARVN were able to regroup and consolidate their positions. The North Vietnamese had also not yet mastered the technique of all-arms co-operations at the tactical level. By 1975, however, they had got it right. In the final 55-day Ho Chi Minh offensive they concentrated their forces for a single overwhelming offensive across the Central Highlands aimed at the lightly defended objective of Ban Me Thuot. There was nothing new about the simultaneous conventional and guerrilla challenge posed by the North Vietnamese. The French had faced it in the Peninsula and the United Nations in Korea. The textbook solution, in both the former cases and in Vietnam, was to deal with the orthodox challenge first. This is undoubtedly right. In Korea, for example, the 'guerrilla' war dispersed once the North Korean conventional army was expelled, and the same would probably have happened in Spain if Wellington had lost. In Vietnam, the guerrillas ceased to be guerrillas when they emerged to mount a conventional attack, and they were now committed to winning the war in conventional, European–American–Soviet terms. The North Vietnamese therefore won the war using an American-style strategy.[114]

The lack of information coming out of Vietnam makes an assessment of Giap's status as a leader in major war difficult. In the late 1960s, before Tet, the Easter invasion and the Ho Chi Minh Offensive, one analyst thought that he displayed few of the qualities which made men like Marlborough, Napoleon, Guderian, or Mao of importance to military affairs. His campaigns showed competence and the ability to learn from mistakes rather than strategic brilliance. His books did not make any new contribution to strategic principles and he made a number of serious misjudgements.[115] But then, many great captains have made mistakes, like Nadir Shah at Baghdad, and the Persian did not write anything on military theory at all. By 1969, after Tet, O'Neill reckoned him 'one of the most skilled practitioners' of military art since 1815, who had shown 'unfailing skill through several years of fighting'.[116] It is now clear that Giap masterminded the subsequent campaign which defeated the South Vietnamese and American armies in Indochina. Of course, had the Americans been sufficiently interested politically to bring their full strength to bear on the ground, the result would have been different, as Giap knew, a reminder that war and politics are inseparable and that it is superiority which can be brought to bear that counts. Giap's higher direction of the war probably gave him little opportunity to demonstrate the skills of a conventional commander on the ground, although it was rumoured that he took direct command in the field during the siege of the American fortress of Khe Sanh in 1968. His writings mostly concern the higher realms of strategy and politics, but certain passages in *The Military Art of People's War* reveal a deep understanding of the nature of land warfare, especially when commenting on American methods:

> Bombing and strafing tactics, aimed at destroying LAF units, resistance bases and the people, have . . . proven ineffective because of poor intelligence and the failure to identify targets accurately. US Air Force bombings and strafings . . . including those of the B52 strategic bombers, have not wiped out one single LAF unit, and have only shattered trees and destroyed empty tunnels, as the imperalists themselves have admitted.
>
> Why have all these been ineffective? Tactics are not separate from strategy, as everyone knows. If strategy becomes defensive and deadlocked, it will affect tactics sharply and adversely. In addition, the failure and deadlock of the various tactics used by the US lies in erroneous tactical thinking. The Americans have believed that they should base their operations on the power of weapons, and have assumed that firepower is their soul. When the use of these bases, weapons and firepower is limited, or fails to become increasingly effective, the tactics become ineffective and themselves fail.[117]

It was rumoured that Giap had died in 1975, but this was subsequently proved false when he was interviewed on television. He could therefore have directed the successful defence of Vietnam against Red China in 1979. The Chinese had no choice but to advance over the same terrain, past Lang Son, as generations of Imperial Chinese generals had done – the same terrain where they faced the Lam Son insurgents in 1426–7. The result was similar, too: an ignominious withdrawal. Giap is clearly one of the great captains of this century. He is probably the only one identifiable after World War Two, with the caveated exceptions of MacArthur and Moshe Dayan of Israel. In order to exercise a defensive influence on events, a great captain must also wield political power.

CONCLUSIONS

The inscrutable and mystical traditions of the east have fascinated many Europeans and Americans, and there is a danger that the oriental in combat may be accorded almost supernatural powers. This was a risk for the British and Commonwealth forces fighting the Japanese in the early stages of the Pacific war, while Japanese success in business has made it fashionable in those circles (as in warfare and military organization) to imitate the pre-eminent power. The works of Sun Tzu and Miyamoto Musashi (the latter writing about sword-play rather than strategy in its wider sense)[118] have become fashionable dinner-party conversation pieces in the world of finance and business. The fall of South Vietnam, six years after the Americans began the policy of 'Vietmanization' has been attributed to the fact that American strategists played two-dimensional chess, the orientals three-dimensional *go*.[119] In the author's view, this is esoteric rubbish. The Americans killed North Vietnamese in their hundreds of thousands, and in spite of problems, were much more effective, unit for unit. The nation was just not prepared to pay the price of victory.

It is however possible to identify two broadly distinct 'Asiatic' patterns of development in warfare. The first evolved in the steppe, *en rase campagne*, as the French would say. Certainly in the pre-industrial period, the great military commanders of Asia placed greater stress on the firepower mobility elements of the firepower-mobility-protection triangle (Chapter 1). In contrast, the Europeans stressed firepower-protection more. Harold Lamb's verdict on a hypothetical encounter between the infantry-defensive system of the English and the cavalry-offensive system of the Mongols is justified. The Black Prince, commanding the English at Crécy, 'would have been as utterly helpless as his European cousins . . . utterly unable to prevail against the rapidly manoeuvring Mongols, led by generals such as Mangu and Kaidu – *veterans of a lifetime of war on two continents*.[120]

This brings out two points in common with the other branch of Asiatic development: the branch which evolved in the complex, teeming, agricultural counterpane of China and Vietnam. The prime characteristic of the latter is integration of guerrilla and large-scale, 'orthodox' warfare. The first point in common is timescale. Many of the great Asian conflicts examined have been extremely protracted, and so have some individual campaigns although the Mongols, for example, acted with devastating swiftness when they could. The Mongols planned to subdue Europe in 18 years, the Taiping Rebellion lasted 13, and the Chinese Communist campaign against Chiang, the Japanese and Chiang again lasted for nearly 20. Giap did not approach the conquest of South Vietnam with indecent haste. European soldiers and scholars alike are mesmerized by swift campaigns with decisive results: that is probably why the word 'Blitzkrieg' is so popular.

The second is what Liddell Hart called the indirect approach (a phrase coined by Sun Tzu, of course).[121] The Mongols would certainly not have charged the front of the English defensive positions at Crécy and Agincourt as foolishly as the French did. The strategy used by the NVA and the Vietcong in Vietnam similarly aimed to avoid presenting the enemy (Americans ARUN), with a target. One of the most difficult problems for the Americans in Vietnam was to find the enemy. This applied in the specialized context of underground warfare, the Vietnamese digging complex and ingenious defensive systems to escape detection and heavy bombing, and also on the surface.[122] The enemy was hard to find because the war was fought 'on a battlefield that has neither a front nor a rear. The enemy himself is the only valid point of reference, and it is he who dictates the limits and the dimensions of the battlefield'.[123] Even in the Middle Ages, the chronicler Fra Carpini noted that 'the Tartars fight more by strategem than by sheer force'.[124] As Sun Tzu noted, to subdue the enemy without fighting was the acme of skill, and Genghis Khan, the 'perfect warrior', did not fight if he did not have to.[125] The transition and interplay between guerrilla and regular forces was a particular feature of the Vietnam War. The Americans soon realized that the NVA and the VC were different enemies who had to be met with different tactics. If an American unit faced a transition in the enemy from NVA to VC, the adjustments in tactics required soon became apparent.[126]

The philosophical traditions of Asia have contributed to a more blurred view of the division between war and peace than in European thinking. Like the Buddhist prayer wheel, the circle of comparative peace, guerrilla war, mid-intensity war, major war, revolves and returns to its starting point, and revolves again. Smooth transition from one level of conflict to another is part of the East Asian approach to war. This tradition has thus provided fertile ground for theories and practice of 'People's War' and 'Revolutionary War' to germinate. The interplay between the different levels of conflict is apparent from the careers of Mao and Giap. The ebb and flow of war, terrorism and diplomacy recalls Sun Tzu's simile, 'an army may be likened to water, for just as flowing water avoids the heights and hastens to the lowlands, so an army avoids strength and strikes weakness'.[127] In a world where clear-cut, major military operations are looking less and less likely, and where conflict is more likely to take the form of 'low intensity' or 'counter-insurgency' operations, where violence at a certain level short of fully fledged warfare has become almost endemic, the history of the art of war in Asia is more relevant than ever. However, the principles of war apply at every level, whether in a guerrilla struggle or in large-scale conventional operations. The Asiatic traditions underline this, stress the interdependence of different levels of conflict, and at every level Asia has produced some of the most successful and brilliant military commanders in history.

Chapter seven

Some conclusions and prognoses

Throughout this book, the author has demonstrated that the intelligent study of military history is of practical value for today and the future. This does not mean taking things out of their historical context, or making sweeping and absurd parallels between events and conditions in widely differing eras.[1] For that reason, accuracy and technical detail are important: how and why commanders reacted as they did, given the enemy they faced, the technics and the information they had, *at that time*. The author has chosen some examples which may be unfamiliar, but which demonstrate key principles and ideas as well as, or better than, more familiar ones, especially with regard to the likely location of, and participants in, future conflict. Where others have made sensitive and intelligent use of historical experience – for example, MacArthur's analysis of Genghis Khan, Kishmishev's analysis of Nadir Shah's campaigns, or Mao's of the Taiping Rebellion – that is of interest and value in itself.

A broad view of history naturally leads one to look into the future; a prognosis, that is, a prediction based on known facts. The author will confine himself to but six conclusions from the study of warfare, all of which have some bearing on the future, as well as on an understanding of the past.

1 IT WORKS!

The first observation is prompted by a genuine, though unreferenced account of a meeting with a Soviet Marshal who had been in a senior headquarters when the Germans launched their attack on the Soviet Union on 22 June 1941. At first, the Marshal recalled, there was total confusion and things were 'pretty bad'. But after a few hours, things 'lightened a little'.

' "Lightened". What do you mean, "lightened"?'

Realization of what was happening would surely have made the situation appear even more dire.

'Our troops were fighting.'

The Marshal did not mean to cast aspersions on the courage or professional competence of his subordinates. What he meant was that *the military machine was actually functioning*. In spite of the complexity, the friction and the inertia that attend the co-ordinated movement of huge bodies of individuals, a creaking machine, with a number of key parts removed, that had stood largely idle, as far as real war was concerned, for twenty years, was groaning, grinding into action, against one that had been finely tuned, and running, for two years.

Similarly, take the Falklands conflict in 1982. The British Army and Royal Marines were a most professional force, and the Army deployed some of its best units. But is it not almost incredible that after years of comparative peace, with no active service against serious opposition in a mid-intensity war, you can take a brigade, or a component of that brigade, a battalion, put it down in an environment where it has not expected to fight, switch it on, *and it works!* That this was possible underlines the high degree of professional expertise, professional studies, training and discipline of all ranks. They had, of course, practised battalion attacks, even the occasional brigade attack, but the gulf between that and real warfare is still stupendous. On exercise, cold and physical and mental tiredness are the worst enemies. In combat, men seem to forget the first, and the second is not so much an enemy in itself as a constant debilitating, dragging impediment which makes them more likely to become a casualty, facing searing pain and death. Unlike a peace-time exercise, this time the day and approximate time of 'endex' is not known. The future is a horrific, yawning, potentially bottomless void. Even under these circumstances, the professional soldier, at the flick of a switch, the nod of a head, gets on with it, according to the peace-time model.

2 LARGE-SCALE WARFARE HAS PRICED ITSELF OUT OF THE MARKET

This brings us to the second point. Major war, involving total commitment, between developed nations has become such a staggeringly costly and complex process that it is almost impossible to imagine it being used as a political instrument in future. Recent hostilities between such nations have been strikingly limited in their overall scope if (like Frederick the Great's battles) extremely savage within the agreed bounds. The Anglo-Argentinian conflict of 1982 is once again a case in point. While conventional forces fought vicious actions on the approach to Port Stanley, the security alert state in the Ministry of Defence's Main Building in Whitehall remained at black. Nobody expected an Argentinian post-graduate physics student to walk in with a home-made atom bomb, obviously. Another factor in restraining international armed conflict has probably been the United Nations. Unlike the League of Nations which preceded it, the United Nations has some teeth (as demonstrated in Korea) and exercises a degree of control over potentially delinquent states. The achievement of a ceasefire in the apparently interminable Gulf War, in 1988, was a major achievement.

Modern nations no longer accept war as a relatively natural aspect of international relations. The mere fact that earlier this century the United States had a 'War Department', the United Kingdom a 'War Office' and Russia a 'War Ministry', but that they all now have Departments or Ministries of Defence underlines this. It is not just a reflection of the deplorable modern tendency to substitute anodyne euphemisms for calling a spade a spade. It does reflect a profound reappraisal of the role of armed conflict in international relations.

This brings us again to the paradox that war*fare* (that is, the conduct of military operations) naturally strives for extremes, while *war*, as an activity in pursuit of some political objective tends, inherently, to be limited. The example of Frederick the Great's wars is crucial here, and they were no doubt at the forefront of Clausewitz's mind when he wrestled with this paradox in Book 1, Chapter 1, of *On War*.[2]

If it begins to seem that war in its most extreme sense between the pre-eminent powers is becoming less and less likely, there remains the problem of the 'lunatic with an axe'. There is no effective way of controlling him other than by force. World War Two arguably represents an example of a war of this type. The western democracies and the Soviet Union (the latter militarized to an astonishing degree) all crept, like armadillos, unwillingly to war. Eventually, even deeply held political beliefs were subordinated to the need to destroy the delinquents, and the western democracies employed the newest destructive technology in the crudest way to obliterate Nazi Germany and Imperial Japan.

3 LARGE-SCALE WARFARE AS A FINITE PHASE IN HUMAN EVOLUTION

The third point is related: the position of warfare as part of the evolution of human societies in general. Warfare, it seems, was primarily a product of agricultural or, in the case of the Asiatic style of war, nomadic herding, societies. They found it a useful political tool: otherwise, it would not have continued. The moment industrial societies developed, in the nineteenth century, warnings of the potentially disastrous effect of major conflict employing new, industrial technologies became widespread. Nobody took up the military instrument lightly, but professional soldiers were perhaps not as effective as they might have been in persuading their political masters that the military instrument was now excessively dangerous and liable to run out of control. World War One was, perhaps, the last attempt by responsible major powers to use war as a rational tool of politics. When it ended, it had destroyed many of the political regimes which had brought it into being.

War, as a product of the agricultural era, survived into the industrial, but to a crescendo of protests against its use. There were those, like Norman Angell and Ivan Bliokh, who thought that the complex structure of international trade and interdependence had made war between the most developed nations impossible by the beginning of the twentieth century: it lingered on to the middle of that century, but at horrendous cost.[3] For post-industrial economies, where autarchic production has given way to far more sensitive international markets and the satisfaction, not of basic economic needs but of artificially stimulated desires, war really has become unnecessary and obsolete, with the exception of the eventualities described above.

Are the vast, baroque artillery parks and tank and air fleets of NATO and the Warsaw Pact therefore anachronistic legacies of an age which has truly passed? Are nuclear weapons, unusable in any imaginable, rational, political context, bizarre mutants, examples of warfare's own inherent tendency to the extreme got out of hand (an eventuality which Clausewitz would, contrary to the view of many of his so-called interpreters have abhorred)? There are some good grounds for thinking so. Warfare appeared in human history at about the same time as organized religion and there are some parallels that can be drawn. Warfare, like organized, institutionalized religion, is no longer a regular aspect of intercourse between the most developed states. The role of religion, like that of warfare, has become more low-key, concerned (as to a large extent it always was), with helping the needy, the oppressed, and those in the less developed parts of the world. The grandiose ritual and showmanship play an important part in focusing belief and reasserting and stressing the deity's power. But there is far greater tolerance between religions, and the religious complexion of developed nations is far more variegated and complex than would ever have been permitted under Elizabeth I of England, the Spanish Inquisition, or the elders of Salem. So, too, with warfare. The abstruse arguments of NATO (and Warsaw Pact) generals about 'Follow-on-Forces Attack' (FOFA), and whether there will be a 'second echelon' to attack (see Chapter 5) play an important role. But they may be as relevant to the conduct of military operations in future as abstruse and scholastic questions of theology are to the day-to-day work of the many religious organizations doing good work in the world.

Major inter-state warfare is becoming increasingly rare, and increasingly entwined with internal disputes. This is fully consistent with the tendency for the phenomenon of warfare to receed to less-developed areas of the globe. The Iran–Iraq war was an anomaly in the sense that it was not directly induced by dissatisfaction at a post-1945 settlement; it was a religious war; it was met, at first, with fanactical enthusiasm. But by the end even the fanatical Iranians had had enough. It belonged, in terms of historical development, to the same era as World War One. Unsurprisingly, the military technique employed, apart from a few high-tech frills, was not dissimilar.

4 FUTURE USES OF WARFARE

The development of major warfare therefore provides us with some surprising crumbs of comfort. Wars are not started by soldiers, but by politicians. Even the thickest politican now knows that warfare is a totally unacceptable and potentially suicidal way of resolving political disputes. It is therefore only likely to be used in the event of a delinquent dictator going berserk. In those circumstances, it will be more effective if all the major powers co-operate, and there are increasing signs that they will be able to. This may include the possibility of East–West military cooperation against a third party. The beast, the lunatic with an axe, may still manifest itself on the planet. The responsible majority of nations will probably perceive the need to retain sufficient force, between them to control what they consider to be a renegade power and prevent it from doing unacceptable damage. This is the course of development which 'conventional' military forces may tend to follow into the twenty-first century.

International 'police actions' of this type must not be confused with 'low-intensity warfare'. They may require the swift and devastating cauterization or amputation of a renegade state which has obtained extremely dangerous resources. They may require large-scale military operations on the classic model, to seize and secure vital areas. In such cases, some more distant military historical precedents may be more applicable than those of the comparatively recent past (we may consider World War Two part of the latter). 'Limited war', as shown, may be very vicious with a keen cutting edge. Like Frederick the Great's campaigns, the surgical strike may require the sharpest instruments. Forces will also be operating in an environment where they may not have any recent combat experience, and where there will not be time to gain it. There will not be any room for mistakes. The intelligent use of past experience – military history – will be vital in avoiding mistakes. There will be no opportunity to re-learn lessons that should have been learned but have been forgotten (as the Russians, in Afghanistan, obviously failed to remember the Finnish War of 1939–40, or to read Kipling).

5 TECHNOLOGY, TECHNICS, AND THE FUTURE FORM OF MILITARY OPERATIONS

Throughout military history, new technics have sometimes made it possible to restore or revive earlier strategic, operational and tactical ideas, though in a different form. The most important case in the future is likely to be the age-old objective of destroying and paralysing the enemy's command and control, preferably without having to fight through a defended front to get at them, as advocated, for example, by Fuller (see Chapter 3). Advanced military forces are dependent on computers, radio and other communications, and satellites, for reconnaissance, navigation, and communications. Attacking the enemy's 'brain and stomach' need no longer depend on tanks racing round a flank, or aircraft pounding headquarters and industrial centres from above. The enemy's brain and nerve system can be seared and paralysed by jamming, and various types of electromagnetic weapons. Electronic warfare, and other 'soft kill' weapons are likely to usurp the position envisaged for tanks and aircraft in much of the 1930s military theory. Low-frequency weapons and the application of bioelectronics may severely reduce the effectiveness and alertness of enemy forces, commanders, and political leaders.

But there are two problems. The first is that, as always, the enemy is likely to have something similar, although the disparity between the most developed and less developed states may become more decisive than it has been previously. Second, tanks and aircraft were not much use for occupying and controlling terrain. Electromagnetic weaponry of the type described is of even less use. Having burned out the enemy's circuits, blinded him with lasers, deafened him in the ether, you still have to have forces capable of occupying his territory and forcing him to 'do your will'. They may not have to be used. Electromagnetic and other new-generation weapons may fulfil the age-old objective of placing the enemy in a position which, though it may not of itself bring about victory, *ensures that continuation of the struggle by battle is sure to do so*. But forces for occupation and control will have to be available, nevertheless. We may therefore see simply another layer of the latest high tech grafted onto previous layers: when both sides have disabled their C^3I, brains, and nervous systems, they will revert to tanks, then to rifles, and finally to rocks. The author's prognosis begins to sound like a revision of views popular in the 1930s. European demographic problems (fewer people, especially young men) and the increased sophistication of military technics will place more emphasis on small forces. Millions of men in greatcoats with rifles are no longer the principal determinant of military power: even the Russians realize that. Warfare between the superpower blocs is looking even more unlikely. The world is, in any case, reverting to a multi-polar system, with a number of sub-superpower blocs, the Single European Market, India, China, and Japan being the most obvious. The threat to international order will come from smaller nations which feel they have been left out, as well as from other powerful factions capable of purchasing military muscle, like drug barons. Therefore, smaller but more expert forces, armed with exponentially increased firepower, do appear to be an attractive and appropriate option. The 1930s theorists who forecast much smaller professional armies were, perhaps, ahead of their time, but their prognoses may now be fulfilled. Unable to match the expertise and technology of the latter from their own resources, some states may resort to high-tech mercenaries.

Modern economics and technics certainly lend themselves to these much smaller, more potent forces. Bionics, robotics, body armour, and drugs will substantially enhance individuals' performance, and artificial intelligence and small computers will vastly improve C^3I. Substantial reductions of forces in Europe will reduce the relevance of training for large-scale, classical operations. Worldwide challenges, diverse and unexpected, will remain. All operations will become special operations, thus fulfilling the Vietnamese prognosis (Chapter 6).

As part of their training, and as a means of employing the manpower needed to back up the élite, hard-hitting small forces, the military may turn its expertise to the increasing number of ecological and environmental problems. Military men are good at solving large-scale, complicated problems in the open air. They may find this to be an increasingly prominent part of their lives. There is nothing new about this, of course. Troops have been used to bring in the harvest throughout history, and are invariably called in when natural disaster strikes. The huge, if diminishing, pool of trained, fit manpower deployed in armed forces throughout the world could be very useful in putting right some of the damage we have done to the planet.

Whereas in the period from 1945 until very recently, we have seen a pattern of a bi-polar world, with technology 'pushing' – evolving to some extent in its own way, with the evolution of aircraft and missile systems, leading to colossal armaments expenditure, this may now change. A multi-polar world, where national and international priorities pull the development of new technologies to meet specific requirements, is the logical result.

6 VALUE OF THE HISTORIAN

Study of past experience is therefore even more vital in a world where military options – warfare – are likely to be used with increasing precision and brevity (by developed countries, at least) and where the increasingly sensitive and well informed societies thereof will be even less tolerant of mistakes. But what kind of analysis should be used? There have been huge advances since World War Two in mathematical and computer modelling of conflict. Within history itself, there has been a move (encouraged by the spiralling PhD industry), towards increasingly minute analysis of short periods and limited areas. Historians may sometimes be criticized for extrapolating too readily and freely from past to present and future. The true historian's answer to his or her critics is: 'What is your database?' The historian may have to be selective, but he can draw on the experience of everything that humanity has achieved, and that the human mind can comprehend. He should be taken seriously, particularly in a business as grim, as potentially cataclysmic, but as human and timeless, as war.

Another important function of the historian is that he brings to the analysis of military and defence matters an instinctive feel for the flow of time. It takes a historian to recognize how quickly so-called 'current' analysis becomes out of date. Time is measured, above all, in terms of the scope and scale of human life, and especially, of a human working career. Yet many analysts' sense of time seems strangely distorted and flexible. For example, there are still 'authorities' in the defence field who believe that the works of Marshal Sokolovskiy (first published in 1962, with the last edition in 1968) represent, to some extent, current Soviet military thinking.[4] Yet, even if we disregard the well known and documented shifts in Soviet Military Doctrine, as expounded by Gareyev and others, Sokolovskiy's work is a quarter of a century old.[5] That in itself should be enough to make people ask questions. The same people who obstinately believe that Sokolovskiy is in some sense current would, no doubt, be the first to condemn anyone who opposed the work of Fuller, Liddell Hart, Tukhachevskiy and the like in the early 1930s with arguments drawn from the 1904–5 Russo-Japanese War. Such a person would be labelled as an out-and-out reactionary. Yet the time difference, a quarter of a century, is exactly the same.

It is a cliché, and, in the author's view, an erroneous one, that the pace of technological and associated social development has increased at an unprecedented rate since 1945. In fact, most of the things we associate with our daily lives, and the way we think, were around fifty years ago, in the late 1930s: cars, aeroplanes, dates fixed on the telephone, nights at the flicks. Many of the basic life saving medical techniques we know had been discovered. In the military context, for example, the Spanish Civil War was the first in which blood transfusion was regularly used. Go back another fifty years, to the late 1880s, and we really enter another world. The tactical conduct of the combined-arms battle in 1945 differed little from that of 1939. The same cannot be said of 1918 and 1914. The main advances in the conduct of warfare in World War Two concerned the co-ordination of land, air and sea power at the highest levels. It is a sobering thought that the senior generals at the end of World War Two, the beginning of the nuclear age, were all men of the nineteenth century, born into the now enormously alien world of the 1880s: Patton, Montgomery, and so on. When one measures the pace of change in a professional soldier's career, one cannot but baulk at changes witnessed by men like Patton, who spent his early days as an officer as a member of an Indian fighting Army before World War One, or Montgomery, or Rommel. When looking at the evolution of major warfare in historical perspective, there is good reason, perhaps, to believe that these men witnessed the last, and most rapid leap in the development of major land warfare. Or to put it another way, that 'we that are young, shall never see so much, nor live so long'.[6]

Notes

OPENING QUOTATIONS

1 Cited in Lt Gen Baron von Freytag-Loringhoven, *Deductions from the World War*, trans. 'J.E.M.', (Constable, London, 1918), p. 158.
2 Martin Blumenson, *Patton: the Man Behind the Legend, 1885–1945*, (Jonathan Cape, London, 1985), p. 43.
3 Lenin, *The Revolutionary Army and the Revolutionary Government* [June 27 (10 July), 1905], (Progress Publishers, Moscow, 1980), p.10.

INTRODUCTION

1 Napoleon, *Maxims of War* (1831), cited under 'Generalship' in B. B. Heinl, *Dictionary of Military and Naval Quotations,* (US Naval Institute Press, Annapolis, Maryland, 1966), p. 131. This quotation is often confused with 'read and re-read the campaigns of Alexander, Hannibal, Caesar, Gustavus Adolphus, Turenne, Eugene and Frederick. Make them your models. This is the only way to become a great captain', also from *Maxims of War,* cited in Heinl, p. 253. From a modern perspective one might choose other generals: the sentiment remains the same.
2 Miksche was a Czech soldier who left central Europe before World War Two, worked in the USA and Britain and with the French Foreign Legion, and at the Portuguese Staff College. Unfortunately, he had died before I was able to interview him, but I believe his work is the best on military art written in English during and after World War Two. His first book *Attack! A Study of Blitzkrieg Tactics,* (Random House, New York, 1942) is full of insight. On 'Blitzkrieg', and Miksche's subsequent works, see Chapter 4.
3 Richard E. Simpkin, *Race to the Swift: Thoughts on Twenty-First Century Warfare,* (Brassey's, London, 1986) is his crowning work. At the time of his death he was working on a book about command in war, analysing what made great commanders (we corresponded about Genghis Khan) and developing the idea, expounded in *Race to the Swift,* of a quite radical method of training them (essentially, that business was a far better school for command in real war than peace-time soldiering).
4 Theodore Ropp, *War in the Modern World,* (Collier, New York, 1962); Archer Jones, *The Art of War in the Western World,* (Harrap, London, 1987); Ernest R.Dupuy and Trevor N.Dupuy, *The Encyclopaedia of Military History from 3500 BC to the Present* (Macdonald, London, 1970).
5 Chris Bellamy, *The Future of Land Warfare,* (Croom Helm, Beckenham, and St Martin's Press, New York, 1987).

6 For example, the 'Standard of Ur', dating from about 2500 BC in the Babylonian Room of the British Museum.This beautiful box is adorned with battle scenes in which four-wheeled chariots, drawn by onagers, are carefully depicted. These go into battle first, followed by infantry. The warriors wear pointed caps and sheepskin cloaks (perhaps to give a degree of protection), and are armed with javelins carried in quivers fixed to the fronts of the chariots. Assyrian portrayals of their military campaigns are similarly notable for their detail: for example, the accurate anatomical drawing of the muscles in an archer's arm and the way the fingers grip a bowstring.

7 Ferdinand Otto Miksche, 'Is the atomic deterrent a bluff' in John Erickson, (ed.) *The Military Technical Revolution: its Impact on Strategy and Foreign Policy,* (Institute for the Study of the USSR, Pall Mall Press, London, 1966), article pp. 35–61, this ref. p. 49.

8 Caesar, *De Bello Gallico* (*The Gallic War*). Vindolanda is one of the forts on Hadrian's Wall, the southern of the two lines defending the Empire's northernmost frontier. Infrared photographs of the wooden tablets are on display in the museum at the site, near Hexham. See also A. K. Bowman and J. D. Thomas, *Vindolanda: The Latin Writing Tablets,* (Britannia Monograph Series, no. 4, Society for the Promotion of Roman Studies, London, 1983), esp. Chapters 3 and 4.

9 Sinan was a member of the elite corps of Janissaries, raised from selected Christian slaves (*devşirme*), and served as a military engineer where his ingenuity was tested before becoming Suleyman's principal architect. See J. M. Rogers and R. M. Ward, *Süleyman the Magnificent,* (British Museum Publications, London, 1988), pp. 36–7. Leonardo, see A. E. Popham, *The Drawings of Leonardo da Vinci* (Jonathan Cape, London, 1973), esp. pp. 80–4, 'Machinery and architecture'. He cites Leonardo's letter of *c.* 1481 to Lodovico Sforza (a copy of which, not in Leonardo's hand, is in the Codex Atlanticus), in which he asks for employment, mainly as a military engineer. In addition to 'bridges light and strong' and various means of mining and assaulting fortifications, Leonardo could make 'armoured cars, safe and unassailable, which will enter the serried ranks of the enemy with their artillery . . . and behind them the infantry will be able to follow quite unharmed and without any opposition'. This is a startling premonition of the modern tank, and is illustrated by a drawing dating from 1485–8 on the British Museum Department of Prints and Drawings sheet (plate 308). Plate 310 has another contemporary sketch of a machine with scythes revolving in the horizontal plane, dismembering bodies as it goes. Unlike some modern academics, Leonardo was prepared to examine the gruesome mechanics of wars, albeit fancifully and maybe even with tongue-in-cheek. But then, he was a genius.

10 This view in the works of Nietzsche (1844–1900), Spengler (1880–1936) and more recently, the West German W. Picht, *Vom Wesen des Krieges und vom Kriegswesen der Deutschen* (*On the Essence of War and the German Way of War*), (Friedrich Vorwerk Verlag, Stuttgart, 1952), pp. 4, 32, 115.

11 Roger Beaumont and Robin Edmonds, *War in the Next Decade,* (Macmillan, London, and Kentucky University Press, 1975), p. 204.

12 For example, Brian Holden-Reid, and Major General J. J. G. Mackenzie, (ed.), *The British Army and the Operational Level of War,* (Tri-Service Press, London, 1989). Although the book deals with a concept well known to specialists in Soviet and German military affairs for years (see Chapter 3 of the present book), it represented a rather belated recognition of an important issue which the British War Studies establishment had completely missed. In spite of this, not a single original Soviet source is quoted. More seriously, in the

Introduction, on p. 5, there is an allusion to 'new equipment' such as the 'BMP-60'. There are tracked infantry fighting vehicles designated BMP-1 and BMP-2 and wheeled ones BTR-60, -70 and -80. 'BMP-60' is a nonsense. This kind of error is unacceptable and undermines the authority of the rest of the work.

13 Leo Tolstoy, *War and Peace,* trans. Rosemary Edmonds, (2 vols, Penguin, Harmondsworth, 1957), vol. 2, p. 938.

14 Cited in Beaumont and Edmonds, p. 206.

15 See Yezhekel Dror, 'War, violence and futures studies', *Futures,* February, 1974, pp. 2–3; and G. D. Kaye and K. E. Solem, 'Futures studies and conflict', *Futures,* June 1979, pp. 235–8. Number of conflicts from the author's *Future of Land Warfare,* p. 1. Ibn Khaldun (also known as Abd al-Rahman Ibn Muhammad), Charles Issawi, *An Arab Philosophy of History. Selections from the Prolegomena of Ibn Khaldun ofTunis,* 1332–1406, (John Murray, London, 1950), p.7.

16 See, for example, (the approach is general), M. V. Frunze Military Academy, *Proryv podgotovlennoy oborony strelkovymi soyedineniyami* (*Breakthrough of a Perepared Defence by Rifle Formations*), (Voyenizdat, Moscow, 1958), also published as *Der Durchbruch der Schützenverbande durch eine vorbereitete Verteidigung,* (Verlag des Ministeriums für nationale Verteidigung, East Berlin, no. 5082, 1959).

17 Colonel General F. F. Gayvronskiy, *Evolyutsiya voyennogo iskusstva: etapy, tendentsii, printsipy,* (*The Evolution of Military Art: Stages, Trends, Principles*), (Voyenizdat, Moscow, 1987), p. 245: 'military art . . . has trodden a road of many centuries . . . this process did not unfold in a chaotic fashion, nor in a disordered way. Its study and deep reflection permit one to identify firm laws and trends in military art and constantly recall them while constructing forecasts of its further development'.

18 Sergey Gushchev, 'Strategy of science', in Boris Burkov, Benik Bukanazar Yuzbashev, (ed.), *The Year 2017. One Hundred Years after the October Revolution. Past, Present and Future* (Novosti Press Agency, Moscow, 1968), p. 85.

19 General G. I. Pokrovskiy, *Science and Technology in Contemporary War* [a translation of a short book (1956), a pamphlet (1957) and a newspaper article (1957)], Introduction by Raymond L. Garthoff (Atlantic Books, London, 1959), p. viii.

20 Admiral S. M. Gorshkov's *The Sea Power of the State,* (Pergamon, Oxford, 1979) and Marshal V. D. Sokolovskiy, *Soviet Military Strategy,* (ed. and trans.), Harriet Fast Scott and William E. Scott, (Macdonald and Jane's, London, 1975), are English translations of Soviet works of 1976 and 1962–8, respectively. They were of universal relevance and not just as examples of Soviet thinking of their time. I am also thinking particularly of Major General Aleksandr Svechin, *Strategiya,* the previous Soviet volume on *Strategy,* (2nd edn, Voyenizdat, Moscow, 1927), of which no translation exists and is little known even by so-called Soviet specialists, which is a formidable work on strategy and military history by any standards.

21 In 1985–6 China's ground forces were reported to be just below three million strong, but a 25 per cent reduction was underway. *The Military Balance, 1987–88,* (International Institute for Strategic Studies, London (IISS), London, 1987), p. 146, reported that they had indeed been reduced to 2,300,000 and that reductions continued. The 1988–9 *Military Balance* (IISS, London, autumn, 1988) still gave the army 2,300,000 men, and still noted that reductions continued (p. 147). Claire Hollingworth, 'The tasks of China's Defence Minister Qin', *Jane's Defence Weekly,* 23 April, 1988, pp. 788–9, reported that '1.5

million men and women have already been retired or dismissed', since the early 1980s. Soviet Ground Forces were reported to be two million strong in the 1987–8 *Military Balance* but they were also estimated to have 570,000 paramilitary troops (KGB, Border Guards) under arms. The 1988–9 *Military Balance* gave the Ground Forces 1.9 million troops and the same number of paramilitaries (pp. 34, 44). However, the unilateral force reductions announced by Mr Gorbachev on 7 December 1988 amounted to 500,000 personnel of all services (12 per cent), with 240,000 (10 per cent) coming from the Atlantic to Urals area. Assuming a roughly comparable decrease in ground forces alone, this would reduce ground forces (excluding paramilitaries, which might not be affected) to around 1.7 million by November-December, 1990, keeping the Soviet Union in second place behind China: Philip Karber, *Soviet Implementation of the Gorbachev Reductions: Implications for ConventionalArms Control in Europe,* address to House Armed Services Committee, 14 March 1989. Numbers of people under arms are still a rough guide to relative importance of land powers, while recognizing that training and equipment may differ significantly. The next two in terms of manpower strength are India and Vietnam, with armies of 1.2 and 1.1 million respectively (*The Military Balance, 1988–89*), pp. 161, 181. We therefore need to look at Indian and Vietnamese military history.

22 Beaumont and Edmonds, p. 3.

23 ibid. The word was coined by Gaston Bouthoul in 1946.

24 Definitions: *Sovetskaya voyennaya entsiklopediya* (*Soviet Military Encyclopedia -SVE*) (Voyenizdat, Moscow, 1976–80), vol. 2 (1976), pp. 183–8, 211–18.

1 GROUND RULES

1 The author visited the battlefield in December, 1984. Details of battles of Armageddon from Ernest R. Dupuy and Trevor N. Dupuy, *The Encyclopedia of Military History from 3500 BC to the Present*, (Macdonald, London, 1970), pp. 6–7, 988; Captain Jonathan M. House, *Towards Combined Arms Warfare: A Survey of Tactics, Doctrine and Organization in the 20th century*, (Combat Studies Institute, US Army Command and General Staff College, Fort Leavenworth, 1984), pp. 37–9, and local briefing.

2 Carl von Clausewitz, *One War*, (trans. and ed.), Michael Howard and Peter Paret, (Princeton University Press, 1975), p. 87 (bk. 1, ch. 1, section 23 (most important) and section 24: 'War is merely the continuation of policy by other means); p. 75 (bk. 1, ch. 1, section 2), 'act of force'. The original German of the latter is: 'Der Krieg ist also ein Akt der Gewalt, um den Gegner zur Erfüllung unseres Willens zu zwingen', *Vom Kriege*, (Berlin and Leipzig, 1918), p. 1. The previous English edition, trans. Colonels J. J. Graham and F. N. Maude (3 vols, Routledge & Kegan Paul, London, 1962), translate this as 'an act of violence intended to', encapsulating the nastiness of war slightly better.

3 Konrad Lorenz, *On Aggression*, (Bantam, 1971), is the classic work. See also Erich Fromm, *The Anatomy of Human Destructiveness*, (Holt, Reinhart & Winston, New York, 1973).

4 Richard E. Leakey, *The Making of Mankind*, (Abacus, Sphere, London, 1982), pp. 219–37, 242.

5 Fromm, p. 163, for example says that war as an institution was a new invention, like kingdom or bureaucracy, made around 3000 BC. The prehistorian Bernard Campbell reckoned that 'it was not until the development of the temple towns (around 5000 BC) that we find evidence of inflicted death and warfare. This is too recent an event to have had any influence on human nature' (cited in Leakey, p. 242). *Past Worlds: The Times Atlas of Archaeology*, (Times Books,

London, 1988), p. 44, refers to paintings dating from around 10,000 BC showing Mesolithic archers shooting arrows at each other as a (possible) representation of warfare. Even if this is the case (debatable, in the author's view), it does not challenge the conclusion that warfare arrived at a late stage in human development.

Darwinists naturally believed that violence was in some way innate, and that it went back a lot further. The first film ever to portray prehistoric times, D. W. Griffith's *Man's Genesis* (1912) stated that 'soon after man became man, man discovered armed force'. Bruteforce kidnaps the girl Lilywhite, loved by Weakhands, but Weakhands then makes a terrible discovery: the first weapon that can hurt others, and kills Bruteforce with a hand-axe. In one respect the film contains a fundamental truth: throughout military history the 'weaker', or, at any rate, more pacific side has had to resort to more terrible forms of war: witness the British and Americans' adoption of indiscriminate bombing, total war and, ultimately the atom bomb in order to defeat the more martial societies of Germany and Japan.

6 Engels to Marx, 23 May, 1862, in Dona Torr, (trans. and ed.), *Karl Marx and Friedrich Engels, Correspondence 1846–95*, (Martin Lawrence, London, 1934), pp. 243–45. This letter was reproduced almost verbatim in 'The situation in the American theatre of war', in *Die Presse*, 30 May, 1862. *The Civil War in the United States, by Karl Marx and Friedrich Engels*, (Laurence and Wishart, London, 1938), pp. 185–90. The phrase 'war to the knife' was apparently first uttered by José de Palafox, refusing to surrender Zaragoza to the French in 1808; Heinl, p. 344.

7 See Introduction, note 4.

8 For definitions and etymology, see the full edition of the *Oxford English Dictionary*.

9 Clausewitz, *On War*, (ed.) Howard and Paret, p. 128 (bk 2, ch. 1, 'Classifications of the art of war').

10 Cited in Herman H. Hattaway and Archer Jones, *How the North Won: A Military History of the Civil War*, (University of Illinois Press, 1983), Appendix A, 'An introduction to the study of military operations', p. 711.

11 This point is made in Baron N. Medem's strikingly perceptive *Obozreniye izvestneyshikh pravil i sistem strategii* (*Survey of the Best Known Rules and Systems of Strategy*), first printed in 1836, ch. 1, 'O strategii do poyavleniya yeyë v vide nauki' ('On strategy before its appearance as an aspect of science'); L. G. Beskrovny, (ed.), *Russkaya voyenno teoreticheskaya mysl' xix i nachala xx vekov* (*Russian Military-Theoretical Thought of the 19th and Early 20th Centuries*), (Nauka, Moscow, 1960), pp. 99–102.

12 Rogers and Ward, *Süleyman the Magnificent*, pp. 6–8, including citation of G. R. Elton, *Reformation Europe*, (Collins (Fontana), London, 1963).

13 Colonel G. F. R. Henderson, *The Science of War: A Collection of Essays and Lectures, 1892–1903*, (Longmans, London, 1905), p. 1.

14 ibid.

15 Cited in Heinl, p. 186.

16 Clausewitz, *On War*, (ed.) Howard and Paret, p. 86 (bk 1, ch. 1, section 21); Baron Antoine Henri Jomini, *Précis de l'Art de la Guerre*, 2nd edn (the 1st was addressed to the Russian Tsar in a dedication of 6 March, 1837), l-ère partie, (Librairie pour l'Art Militaire, les Arts et les Sciences, Paris, 1855): 'Sur la théorie actuelle de la guerre', p. 26, 'la guerre est un grand drame, dans lequel mille causes morales ou physiques egissent plus ou moins fortement, et qu'on ne saurait reduire a des calculs mathematiques'.

17 Principles of war are collected and analysed in Julian Lider, *Military Theory:*

Concept, Structure, Problems, (Swedish Studies in International Relations, Gower, Aldershot, 1983), pp. 221–3. Key works are J. F. C. Fuller, *Foundations of the Science of War* (Hutchinson, London, 1925); Bruce Keenan III, 'The principles of war: a thesis for change', *US Naval Institute Proceedings*, November, 1967; Don. H. Starry, 'Principles of war' *Military Review*, 9, 1981; André Beaufre, *Introduction to Strategy*, (Faber, London, 1965). Warsaw Pact: Colonel V. Ye Savkin, *Basic Principles of Operational Art and Tactics*, translated from the Russian of 1972 under the auspices of the US Air Force (US Government Printing Office, Washington, 1982), pp. 167–277. These are declaredly the views of 1956–9, but are the latest unclassified sources to which the author had access at the time of writing and are unlikely to have changed much.

18 Clausewitz, *One War*, Paret and Howard translation, p. 361 (bk. 6, ch. 2): 'the defender is better placed to spring surprises by the strength and direction of his own attacks'.

19 Richard Simpkin, *Race to the Swift*, pp. 19–23.

20 Heinl, p. 121, citing Mahan, *Lessons of the War with Spain*.

21 An excellent summary is in Hattaway and Jones, Appendix A, pp. 705–20.

22 Savkin, p. 256: 'The history of military art knows individual cases where by virtue of some special reasons it was unprofitable to complete the encirclement and destruction of the enemy in a specific region.'

23 Jomini, *Précis*, 1855 edn, l-ère partie, ch. 3, article 21, p. 232. Jomini rejected criticism by 'certain German writers' (Clausewitz!) that he confused a 'central position' (more tactical) with 'interior lines': Clausewitz's 'Advice to his Royal Highness the Crown Prince' in 1810, 1811 and 1812, in the edition of *On War* edited and translated by Graham and Maude (London, 1908), vol. 3, p. 212, section 3, 'Strategy: general principles'. In the main body of *On War*, Clausewitz also alludes to interior and exterior lines (Howard and Paret edn, pp. 367–9 (bk. 6, ch. 4). In bk. 2, ch. 2 (pp. 135–6), Clausewitz criticizes those who make 'interior lines' into a 'lopsided, purely geometrical' principle.

24 Cited in Heinl, p. 161.

25 The history of permanent fortification is a massive, specialized subject with an enthusiastic following. The standard work on pre-gunpowder fortifications is still Sidney Toy, *A History of Fortification from 3000 BC to AD 1700*, (Heinemann, London, 1955); see also Toy, *Fortified Cities of India*, (Heinemann, London, 1955); T. E. Lawrence ('of Arabia') was a keen student and wrote his undergraduate thesis on *Crusader Castles*, in 1909 (first general edition, Michael Haag, London, 1986), an exemplary work for those who wish to write scholarly material in the field of military history, as well as an expert study of the subject. On the gunpowder era, Christopher Duffy, *Siege Warfare: The Fortress in the Early Modern World 1494–1660* (Routledge & Kegan Paul, London, 1979). The author's own travels have taken him round most of the fortifications in the UK, the massive pre-World War I Russian coastal fortifications in Finland, the forts of Verdun, and some of the crusader castles in the Middle East.

26 *Chevaux de frise* ('Horses of Friesland') comprise crossed spikes passing through a central beam, and take their name from the siege of Friesland where they were used to plug a breach. Military vocabularly tends to take terms from the language of the dominant military nation at the time. Thus, musket comes from the Spanish *mosquete*, a massive firearm introduced by the Duke of Alva for the bloody fighting in the Netherlands, as do Major and Colonel (from *colunela*, a column), as Spain was the dominant land power in the sixteenth century. During the seventeenth and eighteenth centuries many terms, including

the whole fortress vocabulary came from French (Marshal, Adjutant, Aide-de-Camp, port arms, etc.). In the nineteenth and twentieth centuries German took over (*Schwerpunkt*, storm trooper, Flak, etc.). In recent decades, Russian terms have been adopted, notably *desant*, a useful term for any 'descent' into the enemy rear, however delivered.

27 Alastair Horne, *The Price of Glory: Verdun, 1916*, (Macmillan, London, 1975, first published 1962). Belgian forts, p. 49. Verdun was arguably the 'worst' battle in history (p. 327) and Fort Vaux got the worst of this (pp. 256–7). A visit to the forts makes the impression even more forcefully.

28 At Verdun, casualties in Fort Vaux were about 100 men, with less than 20 actually killed. Four attacking German battalions expended over 2740 officers and men. As Alistair Horne concludes 'it was hardly surprising that French military thinkers would soon be making some far-reaching deductions about the value of underground forts' (p. 264), see also pp. 337–8. German assessment of Russian fortifications: Colonel E. Lederrey, *Germany's Defeat in the East, 1941–45*, (Official History, War Office, London, 1955); Colonel A. G. Khor'kov, 'Ukreplennye rayony na zapadnykh granitsakh SSSR' ('Fortified Regions on the Western Frontiers of the USSR'), *VIZh*, 12, 1987, pp. 47–54. On revival of fortified regions etc., see Introduction.

29 'Mines', Imperial War Museum, *Martel 4/4* (see ch. 5); 'Hedgehogs', John Keegan, (ed.), *The Times Atlas of the Second World War* (Times Books, London, 1989), pp. 62–63; Kursk, ibid., pp. 124–5; Normandy, ibid., pp. 152–3 and subsequent campaigns in east and west.

30 HQ USAFE and Eighth US Army (Rear), *Japanese Studies on Manchuria* (13 vols, Department of the Army, Office of the Chief of Military History, 1956), vol. II, *Study of Strategical and Tactical Peculiarities of Far Eastern Russia and Soviet Far Eastern forces*.

31 Recent experience and the future role of fortifications are explored in more detail in the author's *The Future of Land Warfare*, pp. 8–9, 288–90.

32 Sun Tzu, *Art of War*, ch. 1, 'Estimates', 17, trans. Griffith, p. 66. Aphorisms continue to 1.26, p. 70.

33 Barton Whaley, 'Toward a general theory of deception', *Journal of Strategic Studies, Special issue on Military Deception and Strategic Surprise*, March, 1982, pp. 178–92. See also the other articles, in particular Whaley's 'Deception and Misperception' (pp. 3–39), Michael I. Handel, 'Intelligence and Deception' (pp. 122–54). Also Whaley, *Strategem: Deception and Surprise in War* (MIT Center for International Studies, Cambridge, Mass., 1969). Pioneering work in the field was done by the artist who designed World War One camouflage, Solomon J. Solomon, *Strategic Camouflage* (London, Murray, 1920).

34 Whaley, 'Toward a general theory of deception' uses the example of the German attack on Liège in 1914, when the Germans deduced from prisoners taken that they faced 5 corps, whereas in fact they only faced 6 brigades drawn from those 5 corps.

2 TECHNICS AND WARFARE

1 The *Battle Studies* of Colonel du Picq (1821–70), killed in the Franco-Prussian War, are a remarkable examination of the microcosm of combat, beginning an important if patchy tradition of military writing including, for example, John Keegan's *The Face of Battle*, (Cape, London, 1976).

2 See for example Lewis Mumford, *Technics and Civilization* (Routledge, London, 1934). Many insights into this subject are in John F. C. Fuller, *Machine Warfare: An Enquiry into the Influences of Mechanics on the Art of War*, (Hutchinson, London, 1942).

3 Colonel E. L. Cordonnier, *The Japanese in Manchuria* (1911), translated by Captain C. F. Atkinson, (London University OTC, London, 1914), vol. 2, part 3, p. 258: 'Technics are only too attractive to officers of the scientific arms, who are sometimes wholly absorbed by them. . . . Now tactics have reasserted their rights.'

4 Michael Simpkins and Donald Embleton, *The Roman Army from Caesar to Trajan* (revised edn), (Osprey, London, 1984), p. 10, and Donald R. Morris, *The Washing of the Spears*, (Cardinal, London, 1973), pp. 47, 51.

5 *De Regis Misericordia* (On the Duties of Kings) consulted by the author in Christ Church, Oxford. It was prepared for the young King Edward III of England. The picture of the gun (firing an arrow) is right at the end, after a series of pictures of catapults. It is not thought that the pictures were added after the text. Florentine manuscript, Carlo M. Cipolla, *Guns and Sails in the early Phase of European Expansion, 1400–1700* (Collins, London, 1965), p. 90, which includes a photograph of the text. Any earlier references are dubious, and may be connected with the use of 'Greek Fire' (a sort of medieval napalm) hurled from torsion or counterpoise engines. All the evidence indicates that guns were a western European invention (*not* Chinese), and spread from west to east.

6 Cipolla, *Guns and Sails* . . ., p. 90; the author's 'The Firebird and the Bear: 600 Years of the Russian Artillery', *History Today*, September 1982, pp. 16–20.

7 Significantly, guns began to play a role in more open warfare in concert with the *Wagenburg* – a wild-west style circle of wagons or 'wagon fort' – used by the Hussites in fifteenth-century Bohemia and defended against the expensively armoured chivalry of the Holy Roman Empire with handguns, light artillery and crossbows. Niccolò Machiavelli, the renaissance political and military theorist, did not rate artillery in open warfare at all: *The Art of War* (translation of *Arte della Guerra*), (revised edn, of the Ellis-Farnworth translation, with Introduction by Neal Wood), (Bobbs-Merrill, Indianapolis, New York and Kansas, 1965), pp. xxxiii, 96; Sir Charles Oman, *History of the Art of War in the Sixteenth Century*, (Methuen, London, 1937), pp. 93–4.

8 Public Record Office (PRO), *Defe II 1251 TWC (45)*, 24 October, 1945, Memorandum by Sir George Thompson, 'Effect of atomic bombs on warfare in the next few years', p. 4; PRO *Defe II, 1252, Examination of the Possible Development of Weapons and Methods of War, TWC (46)* 3 (Revised), 30 January, 1946, part II, 'Effects on warfare', p. 8, and part I, 'Matters of fact relating to atomic energy', section B (e), 'Effect on army targets'.

9 Clausewitz, *On War,* bk. 3, ch. 8, Paret and Howard translation, p. 195.

10 ibid., bk. 2, ch. 1, p. 127.

11 Jomini, *Précis*, (Dedication dated 6 March, 1837), trans. Captain G. H. Mendell and Captain W. P. Craighill, (Lippincott, Philadelphia and Trubner, London, 1879), ch. 2, article 13, pp. 48–9.

12 ibid., Appendix 2, p. 355. The Minié rifle utilized a simple principle. For centuries the problem of imparting spin to a ball loaded from the muzzle had been all but insoluble: in order to make the rifling bite the ball it would have to be hammered down the barrel; slow, arduous, and time-consuming. The Minié 'ball' (in fact, more like a modern bullet in shape) had a cavity at the base. It could be dropped down the barrel, as with a smoothbore but when the weapon was fired the explosion also forced the edges of the cavity outward, so that they engaged the rifling as the bullet travelled back up the barrel.

13 ibid., p. 360.

14 Cipolla, *Guns and Sails* . . ., p. 140. See also Geoffrey Parker, *The Military Revolution. Military Innovation and the Rise of the West, 1500–1800*,

(Cambridge University Press, Cambridge, 1988).

15 The ability to envision a different future, and to make plans to change the present, is an important advantage which appears to be the preserve of advanced (and, in particular, western) civilizations. Foresight applies at all levels of war, from the immediate foresight required of an operational commander to long-term planning by foreign and defence ministries. In this respect, western political regimes which have a life of perhaps five years and whose horizon lies with the next election may be at a serious disadvantage compared with non-democratic regimes.

16 *SVE*, vol. 4, 1977, p. 15.

17 'Past worlds', *The Times Atlas of Archaeology* (Times Books, London, 1988), p. 44.

18 Oman, *History of the Art of War in the Sixteenth Century*, pp. 382–4, citing Sir Humphrey Barwyck: 'if he [the archer] have not his three meals a day, as is his custom at home, nor lies warm at night, he presently waxes benumbed and feeble'.

19 Rogers and Ward, *Süleyman the Magnificent*, p. 14.

20 Mario Modiano, 'Muscle powered missile hits the bull's eye', *Sunday Times*, 2 August, 1987, p. 29, on the work of Professor John Morrison and John Coates.

21 ibid., and Rogers and Ward, *Süleyman . . .*, p. 14. According to a 1655 observation, the Turks crewed their galleys with three categories of oarsmen: slaves (the most privileged and professional), convicts and 'men brought from Asia'. 'Were it not for the aid of the slaves, who are accustomed to the sea, it [The Turkish Navy] would surely be very weak and almost useless': Cipolla, *Guns and Sails . . .*, pp. 161–2.

22 Christopher Duffy, *Frederick the Great: A Military Life* (Routledge, London, 1988), p. 3.

23 Ian V. Hogg, *German Artillery of World War II*, (Arms and Armour Press, London, 1975), pp. 162–72, esp. p. 170.

24 Trevor N. Dupuy, *The Evolution of Weapons and Warfare*, (Jane's, London, 1982), p. 303.

25 See Sir Charles Oman, *History of the Art of War in the Middle Ages*, (Methuen Reprints, London, 1978).

26 Cited in Heinl, p. 105.

27 Patrick Turnbull, *Solferino: the Birth of a Nation*, (Robert Hale, London, 1985), p. 144.

28 General H. Langlois, *Lessons from Two Recent Wars (The Russo-Turkish and South African wars)*, (General Staff, War Office, London, 1909), (translated from the French of 1904), pp. 36, 39.

29 Trevor N. Dupuy, 'Understanding war from a historical perspective', *Marine Corps Gazette*, June, 1985, pp. 53–8. See also *The Evolution of Weapons and Warfare*, pp. 307–10.

30 See, for example, Geoffrey Jukes, 'The strategic nuclear balance to 1983', *Futures*, December, 1970, p. 364. Tukhachevskiy and stratospheric bombers: 'Novye voprosy voyny' ('New questions of war', *Voyenno-istoricheskiy zhurnal* (*Military-Historical Journal (VIZh)*) 2, 1963, pp. 62–77, this reference, p. 65. The article was published after Tukhachevskiy's return to favour, 30 years after it was written (1931–2) and 25 after the announcement of his death.

31 Thomas S. Kuhn, *The Structure of Scientific Revolutions*, (2nd edn, enlarged, *Foundations of the Unity of Science*, vol. 2, no. 2, *International Encyclopedia of Applied Science*, (University of Chicago, 1970), (1st edn 1962).

32 Correlli Barnett, 'The War that Never was', *Futures*, August, 1975, reproduced in Ralph Jones, *Readings from Futures, 1974–80*, (Westbury, Guildford, 1981),

pp. 232–7, this reference p. 232.

33 See, for example, Harriet Fast Scott and William F. Scott, *The Armed Forces of the USSR*, (Westview, Boulder, Colorado, 1979), p. 47. 'The "Revolution in Military Affairs" was selected as the slogan . . . to explain the changes in warfare . . . that resulted from the breakthrough in nuclear weapons and ballistic missiles'. That is a Kuhnian paradigm *par excellence*.

34 Major General P. B. Hughes, *Open Fire: Artillery Tactics from Marlborough to Wellington*, (Antony Bird, Chichester, 1983), p. 13.

35 The author's 'The Russian artillery and the origins of indirect fire', *Army Quarterly and Defence Journal*, vol. 112, nos. 2 and 3 (April and July, 1982), pp. 211–22 and 330–37, respectively, explores this key issue in detail. See also *Red God of War: Soviet Artillery and Rocket Forces*, (Brassey's London, 1986), pp. 28–33 for illustrations.

36 Kuhn, *The Structure of Scientific Revolutions*, p. 99.

37 ibid., Ch. 8 'The response to crisis', in particular, pp. 86–9.

38 See Michael Salewski, *Zeitgeist und Zeitmaschine: Science Fiction und Geschichte* (*The Spirit of the Time and the Time Machine: Science Fiction and History*), (DTV, München, 1986), p. 170–71.

39 General Baron Dr Hugo von Freytag-Loringhoven, 'Wandlungen im operativen und taktischen Verfahren Napoleons verglichen mit denjenigen des Weltkriegs' ('Changes in Napoleon's operational and tactical methods compared with those of the [First] World War'), *Wissen und Wehr* (Military Thought), (Berlin, 1921), pp. 336–8.

40 Quote, Roger Beaumont, 'Nuclear warfare: the illusion of accurate forecasting', *Futures*, February, 1977, pp. 53–8, this p. 53. 'Mind-sets', Carl G. Jacobsen, (ed.), *The Uncertain Course: New Weapons, Strategies and Mind-Sets*, (SIPRI/ Oxford University Press, 1987).

41 Philip A. G. Sabin, 'World War Three: a historical mirage?', *Futures*, August, 1983, pp. 272–80, this p. 273.

42 ibid., p. 273.

43 Colonel General M. A. Gareyev, Lecture to the Royal United Services Institute, 18 October 1988; on recent Pentagon terminology and reaction, I am indebted to my Edinburgh colleague, Lieutenant Colonel (retd) John Hines.

44 See Sabin, 'World War Three . . .', p. 273. For this and other World War Two manoeuvres, John Keegan, (ed.), *The Times Atlas of the Second World War*, (Times Books, London, 1989).

45 The accepted word 'vulnerability' means 'ability to be wounded', and is therefore an accurate and acceptable gauge of a weapons system's chances of surviving – or not. 'Survivability' means, one assumes, 'ability to be survived'. A weapon that 'can be survived' is presumably pretty useless.

46 Trevor N. Dupy, *The Evolution of Weapons and Warfare*, p. 312.

47 See the author's *The Future of Land Warfare*, ch. 8.

48 Captain Jonathan M. House, *Towards Combined Arms Warfare: A Survey of Tactics, Doctrine and Organization in the 20th Century*, (Combat Studies Institute, US Army Command and General Staff College, Fort Leavenworth, 1984), p. 57.

49 'And finally, no war is any longer possible for Prussia-Germany except a world war and a world war of an extension and violence hitherto undreamt of!' (Engels to Sorge, London, 7 January, 1888; 'From the moment warfare became a branch of the *grande industrie* (ironclad ships, rifled artillery, quickfiring and repeating cannons [machine guns?], repeating rifles, steel covered bullets, smokeless powder, etc.), *la grande industrie*, without which all these things cannot be made, became a political necessity' (Engels to Danielson, London,

22 September, 1892), cited in *Karl Marx and Friedrich Engels correspondence*, (ed.) Torr, pp. 455, 498.

50 After the Germans had fired off most of their nitrate stocks in the 1914 invasion of France, they summond the chemist Fritz Haber to help devise substitutes. After their check on the Marne, they asked Haber to investigate the possibility of gas war. I. F. Clarke, 'Methods of prediction 1918–39', *Futures*, December 1970, pp. 375–9, this on p. 376.

51 A well-known example of a major power suffering a 'technology gap' is the case of the Soviet Union and heavy long-range bombers after World War Two. US B-29 Superfortresses were copied faithfully to produce the TU-4, a famous piece of 'reverse engineering' to fill the gap. This was no mean feat. In the process, Soviet industry learned much about high-powered engines, pressurized cabins, remotely controlled armament and other technologies employed in the B-29, a most advanced aircraft for its time. The process of *perestroika* going on at the time of writing may owe not a little to the perceived need to fill other technology gaps in vital strategic areas, notably the mass production of high-tech electronics, microprocessors, and so on. B-29s, Vadlav Nemecek, *The History of Soviet Aircraft from 1918* (translated from the Czech *Sovetska Letadla*), (Collins, London, 1986), pp. 166–9.

3 THE EXPANSION OF THE BATTLEFIELD AND THE OPERATIONAL LEVEL OF WAR, 1800–1918

1 Machiavelli, *The Art of War* (1521). See also his better known work, *The Discourses*. General Henry H. Lloyd, *History of the Late War in Germany* (1766): Lloyd's *Memoir* forms the first part of this.

2 Maurice de Saxe (often listed as Moritz of Saxony), *Les Reveries ou Mémoires sur l'Art de la Guerre de Maurice, Comte de Saxe* (first appeared 1732, new edn, Paris, 1757), translated as *Reveries, or Memoirs Concerning the Art of War*, anonymous translation [by Sir William Fawcett] (Sands, Donaldson, Murray and Cochran, Edinburgh, 1759). Frederick the Great's works are listed chronologically in Jay Luvaas, (ed.), *Frederick the Great and the Art of War*, (Free Press, New York, 1974), p. 152. Of particular importance are *Principes Généraux de la Guerre* (1746), *Des Militarische Testament*, (1768) and *Elements de Castrametrie et de Tactique* (1770). See also Christopher Duffy, *The Army of Frederick the Great* (David and Charles, London, 1974) and *Frederick the Great: A Military Life* (Routledge, London, 1988).

3 Suvorov (1729–1800), being Russian, has been neglected by non-Russian speakers, but was one of the world's greatest Captains and Russia's greatest general. His *Nauka pobezhdat* (*The Science of Winning*), first promulgated *c.* 1795, (Voyenizdat, Moscow, 1980), is a concise gem. See also Philip Longworth, *The Art of Victory: The Life and Achievements of Generalissimo Suvorov*, (Constable, London, 1965).

4 Giuseppe Grassi, compiled and edited, *Opere di Raimondo Montecuccoli* (Stampa di Giuseppe Favali, Torino, 1821), Libro Terzo, *Aforismi applicati alla guerra possible col Turco in Ungheria*, pp. 101–215 and bibliographical note on pp. 301–02. The manuscript containing these aphorisms probably dated from 1685–95 but must have been based on an earlier one as Montecuccoli died in 1680. A geographical survey of Hungary which follows is dated 1673.

5 See for example Herman Kahn, *On Thermonuclear War*, (Princeton University Press, New Jersey and Oxford University Press, London, 1960). Although the title is (probably deliberately) modelled on Clausewitz, it is a work of a quite new school. It is an excellent study and much broader than its title, predicting

the nature of future conflict and related issues. Ferdinand Otto Miksche, *Atomic Weapons and Armies*, (Faber & Faber, London, 1955), is one of the few works to qualify as a recognizable continuation of the pre-nuclear tradition of military writing and thought.

6 This view permeates Simpkin's *Race to the Swift*, which, whilst representing *Thoughts on Twenty-First Century Warfare*, has 'deliberately excluded discussion of nuclear and chemical warfare' (p. xxiii).

7 See Duffy, *Frederick the Great. A Military life*, pp. 296–7.

8 ibid., pp. 264, 313–4; Captain A. F. Becke, *An Introduction to the History of Tactics*, (Hugh Rees, London, 1909), which contains an Appendix on the evolution of column formations, see esp. pp. 71–3; Suvorov, *Nauka pobezhdat*, pp. 23–4.

9 See Duffy, p. 264. In *Elements de Castrametrie et de Tactique* (1770) Frederick forecast that future war would 'be a question of artillery duels and attacking defended positions'.

10 See Becke, *Introduction to the History of Tactics*.

11 Jomini, *Art of War*, trans. Mendell and Craighill, p. 69.

12 Henderson, *The Science of War*, pp. 180–81.

13 Von Freytag-Loringhoven, 'Wandlungen im operativen und taktischen Verfahren Napoleons verglichen mit denjenigen des Weltkrieges' ('Changes in Napoleon's operational and tactical methods compared with those of the Great War'), *Wissen und Wehr* (*Military Thought*), (1921), pp. 333–4; Becke, p. 21. Freytag called this *operativen Umgehung* ('operational encirclement').

14 ibid., p. 334; on Genghis, see Chapter 6.

15 Freytag-Loringhoven, 'Changes . . .', p. 335.

16 ibid., p. 336.

17 See the introductory essays in the Howard and Paret edn (Princeton University Press, Princeton, NJ, 1976).

18 Mendell and Craighill edn, pp. 7–8.

19 P. Geysman, 'Uchrezhdeniye Imperatorskoy voyennoy akademii', ('Establishment of the Imperial Military Academy') *Voyenny Sbornik*, 11, 1908, pp. 81–98 and 12, 1908, pp. 133–56. Jomini's letter, pp. 90–91 of the former, dated 21 March–2 April 1826 (old style). The dedication of *Précis* is dated 6 March, 1837.

20 The connection passed through the late Imperial thinkers, of whom G. A. Leyer is the principal figure (1829–1904). Leyer's emphasis on 'operational lines' represents a strong continuation of the Jominian tradition.

21 Lieutenant Colonel F. E. Whitton, *Moltke* (Makers of the Twentieth Century, Constable, London, 1921), p. 1. On Moltke, see Earle, *Makers of Modern Strategy*; the author also used *Moltke's Tactical Problems from 1858–1882*, Edited by the Prussian General Staff Department for Military History, trans. Karl von Donat, (Hugh Rees, London, 1903) and *The Franco German War of 1870–71*, trans. Archibald Forbes (Harper Bros, London, 1907).

22 Generalleutnant H. von Sandrart, 'Operative Führung über die Gefechtstaktik hinaus' ('Operational command goes beyond combat tactics'), *Europäische Wehrkunde* 9, 1987, pp. 503–5, this ref. p. 504 citing Moltke, *Taktisch-strategische Aufsätze aus den Jahren 1857 bis 1871* (*Tactical-strategic Essays from the Years 1857–1871*), p. 291.

23 See, for example, Charles Dick, 'Soviet operational art: part 1, the fruits of experience', *International Defense Review*, 7, 1988, pp. 755–61. 'Until the mid 1980s, NATO did not even recognize the existence of an intermediate level of war between strategy and tactics' (p. 755).

24 Dwight L. Adams and Clayton R. Newell, 'Operational Art in the Joint and Combined Arms Arenas', *Parameters* (Journal of the US Army War College),

June, 1988, pp. 33–9, esp. the former, 'The Army has always recognized three levels of war, even though it may not have always used that terminology.' Absolutely.

25 John F. C. Fuller, *On Future Warfare*, (Sifton Praed, London, 1928), p. 104.
26 John F. C. Fuller, *The Foundations of the Science of War*, (Hutchinson, London, 1926), pp. 88–9.
27 Sun Tzu, *Art of War*, trans. Griffith, p. 77, (ch. 3, 'Offensive strategy', p. 4). It may be significant that in the Griffith version, on the following page, it is recorded that: 'All the generals said: "This is beyond our comprehension".'
28 Simpkin, *Race to the Swift*, p. 24.
29 *Soviet Military Encyclopedia*, vol. 1 (1976), p. 62.
30 Haig: G. A. B. Dewar and Lieutenant Colonel J. H. Boraston, *Sir Douglas Haig's Command, December 19, 1915 to November 11, 1918*, (2 vols, Constable, London, 1922), vol. 2, pp. 354–6. The linking of combats in an operation and role of operations as leaps along a strategic path was well put by Svechin in his *Strategy* (2nd edn, 1927):

'Normally, this way to a final aim is divided into a series of operations; the latter are divided in time by pauses of greater or lesser significance. . . . We define the operation as that act of war during which the exertion of forces is directed within a defined region of the theatre of strategic military action (TVD) without a break, towards the attainment of a defined intermediate aim' (pp. 14–15).

31 Sigismund Wilhelm Lorenz von Schlichting, *Taktische un Strategische Grundsätze der Gegenwart*, (1897–9).
32 Dr Jacob W. Kipp, *Mass, Mobility and the Red Army's Road to Operational Art, 1918–1936*, (Soviet Army Studies Office, Fort Leavenworth, Kansas, 1988), p. 5; quote from Boris M. Shaposhnikov, *Vospominaniya: Voyenno-nauchnye Trudy* (*Memoirs: Military Scientific Works*), (Voyenizdat, Moscow, 1974), p. 144.
33 Infantry General Dr Baron Hugo von Freytag-Loringhoven, *Deductions from the World War*, translated from the German of 1917, introduction by 'J. E. M.', (Constable, London, 1918), p. iii.
34 Von Freytag-Loringhoven, *Generalship in the World War* (translated from the German of 1920, US Army War College, Washington DC, 1934), p. 34.
35 Von Freytag-Loringhoven, 'Changes . . .', p. 332.
36 A. Golubev, *M. V. Frunze o kharaktere buduschchey voyny* (*M. V. Frunze on the Character of Future War*), (Voyenizdat, Moscow, 1931), p. 8.
37 B. M. Shaposhnikov, ' "Konnye massy" na flange armii. Ocherk deystviy ruskkoy konnitsy na levom beregu Visly v avguste 1914 g' (' "Cavalry masses" on the Army's Flank. Sketch of the actions of the Russian Cavalry on the left bank of the Vistula in August 1914'), *Voyennaya nauka i Revolyutsiya* (*Military Science and the Revolution*), (a military-scientific journal), (Moscow, 1921), bk 1, July–August, pp. 95–126, this on p. 96.
38 Kipp, *Mass, Mobility* . . ., p. 17. The 1st edition of Svechin's *Strategy* appeared in 1926; see 2nd edition (1927), pp. 14–15; adoption from late 1920s, *Soviet Military Encyclopedia*, vol. 6, (1979), p. 55 and vol. 7 (1979), p. 264.
39 General Sir James Marshall-Cornwall, *Grant as Military Commander*, (Batsford, London, 1970), pp. 7–8; Hattaway and Jones, *How the North Won*.
40 *FM 100-5 Operations* (Department of the Army, Washington DC, 20 August, 1982), pp. 8–1 to 8–3.
41 Dupuy, *The Evolution* . . ., pp. 201–2.
42 ibid., p. 201.
43 Sergeant Harry Bell, 'Cavalry raids and the lessons they teach us', (translated

from 'Employment of cavalry in independent action against the flank and rear of a hostile army', by Lieutenant Colonel Wenniger, Bavarian Cavalry, *Austrian Cavalry Monthly*, April 1908), *Journal of the US Cavalry Association*, July, 1908, pp. 142–52, this on pp. 142–3.

44 Hattaway and Jones, *How the North Won*, raids and penetrations, pp. 82–3, quote p. 250.

45 ibid., pp. 323–4, 721–2.

46 Wenniger, trans. Bell, 'Cavalry raids . . .', p. 144.

47 Captain George T. Raach, 'Raid: an historical example' (Grierson's Raid of April, 1863), *Armor*, May–June, 1973, pp. 8–13; D. Alexander Brown, *Grierson's Raid* (University of Illinois Press, Urbana, 1954) and Edward G. Longacre, *Mounted Raids of the Civil War* (A. S. Barnes, Cranbury, NJ, 1975); Hattaway and Jones, pp. 366–7, 489–92.

48 Hattaway and Jones, pp. 233, 417, 487.

49 ibid., pp. 700–1.

50 Whitton, Moltke, pp. 70–1; Edwin A. Pratt, *The Rise of Rail Power in War and Conquest, 1833–1914* (P. S. King & Son, London, 1915), is a first-class and most valuable work on this key issue. The Franco-Austrian War of 1859 saw 'the first employment of railways in close connection with vast military operations'. Both the French and Italians and their opponents, the Austrians, tore up lines and destroyed railway bridges: Pratt, pp. 9, 27. Telegraph, invented, Gayvoronskiy, *Evolyutsiya* . . . pp. 83–4; Russian move, *Opisaniye voyennykh deystviy Rossiyskikh voysk protiv Vengerskikh myatezhnikov v 1849 godu* (*Description of the Russian Forces' Action against the Hungarian Rebels in 1849*), (Voyennaya tipografiya, St Petersburg, 1851).

51 See, for example, the exchange between A. Petrov and Captain E. A. Martynov in 1894. Martynov's book, *Strategiya v epokhu Napoleona I i v nashe vremya* (*Strategy in the Era of Napoleon I and Now*), (General Staff Press, St Petersburg, 1894), addressed the effect of recent developments in military art on Napoleonic strategy and operational art, which had occupied German theorists, notably von der Goltz, for a decade. Martynov concluded that the mass army would bring about a decline in military art. In 'Zadachi sovremennoy strategii (Po povodu sochineniya kapitana Martynova "Strategiya . . ." ' ('Tasks of modern strategy (with regard to captain Martynov's work *Strategy* . . .'), *Voyenny sbornik* 5, 1894, pp. 35–64, Petrov compared the development of warfare to the development of industry where fewer, but more highly skilled workers were now employed (p. 49). Martynov responded in 'Neskol'ko slov v ob'yasneniye i razvitiye sochineniya "Strategiya . . ." ' ('A few words in explanation and development of the work *Strategy* . . .'), *Voyenny Sbornik* 7, 1894, pp. 21–41, 8, 1894, pp. 232–48. On p. 236 of the latter he argued that state interests required 'not the purity of military art, but military success'.

52 Von Sandrart, '*Operative Führung* . . .', p. 504.

53 For example, Neznamov's comment, 'we did not understand modern war', 'Sovremennaya voyna. Deystviya polevoy armii' (*Modern War: Action of the Field Army*), (2nd edn, Neznamov, St Petersburg, 1912), p. vi. British, Germans, and Americans all produced voluminous reports on their experience with the Russian and Japanese Armies, for example, War Office, General Staff, *Reports of British Observers Attached to the Russian and Japanese Armies in the Field* (3 vols, HMSO, London, 1907); War Department, Office of the Chief of Staff, *US Official Reports on the Russo-Japanese War*, (Washington DC, 1906). For an idea of the war's importance for military thinking, see Philip A. Towle, 'The influence of the Russo-Japanese war on British naval and military thought 1905–14', (PhD thesis, King's College, London, 1973).

54 The literature on the Great War is so vast that no attempt will be made to catalogue it. For the student who wishes to avoid anachronism and hindsight, the official history, Brigadier General J. E. Edmonds and Captain G. C. Wynne, *History of the Great War Based on Official Documents* (by Direction of the Historical Section of the Committee of Imperial Defence, Macmillan, London), published in the 1920s and 1930s, is a good start. On the military technical aspects, John Terraine, *White Heat: The New Warfare, 1914–18*, (Sidgwick & Jackson, London, 1982) and Guy Hartcup, *The War of Invention: Scientific Development, 1914–18*, (Brassey's, London, 1988). John Keegan, *Opening Moves, August, 1914*, (Pan/Ballantine, London, 1973) is an excellent account of the distinctive opening phase; also Gerhard Ritter, *The Schlieffen Plan: Critique of a Myth*, trans. Andrew and Eve Wilson, (Oswald Wolff, London, 1958), (original German edn, Verlag R. Oldenbourg, Munich, 1956).

55 This is used as an example in FM 100–5, p. 10–2. Norman stone, *The Eastern Front, 1914–17*, (Hodder & Stoughton, London, 1975) is probably the best introduction to the war in the east: pp. 61–69 on Tannenberg; Ward Rutherford, *The Ally: the Russian Army in World War I*, (Gordon & Cremonesi, London, 1975), pp. 58–60.

56 Edmonds and Wynne, *History of the Great War . . . France and Belgium, 1915*, (Macmillan, London, 1927), pp. vi-viii, 4–5, notes that after the 'race to the sea', there was a 'period of recuperation' from 23 November to the end of 1914, during which front trenches were gradually connected up. A good account of the formation of the continuous front is in Divisional Commissar M. Galaktionov, *Tempy operatsii* (*The Tempos of Operations*), (Voyenizdat, Moscow, 1937), vol. 1, in which he analyses the 'Shift to positional war' (ch. 7, pp. 299–332).

57 Von Freytag-Loringhoven, 'Changes . . .', p. 336: K.ü.K stands for 'Royal and Imperial'.

58. Rutherford, *The Ally*, pp. 125–26; *Soviet Military Encyclopedia*, vol. 2 (1976), pp. 607–8.

59 Von Freytag-Loringhoven, 'Changes, . . .', p. 338: Rutherford, pp. 126–33.

60 V. D. Grendal, *'Ogon' artillerii* (Artillery Fire), (Voyenizdat, Moscow, 1926), p. 48.

61 House, *Towards Combined Arms Warfare . . .*, p. 33.

62 Edmonds and Wynne, *History of the Great War . . . France and Belgium, 1915* (1927), pp. 176–94. A general history of chemical warfare is Sterling Seagrave, *Yellow Rain: Chemical Warfare, the Deadliest Arms Race*, (Abacus, London, 1981). See also Robert Harris and Jeremy Paxman, *A Higher form of Killing: the Secret History of Gas and Germ Warfare*, (Paladin, London, 1983). Earlier use, David Rosser Owen, 'NBC warfare and anti-NBC protection', *Armada International*, 1, 1984, pp. 78–90.

63 *PRO, PREM 3/89*, Prime Minister to Ismay, D 217/4, 6 July, 1944.

64 House, p. 33.

65 V. F. Kirey, *Artilleriya ataki i oborony* (*Artillery of the Attack and Defence*), (Voyenizdat, Moscow, 1936), p. 72.

66 Timothy T. Lupfer, *The Dynamics of Doctrine: The Changes in German Tactical Doctrine during the First World War*, (Leavenworth Paper no. 4, Combat Studies Institute, Fort Leavenworth, 1981), pp. 38–9, 41, 43.

67 House, pp. 33–4.

68 Lupfer, p. 41.

69 House, pp. 20, 34–5. On artillery see Shelford Bidwell and Dominick Graham. *Fire-power: British Army Weapons and Theories of War, 1904–1945*, (George Allen & Unwin, London, 1982), pp. 102, 111–12.

The effective use of massed artillery depended on air observation and accurate maps. The latter were in short supply in 1914 and not yet gridded: it was necessary to re-survey the entire theatre, and during 1915 12,000 square miles of the western front were mapped and 32 million maps printed. This, and the development of techniques of flash spotting and sound ranging were unglamorous but essential concomitants of the development of large-scale industrial warfare. The range of technologies used to solve these problems became wider and wider (see Chapter 2). The Russians adopted similar techniques in the east, but more slowly: author's *Red God of War*, p. 41. Also Lieutenant Colonel Lucas, General Staff, *L'Evolution des Idées tactiques en France et en Allemagne pendant la Guerre de 1914–18*, (Berger-Levrault, Paris, 1923), and Terraine, *White Heat*.

70 Lupfer, pp. 46–9.
71 Author's *Red God of War*, p. 72.
72 Lupfer, p. 45; Kirey, *Artilleriya ataki* . . .
73 House, pp. 35–36.
74 J. F. C. Fuller, *On Future Warfare*, p. 78.
75 Dewar and Boraston, *Sir Douglas Haig's Command*, vol. I, p. 155.
76 The Amiens operation was widely analysed after the war, and not just by those whose countries had participated. See Fuller, *Decisive Battles* . . .; also Kipp, *Mass, Mobility* . . ., p. 19 and *Soviet Military Encyclopedia*, vol. 1 (1976), p. 169. 'Munitions of mobility', Dewar and Boraston, vol. 2, pp. 44, 45.
77 Dewar and Boraston, vol. 2, pp. 275–6.
78 ibid., p. 44.
79 Winston S. Churchill, *The World Crisis: The Eastern Front* (Thornton Butterworth, London, 1931), p. 83.
80 Von Freytag-Loringhoven, *Deductions from the World War*, p. 45.
81 *The Final Despatch of Sir Douglas Haig*, General Headquarters, British Armies in France, March 21, 1919 (*London Gazette*, April 8, 1919, Supplement dated April 10, 1919), cited in Dewar and Boraston, vol. 2, p. 358.

4 AIR–LAND BATTLE, 1918–77

1 Fuller, *On Future Warfare*, p. 78.
2 ibid., pp. 78–9.
3 Marshal of Poland Jozef Pilsudski, *Year 1920, and its Climax, Battle of Warsaw during the Polish-Soviet War 1919–20, with the Addition of M. Tukhachevsky's March Beyond the Vistula*, (Pilsudski Institute of London and America, London and New York, 1972), pp. 76–81; Norman Davies, 'The Soviet command and the Battle of Warsaw', *Soviet Studies* (Glasgow, April, 1972), pp. 573–85, this reference pp. 574–5; The Battle of Warsaw features in Fuller's *Decisive Battles of the Western World*, vol. 3.
4 Pilsudski, pp. 74–5.
5 Davies, p. 575.
6 *WO 32/13087, Report of the Committee on the Lessons of the Great War* (A 3629), October, 1932, p. 29.
7 House, pp. 37–9; first battle, Dupuy, *The Evolution* . . ., p. 1.
8 Ministère de la Guerre, *Instruction Provisoire sur l'Emploi Tactique des Grandes Unités*, (Paris, 1922), p. 23.
9 Jean Baptiste Estienne, *Etude sur les missions des chars blindées en campagne*, 25 May 1919, cited in Georges Ferré, *Le Defaut de l'Armure*, (Paris, 1948), pp. 34–47.
10 Charles de Gaulle, *Vers l'Armee de Metier*, (Plon, Paris, 1973).

11 Early evolution of indirect fire systems, Colonel H. A. Bethell, *Modern Artillery in the Field*, pp. 283–4, 'massing the fire of dispersed batteries has now become a fine art; 1920s–1930s developments, Janice E. McKenney, *Field Artillery Army Lineage Series*, (Washington, 1980), pp. 266–73; Riley Sunderland, 'Massed Fires and the FDC [Fire Direction Centre], *Army*, 8, 1958, pp. 58–9; on legacy of World War One see also Lieutenant Colonel Pascal Lucas, *L'Evolution des Idées Tactiques en France et en Allemagne Pendant la Guerre de 1914–18*, (Berger-Levrault, Paris, 1923); on the survival of sophisticated indirect fire techniques among the British, Bidwell and Graham, *Firepower* . . ., pp. 197–9, 276. The latter notes that 'the methods worked out with such pain and labour in 1915–17 were carefully preserved by the Royal Artillery. . . . One of the most remarkable and unregarded feats of the artillery of the 8th Army was to produce the Alamein fire-plans without a hitch or mistake after some 18 months of motoring about the desert in independent troops, as it is only too easy to lose the knack of such detailed processes if they are not regularly exercised'; French artillery in 1936, E. O. Miksche, *Attack! A Study of Blitzkrieg Tactics*, (Random House, New York, 1942), p. 169.
12 The phrase is Miksche's, *Attack!*, pp. 73–99.
13 Harris' obituary in *The Times*, 7 April, 1984, p. 10.
14 See Werner von Blomberg, *Visit of the Chief of the Army Board to Russia, August to September, 1928* (Reichswehr Ministerium, Berlin, 1928), trans. W. R. Newby Grant, p. 1; House, *Towards Combined Arms Warfare* . . ., p. 53.
15 General Heinz *Guderian, Panzer Leader*, (with a foreword by B. H. Liddell Hart, trans. Constantine Fitzgibbon, (Michael Joseph, London, 1952), p. 24. Also Kenneth J. Macksey, Guderian, *Panzer General* (Macdonald and Jane's, London, 1975).
16 ibid., pp. 33–4, 53–4. For the establishment of a panzer division in 1935, see Appendix xxiv.
17 See also House, p. 57.
18 Robin Cross, *The Bombers: the Illustrated Story of Offensive Strategy and Tactics in the Twentieth Century*, (Bantam, London, 1987), pp. 81–2. Modern CEP is the *radius* within which 50 per cent of munitions will fall.
19 ibid., pp. 82–3; R. J. Overy, 'From "Uralbomber" to "Amerikabomber": the Luftwaffe and strategic bombing', *Journal of Strategic Studies*, London, September 1978, pp. 154–78.
20 Miksche, *Attack!*, ch. 5, 'the Air Arm over Battle', esp. pp. 74–5.
21 The exhibition in the *Museo del Ejercito* in Madrid brings this home forcefully; a most interesting display of photographs of destroyed and rebuilt bridges, field works, a model of the attack on the Madrid University precinct, and so on.
22 V. K. Triandafillov, *Kharakter operatsii sovremennykh armiy* (*Character of the Operations of Modern Armies*), (3rd edn, Voyenizdat, Moscow, 1936), pp. 11, 37, 39–41.
23 See, for example, Tukhachevskiy's introduction to the Russian translation of J. F. C. Fuller's *The Reformation of War*, '*Reformatsiya voyny*' in M. N. Tukhachevskiy, *Izbrannye proizvedeniya (Collected Works)*, (Voyenizdat, Moscow, 1964), vol. 2, pp. 147–56. The abridged translation of Fuller's book by A. Taube was published by Gosvoyenizdat, Moscow, 1931.
24 J. F. C. Fuller, *The Reformation of War*, (Hutchinson, London, 1923), pp. 136–51.
25 See Nemecek, *Soviet Aircraft* . . ., pp. 124 (Il'ya Muromets), 136 (long-range bombers), 256, 267 (commercial), 298 (agricultural), 356 (arctic).
26 Tukhachevskiy, 'Novye voprosy voyny' ('New questions of war') in *Voyenno-istoricheskiy zhurnal (VIZh)*, 2, 1962, pp. 64–9.

27 John Weeks, with artwork by John Batchelor, *Airborne Soldier*, (Blandford, Poole, 1982), pp. 12–18; Maurice Tugwell, *Airborne to Battle: A History of Airborne Warfare, 1918–1971*, (Kimber, London, 1971), p. 23. Also Philippe de St Croix, *Airborne Operations: An Illustrated Encyclopedia of the Great Battles of Airborne Forces*, (Salamander, London, 1978).

28 V. F. Margelov, I. I. Lisov, Ya P. Samoylenko, V. I. Ivonin, *Sovetskiye vozdushno-desantnye: Voyenno-istorischeskiy ocherk* (*Soviet Airborne Forces: A Military Historical Sketch*), (Voyenizdat, Moscow, 1980), pp. 8–9, 13. On Soviet paratroops see the author's 'Red Star in the west: Marshal Tukhachevskiy and east–west exchanges on the art of war', *RUSI*, December, 1987, pp. 63–73.

29 Colonel N. Ramanichev, '*Razvitiye teorii i praktiki boyevogo primeneniya vozdushno-desantnykh vovsk v mezhvoyenny period* ('Development of the theory and practice of the military employment of airborne forces in the interwar period'), *VIZh* 10, 1982, article pp. 72–7, reference pp. 75–6; misunderstanding of the aim of the operation, *The Times*, 5 March 1936, p. 12.

30 I. M. Maisky, (Soviet ambassador), 'V Londone' ('In London') in N. I. Koritskiy, (ed.), *Marshal Tukhachevskiy, Vospominaniye druzyey i soratnikov* (. . . *Memoirs of Friends and Comrades in Arms*), (Voyenizdat, Moscow, 1965), p. 230.

31 F. O. Miksche, *Paratroops: The History, Organization and Tactical Use of Airborne Formations*, (Faber & Faber, London, 1943), maps on pp. 64 and 98.

32 G. Isserson, 'Vstrechnoye srazheniye budushchego' (The operational meeting engagement of the future'), *Voyennaya mysl' (VM) (Military Thought)*, 7, 1938, pp. 10–26; 'Razvitiye teorii operativnogo iskusstva v 30-e gody' ('Development of Soviet operational art in the 1930s'), 2 parts, *VIZh*, 1, 1965, pp. 36–46 and 3, 1965, pp. 48–61; 'Operativnye perspektivy budushchego' ('Operational perspectives of the future'), *VM* 8, 1938, pp. 14–26, quotation on p. 23 of the latter; People's Commissariat for the Defence of the USSR, *Vremenny polevoy ustav RKKA 1936 (PU-36) (Temporary Red Army Field Service Regulations)*, (Gosvoyenizdat, Moscow, 1937).

33 The British Official History, *History of the Second World War. UK Military Series. Grand Strategy*, (Her Majesty's Stationery Office London): vol. 2, September 1939 to June 1941, by J. R. M. Butler (2nd impression 1971, first published 1957); vol. 3, June 1941 to August 1942, by J. M. A. Gwyer and J. R. M. Butler, (1964); vol. 4, August 1942 to September 1943, by M. E. Howard (1972); *The Mediterranean and Middle East*, (1987), etc.; *Purnell's History of the Second World War*, editor-in-chief B. H. Liddell Hart, editor Barrie Pitt, (Phoebus Publishing/BPC Publishing, London, 1966/1972/1974/1979/1980) (superb); on the impact of science and ingenuity, R. V. Jones, *Most Secret War*, (Coronet, London, 1981); on the Eastern Front, John Erickson, *Stalin's War with Germany*, vol. 1, *The Road to Stalingrad*, (Weidenfeld & Nicholson, London, 1975), vol. 2, *The Road to Berlin*, (Weidenfeld & Nicholson, London, 1983); Brigadier Peter Young, (ed.), cartography by Richard Natkiel, *Atlas of the Second World War*, (Weidenfeld & Nicholson, London, 1973); John Keegan, General Editor, *The Times Atlas of the Second World War*, (Times Books, London, 1989). This includes new scholarship.

34 *Time Magazine*, 25 September, 1939, cover and p. 25. The term first appears as *Blitzkrieger*, and then on p. 25 as *Blitzkrieg*. The word does not appear in the 1933 *Truppenführung (Troop Leadership)* or in the 1940 *Richtlinien für Führung und Einsatz der Panzer Division* (*Guidelines for Leadership and Operation of the Panzer Division*), (OKW, Berlin, 1940). The preface to Miksche's book, *Blitzkrieg*, is dated May 1941, but the introduction does not

discuss its etymology although it does examine *Schwerpunkt* and *aufrollen* in detail (pp. 16–17). Hitler's comment, at the Löwenbraukeller in Munich, Max Domarus, *Hitler: Reden und Proklamationen*, (2 vols, Verlagsdruckeri Schmidt Neustadt . . ., Simhart & Co., Munich, Wurzburg, 1962, 1963), vol. 2, p. 1776.

35 Professor Williamson Murray, 'Forces strategy, Blitzkrieg strategy and the economic difficulties: Nazi strategy in the 1930s, *RUSI* March 1983, pp. 39–43 criticizes the idea of a 'Blitzkrieg Strategy' propounded by Larry Addington, *The Blitzkrieg Era and the German General Staff* (Rutgers University Press, New Brunswick, 1971); Alan Milward, *The German Economy at War*, (University of London, Athlone Press, London, 1965), Burton Klein, *Germany's Economic Preparations for War* (Harvard Economic Studies, vol. 109, Harvard University Press, Cambridge, Mass., 1959). 'Fresh', not new: Barrie Pitt, 'Blitzkrieg! Poland, September 1/October 6, 1939', in *Purnell's History of the Second World War*.
36 Murray, p. 42.
37 House, p. 78.
38 R. J. S. Stolfi, 'Equipment for victory in France in 1940', *History*, vol. 55, no. 183, February, 1970, pp. 1–20, quotation p. 19, *Times Atlas*, pp. 44–5.
39 House, pp. 97–9.
40 Author's *Red God of War*, pp. 49–50, on Soviet procedures; House, p. 79.
41 Imperial War Museum, *Martel 4/4*, Enclosure 64, *Report from British Military Mission to the USSR. Operations up to the end of the year 1943*, p. 9.
42 *Red God of War*, pp. 57–74.
43 Kent R. Greenfield, *Army Ground Forces and the Air–Ground Battle Team, Including Organic Light Aviation*, (AFG Study no. 35, Washington DC, 1948), p. 47.
44 Edgar Snow, 'The Red Army's flying tank', *Saturday Evening Post*, 10 March, 1945), pp. 18–19, 103; debate: '*Boyevye deystviya bombardirovochnoy aviatsii*' ('Military operations of bomber aviation'), *Krasnaya zvezda (KZ) (Red Star)*, 17 April 1938, p. 2; P. Mikhaylov, '*Deystviya shturmovoy aviatsii* ('Action of fighter–ground attack aviation'), *KZ*, 4 April 1938, p. 2; Kombrig P. Ionov, '*Aviatsiya v nastupatel'nykh operatsiyakh*' ('Aviation in offensive operations'), *KZ*, 24 May 1938, p. 2, and subsequent exchanges in *KZ* until Colonel A. Osipenko, '*Aviatsiya na pole boya*' ('Aviation over the battlefield'), *KZ*, 21 August 1938, p. 3.
45 House, p. 133; Bidwell and Graham, *Fire-power*, pp. 185–7, 260–75.
46 Photographs from the Imperial War Museum, London; Bidwell and Graham, p. 289; Ian V. Hogg, *Barrage: the Guns in Action*, (Macdonald, London, 1970), pp. 138–53, esp. pp. 145, 147–8. Most importantly, not a single German telephone wire remained intact after this bombardment, paralysing command and control.
47 Bidwell and Graham, p. 289.
48 House, pp. 133–5.
49 Young and Natkiel, *Atlas of the Second World War, The Times Atlas*, pp. 152–3, p. 226; *Soviet Military Encyclopedia*, vol. 5, pp. 632–4.
50 Young and Natkiel, p. 233; Clausewitz, *On War*, bk. 1, ch. 1 (Paret and Howard edn, p. 75); Sun Tzu, *Art of War*, 3, 'Offensive strategy', trans. Griffith, p. 77.
51 Young and Natkiel, p. 240, *Times Atlas* . . ., pp. 158–9.
52 ibid., p. 242, and exhibits at the airborne museum at Arnhem.
53 Young and Natkiel, pp. 244–5, *Times Atlas* . . ., pp. 160–61.
54 These operations are summarized in C. J. Dick 'Soviet operational art. Part 1: the fruits of experience', *International Defence Review*, 7, 1988, pp. 775–61; otherwise, the relevant entries in the *Soviet Military Encyclopedia* and the

Times Atlas . . .

55 Colonel David Glantz's excellent *August Storm: the Soviet 1945 Strategic Offensive in Manchuria*, (Leavenworth Paper no. 7, US Government Printing Office, 1984).

56 A. G. Kavtaradze, *Voyennye spetsialisty na sluzhbe Respubliki Sovetov, 1917–20 gg.* (*Military Specialists' in the Service of the Soviet Republic*), (Nauka, Moscow, 1988), pp. 225 and penultimate page of photographs between pp. 192 and 193.

57 Army General I. M. Tret'yak, 'Razgrom Kvantunskoy armii na dal'nem vostoke' ('Defeat of the Kwantung Army in the Far East'), *VIZh*, 8, 1985, pp. 9–19, in particular p. 12, (quote).

58 Glantz, *August Storm, the Soviet 1945 Strategic Offensive* . . . p. 3.

59 ibid., pp. 183–7; Glantz, *August Storm, Soviet Tactical and Operational Combat in Manchuria*, (Leavenworth Paper no. 8, US Government Printing Office, 1984), pp. 1–6, *Times Atlas* . . ., pp. 198–9.

60 Martin Blumenson, *Patton. The Man Behind the Legend, 1885–1945*, (Jonathan Cape, London, 1985), p. 186.

61 ibid., p. 185.

62 Michael Howard, 'War and technology' (based on the Roskill Lecture given at Churchill College, Cambridge on 18 February, 1986), *RUSI Journal*, December, 1987, pp. 17–22, this on p. 19.

63 There was a revival of interest in the Korean War in 1988, connected, no doubt, in part with the Seoul Olympics, and marked by an informative TV series. See Jon Halliday and Bruce Cummings, *Korea: the Unknown War*, (Viking/Penguin, London, 1988).

64 House, *Towards Combined-Arms Warfare* . . ., pp. 154–7.

65 ibid., pp. 158–60.

66 Freytag-Loringhoven, *Deductions from the World War*, p. 107.

67 ibid., p. 108.

68 On the US Army, there have been numerous articles in *Military Review* over the past few years. On the Russians, Edinburgh University, *Soviet Military restructuring* . . .

69 Simpkin, Race to the Swift, ch. 7, pp. 117–32.

70 House, pp. 160–63, diagram of the structure of the 1st Cavalry Division on the latter; on later developments, see the author's 'The operational art of the European theatre', *RUSI/Brassey's Defence Yearbook, 1985*, (Brassey's, London, 1985), pp. 249–55.

71 Major General Dave R. Palmer, *Summons of the Trumpet: A History of the Vietnam War from a Military Man's Viewpoint*, (Ballantine, New York, 1978), p. 303.

72 Palmer, pp. 302–9; House, pp. 164–8; Lomperis concludes that the ARVN were routed (pp. 27–8), House, pp. 167–8, that the mission was accomplished.

73 See Palmer and House, ibid.

74 For the opening phase see the author's *The Future of Land Warfare*, pp. 8–17; Colonel A. Barker, *Yom Kippur War*, (Ballantine, Random House, New York, 1974); Chaim Herzog, *The War of Atonement*, (Weidenfeld & Nicholson, London, 1975); ' "Kar", A Personal View of the Yom Kippur War', *British Army Review*, August, 1975, pp. 12–18.

75 Barker, pp. 120–27; Herzog, pp. 208–50; Brigadier Mohammed Ibrahim Nagaty, 'Some lessons of the Ramadan war', *Pakistan Defence Journal*, April, 1975, pp. 8–9, 45; Aron Soffer, 'The wars of Israel in Sinai: topography conquered', *Military Review*, April, 1982, pp. 61–72; Martin van Creveld, *Military Lessons of the Yom Kippur War: Historical Perspectives*, (The

Washington Papers 3, Sage Policy Papers, Beverly Hills, Ca., 1975).

76 Lieutenant Colonel I. E. Mirghani, 'Lessons from the Ogaden war', *British Army Review*, August, 1981, pp. 28–33.

5 CASE STUDY ONE: *CORPS VOLANT* TO OMG: THE PRACTICAL UTILITY OF MILITARY HISTORY

1 This restructuring was the subject of an Edinburgh University study completed in February 1988, *Soviet Military Restructuring: Defensive Sufficiency, Technology and the Revived Corps/Brigade* (Centre for Defence Studies). Remodelling Soviet forces on corps/brigade lines created formations more suited to protracted and fluid conventional operations, made more effective use of existing forces, and eliminated a level of command at a crucial level, enabling Soviet forces to 'get inside the enemy's time loop' in any engagement.

2 An unclassified version of the data expounded at SHAPEX-82 is in John Hines and Philip J. Petersen, 'The Soviet Conventional offensive in Europe', *Military Review*, April 1984, pp. 2–29; see also Philip J. Petersen, 'The modernization of the Soviet Armed Forces', *NATO's Sixteen Nations*, July 1986, pp. 32–8. The author thanks Colonel Hines and Mr Donnelly for their personal recollections and advice.

3 In Soviet parlance strategy and operational art are two levels of military art (see ch. 1). The change of title was unfortunate as it obscured the very element of historical continuity which had made the revelations possible. The typescript was released to Associated Press who in turn disseminated it to western media and TASS.

4 Main English language news reports: 'Soviet "killer punch" war plan', *The Times*, Monday 4 October 1982, p. 4. (another edition printed the same article under the heading 'Russia plans Blitzkrieg while West dithers'); 'Russia said to develop a quick-attack plan', *International Herald Tribune* 4 October 1982, p. 4. The latter cited AP as its direct source and noted that Mr Donnelly's assessment became known 'several days after the International Institute for Strategic Studies (IISS), London reported in its 1982–3 assessment of the military balance that 'NATO forces suffer from a lack of co-ordination and logistic problems', although there was nothing to connect Mr Donnelly's report directly with IISS.

5. Tass in Russian for abroad, 1417 hrs (read in English, 1508 GMT), 4 October 1982, BBC Monitoring Service transcript SWB SU/7149/A1/1, 6 October 1982: 'British military analyst's allegations about Soviet "Blitzkrieg" plans', text of commentary by Vasiliy Kharkov, 'The aim of falsehoods', paras 1 and 4. The latter included the reference to Rogers' statement on FOFA of the previous week. See also 'Non-nuclear defence plan for NATO', *Daily Telegraph*, 29 September 1982, p. 1. Second Russian broadcast, Moscow World Service in English 1500 GMT, 5 October 1982, BBC transcript *SU/7150/A1/7*, 7 October 1982, 'Donnelly's "Blitzkrieg" Theory: another piece of propaganda', excerpts from commentary by Viktor Olin, paras 2, 4, 5. The latter stressed the military problems of implementing the concept, notably keeping the massing of so many troops and vehicles secret, in order to refute Mr Donnelly's analysis. It also suggested that in drawing attention to the unwiedly nature of NATO's nuclear release procedures and their vulnerability to dislocation by, e.g., OMGs, Mr Donnelly was applying pressure to streamline them. Both these condemnations blasted through the ether were aimed at foreign audiences, not at the Soviet people or even the Soviet military. No printed refutation was issued, either in the ordinary press or in any Soviet military journal, and that is

significant.

6 Donnelly's article, 'The Soviet operational manoeuvre group: a new challenge for NATO', *IDR*, 9, 1982, pp. 1177–86; John G. Hines and Philip J. Petersen, 'The Warsaw Pact strategic offensive: the OMG in context' *IDR*, 10, 1983, pp. 1391–5; Charles J. Dick, 'Operational manoeuvre groups: a closer look', *IDR*, 6, 1983, pp. 769–76 and the author's 'Antecedents of the modern Soviet operational manoeuvre group (OMG)', *RUSI Journal*, September, 1984, pp. 50–8. More recent studies: Joseph R. Burniece, 'The operational manoeuvre group: concept versus organisation', *Military Technology (Miltech)*, 10, 1986, pp. 66–79, and Peer H. Lange, *New Options for the Soviet Military Posture in Central Europe: Operational Manoeuvre Groups*? NATO Study *A/1/g/i/-/116*, September 1985. The author does not agree with much of the latter, or with the account of the OMG's exposition, but it does represent a useful bibliography of the literature on the subject up to the time it was completed. Wider studies which nevertheless have a direct bearing are Soviet Army Studies Office (SASO) USACGSC, *The Soviet Conduct of War*. (USACGSC, Fort Leavenworth, 30 March 1987), and Colonel David M. Glantz, *Towards Deep Battle*: *The Soviet Concept of Operational Maneuver*, (US Army War College, Carlisle Barracks, 1985). Eventual Soviet acknowledgement of OMG, Major General Batenin and Colonel General Chervov, to visiting West German MPs, February 1989, cited in Philip A. Karber/BDM, *Soviet Implementation of the Gorbachev Unilateral Military Reductions*: *Implications for Conventional Arms Control in Europe*, House Armed Services Committee, (March, 1989), p. 13.

7 Major Wojciech Michalak, 'Lotnictwo w dzialaniach Rajdowo-Manewrowych Wojsk Ladowych', ('Aviation in raid-manoeuvre operations of the ground forces'), *Przeglad Wojsk Lotniczych i Wojsk Obrony Powietrznej Kraju*, (*Air and Air Defence Force Review*) (*PWL i WOPK*), February 1982, article pp. 5–9, this reference, p. 5, '*operacyjnych grup marszowych*'. The clearest, or, at any rate, most categorical expositions of the OMG revival appeared in Polish journals. This is not because Polish military thinking is (or was) any more advanced than Soviet – on the contrary, all the concepts appear to be Soviet in origin – but because the Poles at this time were prepared to be more forthcoming and tend to leave out much of the dogma and get down to business.

8 SASO, *Soviet Conduct of War*, pp. 63–4. The author has seen no mention of Operational Manoeuvre Groups as such in Soviet open sources. However, the article on 'Army' in *SVE*, vol. 2, p. 255, refers to the tank army as 'the most important means of developing a penetration and conducting operational manoeuvre', while the entry on 'manoeuvre' in *SVE*, vol. 5, p. 144 says that it is carried out by 'large units and formations . . . of all types of forces'. V. Novikov and F. Sverdlov, *Manoeuvre in Modern Warfare*, (Progress, Moscow, 1972), p. 29, stated that 'Operational manoeuvre . . . may take the form of manoeuvre with nuclear strikes delivered by operational–tactical missiles or the army air-force or manoeuvre by operational groups from one sector to another to exploit success or to outflank an enemy group.'

9 The SHAPEX briefing slides appear to have been based on inter alia, that in Michalak's article, *PWL i WOPK*, 2 1982, p. 9. Marshal P. Rotmistrov, in A. A. Grechko, (ed.), *Yaderny vek i voyna* (*War and the Nuclear Age*), (Izvestiya, Moscow, 1964), cited in John Erickson, (ed.), *The Military-Technical Revolution: Its Impact on Strategy and Foreign Policy*, (Institute for Study of the USSR, Pall Mall Press, London, 1966), p. 59.

10 '*V otryve ot glavnykh sil*'. On the interaction with air, the most important source is probably Colonel Pilot Alexander Musial, 'Charakter i znaczenie

operacji powietrznych we wspolczesnych dzialaniach wojennych' ('The character and importance of air operations in modern war'), *PWL i WOPK*, March 1982, pp. 10–12. The interaction with air and probable enhancement of OMGs is addressed by Donnelly, Dick, Hines, and Petersen, in the latter's 'The Soviet conventional offensive in Europe', and so on. Musial in particular stresses the need for integration of all forces, for 'a single concept and plan', ('jednego zamiaru i planu') and the operation's 'combined service character' ('charakter ogolnowojskowy') (p. 12).

11 *SVE*, vol. 6, 1978, p. 373.

12 ibid., 'Zadachi razvitiya uspekha v operatsii vypolnyayut vtorye eshelony fronta (armii)'.

13 The deployment of forces would depend on many factors, and especially on whether the defence was un-, partially- or fully-prepared. See SASO, *Soviet Conduct of War*, pp. 32–9.

14 *SVE*, vol. 6, 1978, p. 282.

15 SASO, *Soviet Conduct of War*, pp. 66–7.

16 *SVE*, vol. 7 1979, p. 94.

17 M. M. Kir'yan, *Fronty nastupali. Po opytu Velikoy Otechest-vennoy voyny. Kratkiy istoriko-teoreticheskiy ocherk*. (*The Fronts on the Offensive. According to the Experience of the Great Patriotic War. A Short Historical-Theoretical Sketch*), (Nauka, Moscow, 1987), pp. 4, 21, 109, 110, 120.

18 ibid. p. 4.

19 V. Mitkevich, 'Kazach'ya lava' ('The Cossack "Lava" '), *Voyenny Sbornik* (*Military Collection*), vol. 204, April 1892, pp. 347–68, esp. p. 359. Captain C. R. Day, 'Cavalry raids: their value and how made', *Journal of the US Cavalry Association* (*JUSCA*), September 1912, pp. 227–38, esp. p. 228.

20 Mitkevich, p. 368. *Teatr voyennykh deystviy* is translated Theatre of Strategic Military Action (TSMA) as for modern Russian. In the late nineteenth century the term, as now, clearly implied the strategic level. For example, the Russian military theorist Leyer (1829–1904) entitled the later edition of his main work *Strategiya – taktika teatra voyennykh deystviy* (*Strategy: the tactics of the TSMA*) (St Petersburg, 1885–98). Mitkevich, writing for publication in 1892, would have known Leyer's work, and was referring to action that was 'strategic' in the parlance of his time.

21 Ye Barsukov, *Russkaya armiya i flot v xviii v* (*The Russian Army and Fleet in the Eighteenth Century*), (Moscow, 1958), pp. 282–3; Duffy, *The Army of Frederick the Great*, p. 194 and *Russia's Military Way to the West*, pp. 114–15. Austrian precedent and French suggestion, Duffy, *Frederick the Great: A Military Life*, pp. 138–9, 208–9.

22 Second Lieutenant W. H. Hay, 'Cavalry raids', *JUSCA*, vol. 4, December 1891, pp. 326–77, quote, p. 365; L. G. Beskrovniy, (ed.), *Pokhod russkoy armii protiv Napoleona v 1813 g.: Sbornik dokumentov*, (*Operations of the Russian Army against Napoleon in 1813: Collected Documents*), (Nauka, Moscow, 1964), esp. p. 100. 'Army Partisan Detachment' 'armeyskiy partizanskiy otryad', in the latter. Biography of later Chernyshev in *SVE*, vol. 8, 1980, p. 461, describes him commanding an 'independent cavalry detachment' ('otdel'ny kavaleriyskiy otryad').

23 Sergeant Harry Bell, 'Cavalry raids and the lessons they teach us' (translated from 'Employment of cavalry in independent action against the flank and rear of a hostile army' by Lieutenant Colonel Wenniger, Bavarian Cavalry, *Austrian Cavalry Monthly*, April 1908, *JUSCA*, July 1908, pp. 142–52, esp. p. 145.

24 Hay, 'Cavalry raids', p. 363. The classic work on the influence of the American Civil War is Luvaas, *The Military Legacy of the Civil War*.

25 N. Glinoyetskiy, 'Inostrannoye voyennoye obozreniye: znacheniye lëgkoy kavalerii v noveyshikh voyn . . .' ('Foreign military review: the significance of light cavalry in most recent wars . . .'), *Voyenny Sbornik (Military Collection)*, no. 4, June 1863, pp. 651–54.
26 V. Dumbadze, *Russia's War Minister* (Simpkin, Marshall & Co. Ltd, London, 1915), p. 51.
27 Luvaas, p. 113.
28 Colonel E. Chenevix Trench, *Cavalry in Modern War*, Military Handbooks, vol. 6, (Kegan Paul, Trench & Co., London, 1884), p. 75.
29 ibid, p. 227. Napoleonic War experience, pp. 192–3.
30 Staff Captain N. Sukhotin, 'Nabeg letuchago otryada za vislu' ('Raid of a flying detachment beyond the vistula'), *Voyenny Sbornik*, 11, 1876, pp. 114–53 and 12, 1876, pp. 365–99, esp. part 2, p. 396. Sukhotin also wrote *Reydi i poiski kavalerii vo vremya amerikanskoy voyny 1861–65 gg.* (*Operational and tactical raids by cavalry during the American War of 1861–65*), (Moscow Military District Staff Press, 1875), especially introduction, p. iii. Sukhotin's article on the 1876 manoeuvres was translated into French, 'Manoeuvres et exercises de la Cavalerie Russe' *Revue Militaire d'Etranger* (*Foreign Military Review*). An earlier exercise of 1875 is described in no. 316, vol. 10, July–December 1876, and Sukhotin's account of the 1876 exercise is reproduced in vol. 11, January-June 1877, nos 333–49. They are also described in George Cardinal von Widdern, *Strategische Kavallerie-Manöver: Studien und Vorschlage angeregt durch die grossen strategischen Manöver der russischen Kavallerie an der Weichsel im Herbst 1876* (*Strategic Cavalry Manoeuvres . . .*), (A Reisewitz, Gera, 1877). The latter contains a better map. Sukhotin gives the dates in the old style used in Russia before 1918, which was 12 days behind the rest of Europe in the nineteenth century and 13 in the early twentieth. The maps with the French and German accounts use the same dates without conversion.

Sukhotin discussed at length the best way of encapsulating the English word 'raid' in Russian (part 1, p. 126). It had its origins in 'to ride', and for this the Cossacks had *pobezhat'* and *probezhat'*, 'to ride through'. To conduct a raid, foray, or incursion to produce a *pogrom* (massacre) or destruction in the enemy rear was a *nabeg*. Finally there was a smaller-scale blow (*nalët*) by cavalry, a *poisk*, often translated reconnaissance raid. Sukhotin therefore described this operational-strategic raid as a *nabeg*, although the Soviet army later adopted the word *reyd* from English.
31 'Single brilliant example': Hay, 'Cavalry raids', p. 369. Obruchev's comments, Military–historical commission of the General Staff, *Osobiye pribavleniye k opisaniyu russko-turetskov voyny 1877–78 gg. na Balkanskom poluostrove, . . . soobrazheniya, kasayushchiyasya plana voyny, (Special Supplement to the Description of the Russo-Turkish War of 1877–78 on the Balkan Peninsula . . . correspondence relating to the war plan)*, (General Staff Military Historical Commission Press, St Petersburg, 1901), pp. 27–39 cover Obruchev's two 'notes' (*zapiski*), quote on p. 29.
32 *Osobiye pribavleniye . . .*, pp. 31–2.
33 ibid., pp. 33 (two armies) 40–69 (Artamonov's notes), esp. p. 43. The need for speed is reiterated on p. 69.
34 *SVE*, vol. 6, 1978, p. 282; 'Deystviya peredovogo otryada Generala Gurko v 1877 godu' ('Action of General Gurko's forward detachment . . .'), *Voyenny Sbornik*, nos 7–100, 1900 vols 254–5; V. T. Novitskiy, (ed.) (sometimes cited as Sytin: Sytin was the publisher), *Voyennaya Entsiklopediya (VE) (Military Encyclopedia*, St Petersburg, 1911–15), vol. 8, p. 543.
35 SVE, vol. 8, p. 543. Gurko (1828–1901) eventually became a Field Marshal. He

wrote no military-theoretical works but had studied those of the Prussian cavalry Major General Carl von Schmidt (1817–75), whose *Instructions for the Training, Leading and Employment of Cavalry* had been published in German in 1876. See the English translation by Captain C. W. Bowdler Bell (HMSO, London, 1881), p. iii.

36 Colonel Epauchin (wrong transliteration: his name was Epanchin), *Operations of General Gurko's Advance Guard in 1877*, (translated from *Voyna 1877–78. Deystviya Gen.-Ad. Gurko, 1895*), trans. H. Havelock, (The Wolseley Series, (ed.) W. H. James, London, 1900), pp. 3–4. F. V. Greene's, *The Russian Army and Its Campaigns in Turkey in 1877–78*, (W. H. Allen, London, 1878), p. 165 gives Gurko 8,000 infantry, 4,000 cavalry and 32 guns. The American Greene's work was considered 'by far the best work in the English language' by the publishers of the British edition of part of that work, published as *The Campaign in Bulgaria*, (Hugh Rees, London, 1903). Epanchin also wrote articles in *Voyenny Sbornik* in 1891–3 and 1897. See the bibliography in V. A. Zolotarev, *Rossiya i Turtsiya. Voyna 1877–78 gg.* (*Russia and Turkey. The War of 1877–78*), (Nauka, Moscow, 1983), p. 211.

37 Epanchin, *Operations* . . ., pp. 4–6; Greene, *The Russian Army* . . . pp. 166–8, *SVE*, vol. 7, p. 187.

38 Epanchin, *Operations* . . ., pp. 4–6.

39 ibid., p. 8.

40 Day, 'Cavalry raids . . .' p. 234 (quote); Greene, *The Russian Army*, p. 167.

41 Epanchin, *Operations* . . . pp. 40, 42–3; Greene, *The Russian Army* . . . p. 167.

42 Epanchin, *Operations* . . . p. 86.

43 ibid., Editor's introduction, p. x, and 106–27.

44 ibid.

45 Ibid., pp. 67–8, 129.

46 N. Epanchin, '*Deystviye peredovogo otryada gen.-ad. Gurko v voynu 1877–78*' ('Action of General Gurko's forward detachment in the war . . .'), *Voyenny Sbornik*, 1896, cited in Zolotarëv, p. 135.

47 Zolotarëv, ibid.

48 Greene, *The Russian Army*, p. 183.

49 War Office, General Staff (UK), *Extracts from General Kuropatkin's Instructions to Commanders of Units of the Russian Army in Manchuria 1904*, (15 April 1904), (HMSO, London, 1905), p. 6.

50 General P. I. Mishchenko, born 1853 was a Cossack by birth and an artillery officer by training. He had fought in the Russo-Turkish war. In 1900, as a Colonel, he was second in command of the force protecting the Chinese eastern railway against Chinese insurgents in the Boxer Rebellion. Wenniger, trans. Bell, pp. 145–6; M. Svechin, *Nabeg konnago otryada General Ad'yutanta Mishchenko na Inkou – organizatsiya i proizvodstva nabega* (*Raid of General Mishchenko's Horse Detachment on Inkou – Organization and Conduct of the Raid*, (Ofitser – vospitatel', St Petersburg, 1907), p. 12. The term *nabeg* – raid in the general sense is used instead of the more specific term used for raids in conjunction with main forces, *reyd.*

51 Svechin, pp. 12–13, 68.

52 ibid., pp. 6–7.

53 ibid., p. 70.

54 The course of the raid is graphically described in Francis McCullagh, *With the Cossacks. Being the story of an Irishman who rode with the Cossacks throughout the Russo-Japanese War*, (Eveleigh Nash, London, 1906). Attempts to prevent reinforcement of Inkou, pp. 178, 179, 183, also Svechin, esp. diagram opp. p.38 and p. 68 for results of raid.

55 Svechin, pp. 65–6; McCullagh, p. 179; Wenniger, trans. Bell, p. 148.
56 Wenniger, trans. Bell, p. 149.
57 *Comptes Rendus publiés par le Rousskii Invalid de Conferences sur la Guerre Russo-Japonais Faites a l'Academie d'Etat Major Nicolas*, (*Accounts of Conferences on the Russo-Japanese War at the Nicholas General Staff Academy, published by the Russkiy Invalid* (*Russian Veteran*), (Charles Lavauzell, Paris, undated), (vol. 7, Offensive of the Manchurian 2nd Army in January, 1905, and General Mishchenko's Mounted Detachment during the 2nd Army Offensive of January 1905), pp. 16–23, 69–115, maps 5 and 6.
58 There was a general efflorescence of work on future large-scale war: see for example, N. P. Mikhnevich, 'Poyavitsya-li milioniya [sic.] armiy v budushchey bol'shoy evropeyskoy voyne?' (Zametki po povodu stat'i A Petrova "K voprosam strategii")' ('Will million-strong armies appear in a future major European War? Remarks on A. Petrov's article "On questions of strategy")'), *Voyenny Sbornik*, vol. 2, 1898, pp. 260–4. The study of particular relevance here is Fedor Gershel'man, 'Kavleriya v sovremennykh voynakh' ('Cavalry in Modern Wars'), *Voyenny Sbornik*, vol. 242 (July and August), 1898, pp. 76–121, 424–41, vol. 243, September 1898, pp. 104–131. These references, pp. 79, 98, 103–7, July 1898.
59. ibid., July 1898, pp. 110–11, 121; August 1898, p. 438; September 1898, pp. 104, 125–6.
60. Service Historique de l'Armée de Terre (Vincennes) (SHAT), *Carton 7N 1472*, Moulin to French War Minister, 28 July 1895. In the previous year (same Carton, 3 April 1894), Moulin had reported (unenthusiastically) on the appearance of Martynov's work, *Strategy* . . . (ch. 3 note 51), which discussed the possibility of a cavalry invasion of East Prussia and noted the importance of railways and their vulnerability. The author thanks Carl Van Dyke for alerting him to this superb collection of material, covering Franco-Russian exchanges on military science and war planning from 1894–1914.
61 *SVE*, vol. 2, pp. 378–19; Yuri Danilov, *Rossiya v mirovoy voyny* (*Russia in the World War*), (Slovo, Berlin, 1924), pp. 98–9. Danilov's book is also available in French, *La Russie dans la Guerre Mondiale. 1914–1917*, (Payot, Paris, 1927), pp. 140–1.
62 Shaposhnikov, ' "Konnye massy" na flange armii', p. 96.
63 Day, 'Cavalry raids', p. 235.
64 ibid.
65 ibid., pp. 235–8.
66 Winston S. Churchill, *The World Crisis*: *The Aftermath* (vol. 5, Thornton Butterworth, London, 1929), ch. 12, 'The Russian Civil War', p. 232.
67 *Soviet Military Encyclopedia* (*SVE*), vol. 6, 1978, p. 373.
68 *SVE*, vol. 4, (1977), p. 305.
69 ibid., vo!. 3, 1977, pp. 6–22.
70 Brevet Captain Hinterhoff, Polish Army, 'General Mamontov's cavalry raid', *The Cavalry Journal*, vol. 25, January to October 1935, pp. 209–22. The opening of Hinterhoff's article is confusing, probably because of translation errors, and initially gives the impression that Mamontov was on the Soviet side (he meant that the Soviet took it very much to heart): the body of the article is accurate and useful; *SVE*, vol 5, pp. 112–13 (Mamontov's raid), vol. 1, p. 616 and vol. 6, pp. 262–3 (Budënny's Cavalry Corps/1st Horse Army details): *Grazhdanskaya voyna i voyennaya interventsiya v SSSR: Entsiklopediya* (*The Civil War and Military Intervention in the USSR: Encyclopedia*), (Sovetskaya entsiklopediya, Moscow, 1983), pp. 341, 444–5. See also General A. I. Denikin, *The Russian Turmoil*: *Memoirs: Military, Social and Political*, (Hutchinson,

London, 1922) and Dmitry V. Lehavich, *White against Red*, (W. W. Norton, New York, 1974). Mamontov (1869–1920) had graduated from the Nicholas Cavalry School in 1890 and fought in World War One, commanding 6th Don Cossack regiment, and after the revolution commanded 4th Cossack corps in the White armies. He had a forceful personality, and this placed him in high regard among the local population (Hinterhoff, p. 213).

71 Hinterhoff, 'General Mamontov's Cavalry raid', pp. 212–13.
72 *SVE*, vol. 5, p. 112.
73 *SVE*, vol. 5 p. 112 gives 6 armoured cars: Hinterhoff pp. 212–13 gives 3 armoured cars and 7 armoured trains (the latter probably a translation error).
74 Denkin's army was now enlarged to 160,000. However, as a proportion of the original 50,000, Mamontov's force would have been about the right size.
75 Hinterhoff, p. 213.
76 ibid., pp. 213–16.
77 ibid., p. 216; *SVE*, vol. 5, p. 112.
78 Hinterhoff, p. 216.
79 ibid., p. 220.
80 ibid., p. 221.
81 ibid., p. 222.
82 S. I. Gusev, *Uroki grazhdanskoy voyny (The Lessons of the Civil War*), (Gosizdat, Moscow, 1921), 2nd edn, p. 22, commenting on the relative effect and cost of Mamontov's raid and that of 5th Red Kuban Cossack Cavalry Division in October 1920. On the latter, see *Grazhdanskaya voyna . . . entsiklopediya* p. 486.
83 Colonel I. Polyakov,interviewing I. Dubinskiy, on Primakov, 'Pervaya sablya chervonnykh kazakov' ('First sabre of the crimson cossacks'), *Krasnaya Zvezda* (*KZ*) (*Red Star*), 29 December 1987, p. 4. Primakov apparently received an inscribed cigarette case from Army Commander Uborevich, in recognition of no less than fourteen 'raids'.
84 *Chervonoye kazachestvo. Sbornik materialov po isotrii chervonogo kazachestva 1918–23 (Crimson Cossacks. A collection of materials on the History of Crimson Cossackdom, 1918–23),* introduction by G. I. Petrovskiy, (Put' prosveshcheniya and Molodoy Rabochiy, Kharkov, 1924), pp. 53–4.
85 Colonel G. Vasilev, 'Reydy krasnykh hazakov', ('Raids of Red Cossacks'), *VV* 2, 1968, pp. 55–6.
86 *Chervonoye kazachestvo* . . . pp. 72–9; Polyakov, '*Pervaya sablya* . . .' col. 4.
87 Vasilev, pp. 55–6. A *tachanka* was a light horse-drawn carriage with a heavy machine gun mounted on it.
88 Gusev, p. 22.
89 *SVE*, vol. 4, pp. 159–60. The breakthrough is referred to as the Zhitomirskiy proryv', part of the Kiev operation, 1920.
90 Captain Kretschmer (no initial), '*Durchbruch der russische Reiter-Armee durch die polnische Front im Frühjahr, 1920*' ('Breakthrough of the Russian Cavalry Army through the Polish Front in Spring, 1920'), *Wissen und Wehr (Military Thought)*, 3, 1921, p. 150.
91 ibid.
92 Kretschmer, p. 158.
93 ibid., pp. 159–60; Lange, p. 19.
94 Kretschmer, pp. 159–60, quote on the latter.
95 ibid., p. 164.
96 ibid., p. 163.
97 ibid.
98 Marshal of Poland Jozef Pilsudski, *Rok 1920, z powodu ksiazki M*

Tukhaczewskiego 'Pochód za Visle', Wydanie III, (Instytut Badania Najnowsze historji Polski, Warszawa, 1931), translated as *Year 1920 and its climax, Battle of Warsaw during the Polish Soviet War, 1920, with the addition of Soviet Marshal Tukhachevskiy's 'March Beyond the Vistula'*, (Pilsudski Institute of America/Pilsudski Institute of London, New York/London, 1972), p. 76 of the latter.

99 ibid.

100 ibid., p. 75.

101 ibid., p. 82.

102 J. F. C. Fuller, *Memoirs of an Unconventional Soldier*, (Ivor, Nicholson & Watson, London, 1936), p. 253.

103 Pilsudski, p. 51 of translation. This is in fact the author's own translation of the Polish original, pp. 69–70.

104 V. Triandafillov, *Kharakter operatsii sovremennykh armiy (Character of the Operations of Modern Armies*), 3rd ed, (Voyenizdat, Moscow, 1937), pp. 204–5.

105 Kretschmer, p. 163.

106 *Chervonoye kazachestvo*, pp. 102–9.

107 ibid., pp. 110–19.

108 Gusev, pp. 22–3.

109 Kretschmer, p. 164.

110 See Gusev, *Uroki grazhdanskoy voyny*, and bibliographies of the 1920s which list 'tekhnicheskiye voyska' as a distinct category.

111 K. Monigetti, *Sovmestnoye deystviye konnitsy i vozdushnogo flota (Combined Action of Cavalry and the Air Force*), (Gosizdat, Moscow and Leningrad, 1928). Monigetti's main historical example was the raid of the Polish corps under Colonel Rummel' from Novograd-Volynsk to Korosten' from 8 to 13 October 1920.

112 Monigetti, pp. 90–3, 101–7; on helicopters and modern OMGs, Michalak, '*Lotnictwo w dzialaniach rajdowo-manewrowych wojsk ladowych*', *PWL i WOPK*, February 1982, pp. 5–9.

113 See, for example, *WO 32/3116, Report of the Committee on the Lessons of the Great War*, (October 1932), pp. 15–19.

114 See in particular Triandafillov, *Kharakter operatsii sovremennykh armiy*; Tukhachevskiy; *Novye voprosy voyny*; and Isserson 'Razvitiye sovetskogo operativnogo iskusstva v 30-e gody', ('Development of Soviet operational art in the 1930s'), 2 parts, *Vizl*, 1, 1965, pp. 36–46 and 3, 1965, pp. 48–61; 'Vstrechnoye srazheniye budushchego' ('The operational meeting engagement of the future'), *Voyennaya Mysl* (*VM*) (*Military Thought*), 7, 1938, pp. 10–26; 'Operativnye perspektivy budushchego' ('Operational perspectives of the future'), *VM* 8, 1938, pp. 14–26.

115 Isserson, 'Razvitiye teorii ...', part 1 (*Vizl* 1, 1965), p. 42.

116 ibid., also 'Vstrechnoye srazheniye budushchego', pp. 21–5.

117 Isserson, 'Operativny perspektivy budushchego', pp. 22, 23; 'Vstrechnoye srazheniye. . . . 'p. 18. ERP: *SVE*, vol. 2 (1976), pp. 575–6 ('Glubokaya operatsiya').

118 *SVE*, vol. 2, 1976, pp. 574–6; biographies, *SVE*, vol. 1, pp. 150–1; vol. 3, p. 299; vol. 8, pp. 164, 656; also Isserson, 'Razvitiye . . .', p. 45, and Polyakov, 'Pervaya sablya . . .'

119 Isserson, 'Razvitiye . . .', p. 45.

120 Isserson, 'Razvitiye . . .' and 'Operativnye perspektivy . . .'; V. Obukhov, 'Mekhanizirovannye chasti v parallel nom presledovanii' ('Mechanized units in parallel pursuit'), *VM*, 8, 1938, pp. 78–92, esp. pp. 79, 87.

121 Opening quotation, Winston S. Churchill, *The Dawn of Liberation. War Speeches by the Rt Hon. Winston S. Churchill CH MP.* Compiled by Charles Eade, (Cassell, London, 1945): 'The hour of our greatest effort is approaching', 26 March 1944, p. 41.

122 Soviet historiography divides the Great Patriotic War into three periods: that of the 'Strategic Defensive' from 22 June 1941 to the counter-offensive at Stalingrad on 19 November 1942; the fundamental shift in the balance, up to the end of 1943; and the complete defeat of Germany during 1944 and 1945. The Soviet offensive against Japanese forces in Manchuria is described as a 'special period': *SVE*, vol. 2, pp. 54–69; Zhilin, p. 165.

123 See, for example, P. Kurochkin, '*Deystviya tankovykh armiy v operativnoy glubine*' ('Action of tank armies in the operational depth'), *VM* 11, 1964, pp. 55–73 and the table on pp. 62–3 which begins with Stalingrad. Martel who returned to Russia during the war also confirmed that Stalingrad was the first real opportunity the Russians had to practise these ideas: Imperial War Museum, *GQM 4/4, Russia Mission*, 15A. III 3, 11–19 May, 1943.

124 I. M. Anan'ev, Tankovye armii v nastuplenii. Po opytu Velikoy Otechestvennoy voyne (*Tank armies in the Offensive. According to the experience of the Great Patriotic War*), (Voyenizdat, Moscow, 1988), p. 42.

125 Colonel B. I. Nevzorov, 'Dostizheniye uspekha pri obshchem prevoskhodstve protivnika v silakh i sredstvakh' ('Attaining success in the presence of a general enemy superiority in forces and means [Men and equipment]') (10th Army in the Moscow counter-offensive), *VIZh*, 12, 1986, pp. 22–9, esp. pp. 25, 27.

126 Marshal of Tank Troops O. A. Losik, (ed.) *Stroyitel'stvo i boyevoye primeneniye Sovetskikh tankovykh voysk v gody Velikoy Otechestvennoy voyny* (*The Composition and Combat Employment of Soviet Tank Forces in the Great Patriotic War*) (Voyenizdat, Moscow, 1979), pp. 114–15.

127 Radzievskiy, *Tankovy Udar*, p. 16.

128 Losik, pp. 115–16.

129 ibid., p. 116.

130 Radzievskiy, p. 24; Losik, pp. 117–18. On 1st Guards Army and 24th Tank Corps, Albert Z. Conner and Robert G. Poirier, *The Red Army Order of Battle in the Great Patriotic War*, (Presidio, Novato, 1985), pp. 11–12, 147–8.

131 Losik, pp. 118–19.

132 *SVE* vol. 7, 1979, pp. 517-521 (Stalingrad), 506–7 (Middle Don), 682–3 (Tatsinskaya). Principal detailed Soviet sources are Marshal SU K. K. Rokossovskiy *Velikaya pobeda na Volge* (*Great Victory on the Volga*), (2 vols, Voyenizdat, Moscow, 1965); Army General D. D. Lelyushenko *Moskva-Stalingrad-Berlin-Praga: Zapiski komandarma* (*Moscow-Stalingrad-Berlin-Prague: an Army Commander's Notebook*), (3rd edn, Nauka, Moscow, 1975). Also 1970 and 1973 editions; also by Lelyushenko '*l-ya i 3-ya gvardeyskiye armii v kontrnastuplenii pol Stalingradom*' ('1st and 3rd Guards Armies in the Stalingrad Counteroffensive'), in A. M. Samsonov, (ed.) *Stalingradskaya epopeya* (*The Stalingrad Epoch*), (Nauka, Moscow, 1968). Most recent and focusing on the relevance of the raid to today is Major General E. V. Porfir'ev, 'Reyd k Tatsinskoy' ('Raid to Tatsinskaya'), *VIZh* 11, 1987, pp. 63–71. Porfir'ev concludes: 'The experience of raiding action in the enemy's deep rear has not lost its significance in modern conditions. Most relevant of all are the high mobility, manoeuvrability, and carrying out military action when severed from the main forces' (p. 71). See also Captain Harold W. Coyle, 'Tatsinskaya and Soviet OMG doctrine', *Armor*, January-February 1985, pp. 33–8.

133 Lelyushenko, in *Stalingradskaya epopeya*, p. 702; Rokossovskiy, *Velikaya pobeda . . .*, pp. 313–15; *SVE*, vol. 7, 1979, pp. 506–7.

134 Rokossovskiy, *Velikaya pobeda . . .*, pp. 317–18, 321, 354; Lelyushenko *Moskva-Stalingrad . . .*, p. 173.
135 Rokossovskiy, *Velikaya pobeda . . .*, pp. 306, 309, 330–1.
136 ibid., pp. 324, 355; Lelyushenko in *Stalingradskaya epopeya . . .*, pp. 702–3.
137 *SVE*, vol. 7, p. 507 and map opposite p. 193; Rokossovskiy, *Velikaya pobeda*, p. 358 confirms the 240 kilometres.
138 *SVE*, vol. 7, p. 682; Rokossovskiy, *Velikaya pobeda*, p. 360; Coyle, pp. 34–5.
139 People's Commissariat of Defence, Department for Utilization of War Experience of the Red Army General Staff, *Sbornik Materialov po izucheniyu opyta voyny*, (*Collection of Materials for the Study of War Experience*), no. 8, August-October 1943, (Voyenizdat, Moscow, 1943), pp. 48–81: 'Nekotorve vyvody po ispolzovaniyu tankovykh i mekhaniziroyannykh korpusov dlyya razvitiya proryva' ('Certain conclusions on utilizing tank and mechanized corps to develop the breakthrough'), this reference, pp. 63–4. The Front Military Soviet ordered the air resupply attempt; Rokossovskiy, *Velikaya pobeda*, p. 365.
140 *Sbornik materialov*, p. 64.
141 N. I. Vasilev, *Tatsinskiy Reyd* (*Tatsinskaya Raid*), (Voyenizdat, Moscow, 1969), pp. 101–14; Lelyushenko, *Moskva-Stalingrad* ... p. 175; Rokossovskiy, p. 364: *SVE*, vol. 7, p. 683.
142 *SVE*, vol. 7, p. 683.
143 Coyle, pp. 37–8.
144 Sun Tzu, *The Art of War*, translated by Samuel B. Griffith, with a foreword by B. H. Liddell Hart, (Oxford University Press, 1971), p. 134.
145 *Sbornik materialov*, p. 52, table. The most strongly reinforced was 4th Tank Corps with 1 Anti-tank artillery regiment, 1 rocket launcher ('Guards Mortar') regiment, 1 AA regiment, 1 howitzer artillery regiment, 2 High Command Reserve gun artillery batteries and 1 motorized sapper company.
146 ibid., pp. 62, 64.
147 ibid., p. 64.
148 ibid., pp. 50 (minefields and control of air), 61 (consolidation by infantry and cavalry), 55–6 (formation of mobile group and merging of corps into tank army).
149 Martel in *IWM GQM 4/4 Russia Mission*, 15A III 3 (11–19 May 1943).
150 Radzievskiy, *Tankovy Udar*, p. 24.
151 ibid., pp. 24–6.
152 I. A. Pliev, *Dorogami voyny* (*Along Paths of War*), (Igr, Ordzhonikidze, 1973), *SVE*, vol. 6 1978, pp. 356–7. Pliev had joined the Red Army in 1922 and attended the Frunze Military Academy (1933) and General Staff Academy (1941). Between 1936 and 1938 he had been adviser to the Mongolian People's Army, significant in view of his later command of an international formation, the Soviet-Mongolian KMG. In 1939 he commanded 6th Cavalry Division participating in the partition of Poland. In December, 1941 he took command of 2nd Guards Cavalry Corps; in April, 1942, 5th Guards Cavalry Corps, and later commanded 3rd and 4th Guards Cavalry Corps also.
153 Pliev, *Dorogami voyny*, pp. 9–14.
154 ibid., esp. pp. 10–11.
155 ibid., p. 12.
156 Martel *LHCMA 1/492*. Martel to Liddell Hart commenting on an article the latter was writing for *John Bull*, 25 February 1949, para. 5.
157 ibid., para. 8.
158 Quote: US War Department, *TM-30-430, Handbook on USSR Military Forces*, November 1945, p. V–83. This manual drew extensively on the records

of the German Eastern Front Intelligence Organization, *Fremde Heere Ost*, as supplied by its former chief, Gehlen, and was compiled by US Officers with extensive combat experience themselves. Liddell Hart had also challenged Martel: *LHCMA 1/492*, Liddell Hart's letter of 7 April, 1949: Martel's reply of 9 September. Martel admitted that he had seen no action after the early part of 1944 but stuck to his opinion, expressed in his 10 October 1950 letter to *The Times*, that the primary task of the Soviet air forces, tanks and artillery was to assist their infantry, and that they lacked sufficient mechanized transport 'to form and maintain mobile armoured forces, except on a very small scale. They count on winning their wars by the advance of their infantry formations on a wide front'. In the light of the copious, detailed and reliable evidence to the contrary, this verges on the ridiculous.

159 Liddell Hart published two articles in *John Bull*: 'How good are Russia's forces?', *John Bull*, week ending 26 March 1949, pp. 7–8, 23, and 'The Russian as a fighting man', 2 April 1949, pp. 8–9, 16. This quotation, p. 16 of the latter. Liddell Hart ignored Martel's comments.

160 Field Marshal Erich von Manstein, *Lost Victories*, edited and translated by Anthony Powell with a foreword by B. H. Liddell Hart (Methuen, London, 1958). p. 295.

161 Kurochkin, '*Deystviya tankovykh armiy v operativnoy glubine*', p. 56.

162 Losik, pp. 119–20.

163 Radzievskiy, *Tankovy udar*, p. 123; see, for example, Marshal of Aviation A. N. Yefimov, 'Opyt ispol'zovaniya soyedineniy vozdushnykh armiy v interesakh podvishnykh grupp frontov pri deystviyakh ikh v operativnoy glubine' ('Experience of employing Air Armies in the interests of Front Mobile Groups in their Action in the Operational Depth') *VIZh*, 8, 1986, pp. 14–21 and Lieutenant General A. A. Sokolov, 'Dostizheniye vysokikh tempov nastupleniya v khode frontovykh operatsii Velikoy Otechestvennoy voyny' ('Attaining High Tempos of the Offensive in the course of Front Operations in the GPW'), *VIZh*, 12, 1985, pp. 8–13, esp. table p. 9.

164 Losik, p. 121. The vast body of literature on Soviet Mobile Operations in this period is documented and classified in Michael Parrish, *USSR in World War II: An Annotated Bibliography of Books published in the USSR 1945–75 (With an Addenda for 1975–80)*, 2 vols (Garland Press, New York, 1981).

165 P. V. Terekhov, Boyevye deystviya tankov na severo-zapade v 1944 g. (*Combat Action of Tanks in the North West in 1944*) (Voyenizdat, Moscow, 1965), pp. 86–7; *SVE*, vol. 7, pp. 265–6.

166 Terekhov, pp. 88–97.

167 ibid., pp. 97–8.

168 ibid., p. 98.

169 ibid., pp. 99–110.

170 Importance of operation, Colonel N. F. Polukhin and Lieutenant Colonel Yu D. Patychenko, 'Material'noye obespecheniye podvizhnykh grupp frontov v Vislo-Oderskoy Operatsii' (Logistic Support of Front Mobile Groups During the Vistula-Oder Operation'), *VIZh*, 1, 1987, pp. 30–6. On the operation in general, *SVE*, vol. 2, pp. 147–50; F. I. Vysotskiy *et al.*, *Gvardeyskaya tankovaya* (*Guards Tank*) (2nd Guards Tank Army, Moscow, 1963); S. I. Mel'nikov, *Marshal Rybalko vospominaniya byvshego chlena voyennogo soveta 3-i gvardeyskoy tankovoy armii (Marshal Rybalko. Memoirs of a Former Member of 3rd Guards Tank Army Military Soviet*) (Izd-vo politicheskoy literaturoy Ukrainy, Kiev, 1980); A. M. Zvartsev, *3-ya gvardeyskaya tankovaya* (3rd Guards Tank [Army]), (Voyenizdat, Moscow, 1982). See also Katukov and Babadzhanyan on 1st Guards Tank Army.

The Belorussian operation is used as an example of the use of Mobile Groups in Richard Armstrong, 'Mobile groups: prologue to OMG', *Parameters, Journal of the US Army War College*, vol. 16, no. 2, summer, 1986, pp. 58–69, *KMGs* tank armies, and movements in *Times Atlas*, pp. 148–9, 174–5.

171 Sokolov, '*Dostizheniye vysokikh tempov . . .*', pp. 8–9, and table.
172 Colonel N. Kireyev, Lieutenant Colonel N. Dobenko, 'Iz opyta boyevogo primeneniya peredovykh otryadov tankovykh (mekhanizirovannykh) korpusov' ('From the experience of employing forward detachments of tank (mechanized) corps in combat'), *VIZh*, 9, 1982, pp. 20–7, this on p. 20.
173 ibid., p. 26; Radzievskiy, *Tankovy udar*, p. 133 (distance ahead of main forces); Polukhin and Patychenko, p. 30; Kurochkin, p. 70 (fuel requirements).
174 Polukhin and Patychenko, pp. 32–3; Dick, 'Soviet operational manoeuvre groups . . .', p. 776.
175 Polukhin and Patychenko, pp. 32–6; Abramov, p. 38 (also mentions employment of local provisions and captured equipment); Radzievskiy, *Tankovy udar*, p. 133.
176 Issa Pliev, *Konets kvantunskoy armii. Zapiski komanduyushchego konno-mekhanizirovannoy gruppoy sovetsko-mongol'skikh voysk (End of the Kwantung Army). (Notes of the Commander of the Soviet-Mongolian Cavalry-Mechanized Group)*, (second edn, 'Ir', Ordzhonikidze, 1969), p. 10; KMG and movement illustrated in *Times Atlas*, pp. 198–9.
177 ibid., pp. 33–4.
178 Glantz, *August Storm*: *The Soviet 1945 Strategic Offensive* . . . pp. 81, 106.
179 Pliev, Dorogami voyny, pp. 431, 444, 451; *Konets kvantunskoy*, pp. 95, 109, 117. Pliev might have reflected that Genghis Khan had camped at Dolonnor (Tolun) after a hard campaign against the Chinese, in 1214: Gale, p. 77.
180 Glantz, p.106; Pliev, *Konets kvantunskoy*, p. 187.
181 Sun Tzu, *Art of War*, II, 'Waging war' trans. Griffith, p. 73.
182 See author's *The Future of Land Warfare*, pp. 168–9.
183 Pliev, *Konets kvantunskoy*, p. 18.
184 ESECS, *Strengthening Conventional Deterrence in Europe*, pp. 16–17.
185 Simpkin, *Race to the Swift*, p. 44.
186 ibid.
187 'SZA' '*Operacyjne grup*', p. 4; Lieutenant Colonel Ryszard Konopka, 'Wprowadzenie oddzialu wydzelonego do walki' ('The commitment of forward detachments to battle'), *PWL*, 5, 1982, pp. 21–4.
188 ibid.
189 Lange, p. 19.
190 Colonel N. P. Korol'kov, *Gubokiy Reyd* (*Deep Raid*) (1st Tank Army in the Vinnitsa Breakthrough, 1943–4), (Voyenizdat, Moscow, 1967), on yet another OMG precedent. This incident, pp. 17–19.
191 US Secretary of Defence, *Soviet Military Power 1987* (US Government Printing Office, Washington DC, 1987), p. 71.
192 See Brian Holden Reid, 'J. F. C. Fuller's Theory of mechanized warfare', *Journal of Strategic Studies*, December 1978, pp. 295–312, this on p. 303.
193 Sun Tzu, *Art of War*, VII, 'Manoeuvre' trans. Griffith, p. 106.

6 CASE STUDY TWO: DON'T GET INVOLVED IN A LAND WAR IN ASIA

1 Sun Tzu, *Art of War*, trans. Griffith, 1963, p. 11.
2 '(Vitet) miles sagittas et celeram fugam Parthi', Horace, *Odes*, bk. II, 13, v. 17.
3 The Parthians looked both east and west, evincing both European and oriental characteristics. Their shooting backwards while on the run became a cliché

among Augustan poets; 'and the Parthian bold with his retreating horses' (Horace, *Odes*, bk. 1, 19, v. 12) and 'The Parthian puts his trust in flight and arrows in retreat' (Vergil, *Georgics*, III, 31). My thanks to David Ball, formerly classicist at Merton, for these references.

4 Appearance of the horse in war: Martin Windrow and Richard Hook *The Horse Soldier*, (Oxford University Press, Oxford, 1986), pp. 4–5; Miklos Jankovich, *They Rode into Europe*, (Harrap, London, 1971), pp. 5–7, 12–29; Dr E. V. Chernenko, *The Scythians, 700–300 BC*, (Osprey Men-at-Arms Series, London, 1983), p. 4. On Huns, Avars etc., and keleti harcmodor, F. Rubin's letter in *RUSI, Journal of the Royal United Services Institute for Defence Studies*, Vol. 128, no. 2, June 1983, p. 80, in response to the author's article 'Heirs of Genghis Khan, and the influence of the Tartar Mongols on the Imperial Russian and Soviet armies, *RUSI,* vol. 128, no. 1, March 1983, pp. 52–60.

5 On pursuit, see ch. 5 in particular.

6 General Sir Richard Gale, *Kings at Arms*: *the Use and Abuse of Power in the Great Kingdoms of the East*, (Hutchinson, London 1971), pp. 57–88, birth on p. 60; S. R. Turnbull and Angus McBride, *The Mongols* (Osprey Men-at-Arms Series, London, 1980), author's 'Heirs of Genghis Khan'; Harold Lamb, *Genghis Khan, the Emperor of all Men*, (Thornton Butterworth, London, 1928).

7 Gale, *Kings at Arms. . .*, p. 73.

8 On Giap, Robert O'Neill, *General Giap, Politician and Strategist* (Cassell, Victoria, 1969), and *The Strategy of General Giap since 1964*, (Canberra Papers on Strategy and Defence, Australian National University Press, Canberra, 1969).

9 Gale, *Kings at Arms. . .*, p. 75 (Peking); on Napoleon, see ch. 5.

10 'Mongol Strategy', Proceedings of the fourth East Asian Asiatic Conference, (ed.) Ch'en Chieh-hsien, (Taiwan, 1975), reproduced in Denis Sinor (ed.), *Inner Asia and its Contacts with Medieval Europe*, (Variorum reprints, London, 1977), pp. 244–7. There are cryptic references to Mongol forces apparently tarrying in an area because it was 'not yet time' and so on whilst on one occasion a subordinate general *rebuked Genghis Khan for being late*, which would surely have been an unwise move unless timing was of crucial operational and strategic significance.

11 Sun Tzu, *Art of War*, (ed.) and trans. James Clavell, (Hodder & Stoughton, London, 1981), p. 34. The Samuel B. Griffith translation, (Oxford University Press, Oxford, 1963), renders this as 'a skilled commander seeks victory from the situation and does not demand it of his subordinates. . . . Experts in war depend especially on opportunity and expediency. They do not place the burden of accomplishment on their men alone', (p. 93, ch. 5, 'Energy', 21).

12 Dupuy and Dupuy, *Encyclopedia*, p. 341. Liddell Hart, in 'Jenghiz Khan and Sabutai' in *Great Captains Unveiled* (William Blackwood & Sons Ltd, Edinburgh and London, 1927), p. 31, says 'weapon power multiplied by . . . mobility'.

13 'Supernatural speed!': 'An attack may lack ingenuity, but it must be delivered with supernatural speed', Sun Tzu, trans. Griffith, p. 73 ch. 'Waging war', 6. On the Israeli study, author's *The Future of Land Warfare*, pp. 227–9; General M. I. Ivanin, *Opisaniyie zimnyago pokhoda v Khivu 1839–40 gg. (Description of the Winter Expedition to Khiva in 1839–40)*, (St Petersburg, 1874), p. 6.; Turnbull and McBride, p. 18.

14 Dupuy and Dupuy, p. 348.

15 On the need for dispersal in nuclear conditions, Savkin, Basic Principles. . .', p. 170.

16 'Fire' to prepare assault, Dupuy and Dupuy, p. 349; Liddell Hart, p. 28; pursuit, *tulughma*, Turnbull and McBride, pp. 23, 26.
17 Compare Sun Tzu, ed. and trans. Griffith, p. 78: 'the worst policy is to attack cities' ch. 3, 'Offensive strategy', 7; 'Know your enemy. . .', p. 84, ch. 3, 31. Turnbull and McBride, p. 28; Dupuy and Dupuy, pp. 342, 344; '18 years', J. A. Boyle (ed.) *The Mongol World Empire* (collected essays, Variorum, London, (1977), article 5, p. 339; 'Chins', Gale, p. 75.
18 Douglas MacArthur, *MacArthur on War*, (ed.), Frank C. Waldorp, (Duell, Sloan and Pearce, New York, 1942), pp. 305–6.
19 Tamerlane is the European rendering of Timur-i-Lang, Timur the Limper. He was called simply Timur until his foot was injured by an arrow. Asian historians speak of him as Amur Timur Gurigan – Lord Timur the Splendid, and only call him Timur-i-Lang when speaking pejoratively. It is a sobering thought that if Temujin (Genghis) and Timur had been British, they would both have been called Smith. Harold Lamb, *Tamerlane the Earth Shaker*, (Thornton Butterworth, London, 1928), p. 25.
20 Gale, pp. 128–9.
21 ibid., p. 134: Dupuy, 'The Evolution. . .', p. 201; Lamb *Tamerlane*. . . , pp. 110–125.
22 Muhsin Mahdi, *Ibn Khaldun's Philosophy of History: A Study in the Philosophical Foundation of Science of Culture*, (George Allen & Unwin, London, 1957), p. 59.
23 Gale, p. 75. On the two opposed command styles, see Martin van Creveld, *Command in War* (Harvard University Press, Cambridge, Mass., 1985), pp. 203–31.
24 Lamb, *Tamerlane the Earth Shaker*, p. 254. See also Lamb, *Genghis Khan, the Emperor of all Men*, p. 13. He compares Genghis Khan and his generals very favourably with Napoleon, who 'abandoned one army in Egypt, left the remnant of another in the snow of Russia, and finally strutted into the debacle of Waterloo'.
25 General Palmer, *Summons of the Trumpet*. . . , p. 303 (see ch. 4).
26 Douglas Pike, *PAVN: People's Army of Vietnam*, (Brassey's London, 1986), pp. 9–38.
27 Phan Huy Le, Bui Dang Dung, Phan Dai Doan, Phan Thi Tam, Tran Ba Chi, *Our Military Traditions*, (Vietnamese Studies Series, no. 55, Foreign Languages Publishing House, Hanoi, 14th Year (1979)), p. 156.
28 Samuel B. Griffith, trans. and ed., *Mao Tse-tung on Guerrilla Warfare*, (Praeger, New York and Washington, 1961), p. 20 and Mao's *Yu Chi Chan* (*Guerrilla Warfare*), p. 42.
29 Peter Paret and John W. Shy, *Guerrillas in the 1960s*, (Pall Mall, London, 1962), pp. 6–11: 'guerrilla strategy was not the preferred strategy: it was the only strategy that remained available. . . . In Spain, as throughout history, guerrilla warfare was the weapon of the militarily weak . . . even victorious guerrilla leaders have rarely argued that guerrilla operations can succeed without the eventual aid of regular forces'.
30 Phan Huy Le *et al.*, p. 155.
31 ibid.
32 ibid., p. 156. Clausewitz, the first westerner to analyse irregular operations in warfare, held the same views. In Bk 6, ch. 26 of *On War*, 'The people in arms', Paret and Howard pp. 479–87, he notes the influence of rough and inaccessible terrain, the need for operations to be protracted, and that 'militia and bands of armed civilians cannot be employed against the enemy main force', not 'to

pulverize the core but to nibble at the shell and around the edges' (pp. 480–1).

33 Vo Nguyen Giap, *The Party's Military Line is the Invincible Banner of People's War in our Country*, (People's Army Publishing House, Hanoi, 1973); quote, Phan Huy Le, p. 157.
34 Phan Huy Le, p. 158.
35 See maps, ibid., pp. 8, 32, 58, 82; quote, p. 24.
36 ibid., p. 10.
37 Pike, pp. 9–10.
38 Phan Huy Le, p. 12.
39 ibid., pp., 15–56.
40 From the French 'point-blanc' – 'white mark', or the bull's eye. Technically, 'point blank range' means close enough to aim directly at the mark and not to have to compensate for fall of shot. With small-calibre, flat trajectory rifles introduced from the 1880s, 'point blank range' was considerably further away.
41 Phan Huy Le, pp. 105–20.
42 ibid., p. 60. Tay Do, Dien Chau, Nghe An, Tan Binh and Thuan Hoa. Many Vietnamese villages have the same name (there are five Tan Binhs listed in the US Board On Geographical Names *Gazetteer*) and the author has been unable to locate them all for certain.
43 ibid., p. 69. Nhan Muc bridge was 'deep in the area under Ming control', but appears to have been on the way back to Hanoi. There are Nhan Mucs at 21°26'N, 105°21'E and at 22°02'N, 105°01'E, the last well to the north.
44 Literally 'little log of firewood' in Finnish, which explains the concept graphically.
45 The Vietnamese, like the Russians, see 'left' and 'right' of a river from the source, so on a southward-flowing river the 'right' bank is the *west* side, as in 'Right Bank Ukraine' – the Ukraine west of the Dnepr.
46 Phan Huy Le, p. 79.
47 ibid., pp. 81–104.
48 Vietnamese military philosopher Nguyen Trai, cited in Phan Huy Le, p. 159.
49 Phan Huy Le, interpreting San Tzu, p. 159.
50 ibid., pp. 165–6.
51 Colonel Mohammed Yahya Effendi, 'The north-western routes and the invasions of the Indian sub-continent: a historical study in modern perspective', *Pakistan Army Journal*, June 1987, pp. 2–14.
52 Gale, *Kings at Arms*, pp. 155–80, on Babur (Battle of Panipat, p. 177).
53 ibid., pp. 181–9; Laurence Binyon, *Akbar*, (Peter Davies, London, 1932), esp. pp. 67–70, 79–85.
54 Laurence Lockhart, *Nadir Shah: A Critical Study Based Mainly Upon Contemporary Sources*, (Luzac, London, 1938), esp. pp. 1, 266–7.
55 Lieutenant General O Kishmishev, *Pokhody Nadir Shakha v Gerat, Kandagar, Indiyu, i sobytiya v Persii posle yego smerti (Nadir Shah's Expedition to Herat, Kandahar and India, and Events in Persia after his Death),* (Military Historical Department attached to the Staff of the Caucasian Military District, Tiflis, 1889). The Russian Asiatic expert, General Ivanin, had recommended the study of the campaigns of Genghis Khan, Tamerlane and Nadir Shah in 1875, and this was obviously taken up. See Bellamy, 'Heirs of Genghis Khan…', footnote 72.
56 Kishmishev, pp. 176–7.
57 Lockhart, p. 267.
58 Lockhart, p. 269.
59 Laurence Lockhart, 'The Navy and Nadir Shah', *Proceedings of the Iran Society*, London, 1936, pp. 3–18.

60 *The Times*, 30 August 1853, cited in S. Y. Teng, *The Taiping Rebellion and the Western Powers: A Comprehensive Survey*, (Oxford University Press, 1971), p. 1. According to Teng, almost 2,000 articles and books had been written about the rebellion, although in the author's experience few western military historians and analysts had ever heard of it! Sources are S. Y. Teng, *Historiography of the Taiping Rebellion* (1962) and *New Light on the History of the Taiping rebellion* (1966); Compilation Group for the 'History of Modern China' Series, *The Taiping Revolution*, (Foreign Languages Press, Peking, 1976); Lin-Le (Augustus F. Lindley), *Ti-Ping tien Kwoh: the History of the Ti-Ping Revolution, including a Narrative of the Author's Personal Adventures*, (2 vols, Day, London, 1866) (a wonderful read, though Lindley is somewhat biased); Irish University Press Area Studies Series. British Parliamentary Papers. *China 32. Correspondence, Memorials, Orders in Council and other Papers Respecting the Taiping Rebellion in China 1852–64*, (Irish University Press, Shannon, 1971); Karl Marx (and F Engels), *Marx on China, 1853–56. Articles from the New York Daily Tribune*, Introduction and Notes by Dona Torr, (Lawrence and Wishart, London, 1951); A. E. Hake, *Events in the Taeping Rebellion. Being Reprints of MSS Copied by General Gordon*, (W. H. Allen, London, 1891); Captain T. W. Blakiston, *Five Months on the Yang-Tze . . . and Notices of the Present Rebellions in China*, (London, 1862); Commander L. Brind, *The Taeping Rebellion in China. A Narrative of its Rise and Progress* (London, 1862); J. C. Chester, *Chinese Sources for the Taiping Rebellion, 1850–64*, (1963), Field-Marshal G. J. Wolseley, *Narrative of the War with China in 1860; to which is Added an Account of a Short Residence with the Taiping Rebels at Nankin*, (1862).

61 Tang, p. vii.

62 ibid.

63 G. D. Kaye, D. A. Grant, E. J. Edmond, *Major Armed Conflict: A Compendium of Interstate and Intrastate Conflict, 1720 to 1985* (Operational Research and Analysis Establishment, Department of National Defence, Ottawa, 1985), ORAE Report R95, p. B48; classes of conflict listed in author's *The Future of Land Warfare*, p. 2.

64 Teng, p. 411.

65 Kaye, Grant, Edmond, *Major Armed Conflict*, p. 30; *The Future of Land Warfare*, p. 2.

66 Teng, pp. 411–2.

67 ibid., p. 1.

68 *Mao Tse-Tung on Guerrilla Warfare*, (ed.) Griffith, p. 62.

69 Compilation Group, *The Taiping Revolution*, pp. 1–26; chronology in *Marx on China*, p. xix; Teng, p. 6, Women and feet, p. 109.

70 Compilation Group, *The Taiping Revolution*, p. 27; *Mao Tse-Tung on Guerrilla Warfare*, p. 92:

Mao (1937)	*Taipings (1851)*
All actions are subject to command	Obey decrees and orders
Do not steal from the people	Separate men's and women's regiments/camps
Be neither selfish nor unjust	Forbids slightest violation of the people's interests
	Be selfless, friendly, obey chiefs
	Be co-operative and never retreat in battle

71 Lin-Le, vol. 1, p. 89.
72 *Mao Tse-Tung on Guerrilla Warfare*, (ed.) Griffith, p. 87.
73 *The Taiping Revolution*, pp. 31–2; Lin Le, p. 86.
74 *The Taiping Revolution*, pp. 38–9 gives 13,125 but elsewhere (p. 50) indicates 13,156; Teng, p. 109 gives the latter.
75 See ch. 4. There were indications that the 1980s Soviet corps, like the pentomic division, might be based around 'fives'.
76 *The Taiping Revolution*, p. 38; Manchu Governor of Kwangsi to Consul Meadows cited in Lin-Le, vol. 1, p. 81. The Manchu Governor says an army had 13,270 men. He assessed that 'Hung sui-tshuen [Hung Hsiu-chuan, the Taiping leader] practised 'the ancient military arts' and 'the tactics of Sun-Pin (an ancient Chinese warrior and celebrated tactician)'.
77 Teng, p. 109; *The Taiping Revolution*, p. 39. See also Lin-Le's (Lindley's) comments, vol. 2, p. 472.
78 *The Taiping Revolution*, p. 44.
79 Lin-Le, p. 137.
80 Teng, *The Taiping Rebellion and the Western Powers*, p. 125.
81 *The Taiping Revolution*, pp. 55, 66–71.
82 ibid., pp. 55, 69; Teng, pp. 127–34.
83 *The Taiping Revolution*, pp. 59–60; Teng, pp. 134–8, 329.
84 Teng, pp. 143–8, Laai Yi-faai, 'River strategy: a phase of the Taipings' military development', *Oriens*, ii 1952, p. 329.
85 *The Taiping Revolution*, pp. 78–81.
86 ibid., p. 99, 131; Marx, (ed.) Torr, pp. xxii-xxiii (the 'Ever victorious army', first commanded by an American, then Gordon). Lin-Le, vol. 2, pp. 502–3 gives an example. Wong-ku-dza was an entrenched camp with a garrison of about 4,000, comprising: 'some ten or twelve stockades, each surrounded by a ditch, yet communicating with the others . . . having taken up a position fairly within range of their Enfield rifles and artillery but safely out of range of the useless gingalls and matchlocks of the Ti-Pings, the 'foreign brethren' opened a murderous fire upon the line of entrenchments'.
87 *The Taiping Revolution*, pp. 111–12, 117, 140.
88 ibid., pp. 140, 147, 149–57.
89 ibid., pp. 164, 167.
90 Teng, pp. 326–31, using Li Shun's study of the number of combatants.
91 Lin-Le, vol. I, pp. 78–9
92 Teng, p. 332.
93 'Persia–China', from *New York Daily Tribune* 5 June 1867, written by Engels, in *Marx on China* (ed.) Torr, pp. 45, 50.
94 Consul Meadows to Lord John Russell, 19 February 1861 in Irish University Press, *China 32*, p. 114.
95 Hake, *Events in the Taeping rebellion*, p. 508.
96 On Taiping shortages and attempts to overcome, Teng, p. 335; quote from Lin-Le, vol. 1, p. 252.
97 *Mao Tse-Tung on Guerrilla Warfare*, p. 82–3.
98 Lin-Le, vol. 1, p. 236.
99 *Mao Tse-Tung on Guerrilla Warfare*, (ed.) Griffith, pp. 12–19 (biography and Long March). The underlying philosophy of this was 'conservation of one's own strength: destruction of enemy strength' (Mao, p. 95). On influences: pp. 37 (Griffith on Sun-Tzu), 46 (Lenin), 48 (S. I. Gusev, *Lessons of Civil War* (*Uroki grazhdanskoy voyny*), General Staff, Ukraine, 1918, revised, GIZ, Moscow, 1921), 49 (Clausewitz).
100 ibid., pp. 47–8.

101 ibid., p. 65.
102 ibid., pp. 105–7.
103 ibid. p. 107.
104 Robert O'Neill, *General Giap, Politician and Strategist*, (Cassell, Victoria, 1969), pp. 84–6; Harry G. Summers, Jr, *On Strategy: A Critical Analysis of the Vietnam War*, (Presidio, Novato, Ca., 1982), pp. 75–6.
105 Sun Tzu, *The Art of War*, ch. 1 'Estimates', 20 (ed. Griffith, p. 66).
106 Admiral of the Fleet of the Soviet Union, Sergey M. Gorshkov (d. 1988) *The Sea Power of the State*, (Pergamon, Oxford, 1979), pp. 221—2.
107 O'Neill, *General Giap* . . . pp. 143–7.
108 ibid., pp. 150–1.
109 ibid., pp. 152, 156; Bernard B. Fall, *Hell in a Very Small Place: The Story of Dien Bien Phu*, (Pall Mall Press, London, 1967), pp. 134–47.
110 O'Neill, p. 155.
111 Summers, p. 133.
112 ibid., pp. 154–6.
113 Timothy J. Lomperis, 'Giap's Dream, Westmoreland's Nightmare', *Parameters*, Journal of the US Army War College, June 1988, pp. 18–32, esp. 19–22.
114 ibid., esp. pp. 28–30. Lomperis concludes that we can agree with Sun Tzu that '"what is of supreme importance in war is to attack the enemy's strategy" ' [III, 'Offensive strategy', 4, (ed.) Griffith, p. 77]. 'To this eternal verity General Giap can legitimately add the postscript that it is doubly clever to steal it' (Lomperis, p. 30).
115 O'Neill, *General Giap* . . . pp. 195, 203.
116 O'Neill, *The Strategy of General Giap Since 1964*, (Canberra Papers on Strategy and Defence, Australian National University, Canberra, 1969), p. 19.
117 Russell Steth, *The Military Art of People's War. Selected Writings of Vo Nguyen Giap* (Monthly Review Press, New York and London, 1970), pp. 299–300.
118 Miyamoto Musashi, *A Book of Five Rings: The Classic Guide to Strategy*, trans. Victor Harris, (Overlook, New York, 1974), 'Japan's answer to the Harvard MBA', etc.
119 Scott A. Boorman, *The Protracted Game*, (Oxford University Press, Oxford, 1969). This is a skilled and interesting analysis, but the present author inclines to more pragmatic explanations.
120 Lamb, *Tamerlane the Earth Shaker*, pp. 241–3 and *Genghis Khan, The Emperor of all Men*, p. 234.
121 Sun Tzu, *Art of War*, I, 'Estimates' 6 and VII, 'Manoeuvre', 16, trans. Griffith, pp. 64, 106.
122 Albert N. Garland, (ed.), foreword by General William C. Westmoreland, *Infantry in Vietnam. Small Unit Actions in the Early Days, 1965–66*, (Battery Press, Nashville, 1982), esp. pp. 229–39; *Infantry* Magazine (ed.), Reflections by Garland, foreword by Westmoreland, *A Distant Challenge. The US Infantry in Vietnam, 1967–72*, (Battery Press, Nashville, 1983).
123 Garland, *Infantry in Vietnam*, p. 15.
124 Lamb, *Genghis Khan* . . ., p. 236.
125 Sun Tzu, *Art of War*, III, 'Offensive Strategy', trans. Griffith, p. 77.
126 Captain Anthony V. Neglia, 'NVA and VC . . .' in Garland, *A Distant Challenge*, pp. 173–80.
127 Sun Tzu, *Art of War*, VI, 'Weaknesses and strengths', 27, trans. Griffith, p. 101.

7 SOME CONCLUSIONS AND PROGNOSES

1 The use of historical parallels has become temporarily fashionable in the US Army, but is not always done subtly. Examples of the use of historical parallels are Captain Dana H. Pittard, 'Genghis Khan and 13th Century AirLand Battle', *Military Review*, July 1986, pp. 18–27; Captain Hilario Ochoa 'Operation Michael: the Seeds of AirLand Battle', *Armor*, January-February 1988, pp. 40–43, in which the Hutier tactics of World War One (ch. 3) are seen as having similar offensive concepts to those contained in AirLane Battle Doctrine, but the limitations on those tactics in World War One are not mentioned; Lieutenant Colonel A. Galloway, 'Who Influenced Whom', *Military Review*, March, 1986, pp. 46–51, which reveals that the doctrine was partially influenced by Jomini, Clausewitz, and Sun Tzu; and Major S. Agersinger, 'Karl von Clausewitz: Analysis of FM 100—5', *Military Review*, February 1986, pp. 68–75, in which Clausewitz criticizes the terminology used in the latter in a letter to the journal - an attractive and original device. The present author certainly agrees that the terminology used in much modern military analysis is execrable.

2 Clausewitz, *On War,* 1, ch. 1, 5 ('The maximum exertion of strength'); 6 ('Modifications in practice'); 11 ('The political object comes to the fore again'), Paret and Howard edn, pp. 77–81.

3 Norman Angell, *The Great Illusion,* (London, 1914); Ivan S. Bliokh (his name is often incorrectly transliterated as Bloch), *Pudushchaya voyna v tekhnicheskom, ekonomicheskom i politicheskom otnosheniyakh (Future War in its Technical, Economic and Political Aspects),* 6 vols, St Petersburg, 1898). The *General Conclusions* to the latter were published in English in 1899 as *Is War Now Impossible?*. With the benefit of hindsight, that was a prescient translation, although at the time it was quite inappropriate.

4 The author's discussions with academics in the filed have confirmed this; see also, for example, General Wallace H. Nutting, 'From my bookshelf', *Military Review,* July 1988, pp. 91–2.

5 Colonel General Makhmut A. Gareyev, *M. V. Frunze: Voyenny teoretik* (*M. V. Frunze: Military Theorist*), (Voyenizdat, Moscow, 1985), p. 239

6 William Shakespeare, *King Lear* Act V, Scene 3.

General select bibliography

Sources for the special case studies are normally omitted, but are cited in full in the notes

BOOKS AND FULL-LENGTH STUDIES

Addington, Larry, *The Blitzkrieg Era and the German General Staff*, Rutgers University Press, New Brunswick, 1971.

Angell, Norman, *The Great Illusion: a Study of the Relation of Military Power to National Advantage* (Heinemann, London, 1913), (first published as *Europe's Optical Illusion*, 1909).

Auvergne, Edmund B., *The Prodigious Marshal. Being the Life and Extraordinary Adventures of Maurice de Saxe, Marshal of France*, Selwyn & Blount, London, 1930.

Bailey, J. B. A., *Field Artillery and Firepower* (The Military Press, Oxford, 1989).

Barker, Colonel A., *Yom Kippur War*, Ballantine, Random House, New York, 1974.

Beaufre, André, *Introduction to Strategy*, Faber, London, 1965.

Beaumont, Roger, and Edmonds, Martin, *War in the Next Decade*, Macmillan, London and Kentucky University Press, 1975.

Becke, Captain, A. F., *An Introduction to the History of Tactics*, Hugh Rees, London, 1909.

Bellamy, Christopher D., *Red God of War: Soviet Artillery and Rocket Forces*, Brassey's, London, 1986.

The Future of Land Warfare, Croom Helm, Beckenham, 1987.

Bethell, Colonel H. A., *Modern Artillery in the Field*, (Cattermole, F. J. Woolwich, 1910) with amendments and additions to 1 March 1912.

Bidwell, Shelford and Graham, Dominick, *Fire-power. British Army Weapons and Theories of War, 1904–1945*, George Allen & Unwin, London, 1982.

Binyon, Laurence, *Akbar*, Peter Davies, London, 1932.

Bliokh, Ivan, *Budushchaya voyna v tekhnicheskom, ekonomicheskom i politicheskom otnosheniyakh*, (*Future War in Its Technical, Economic and Political Aspects*), 6 vols, St Petersburg, 1898.

General Conclusions published as *Is War Now Impossible?* London, 1899.

Blomberg, Werner von, *Visit of the Chief of the Army Board to Russia, August to September 1928*, Reichswehr Ministerium, Berlin, 1928, trans. W. R. Newby-Grant, RMA Sandhurst.

Blumenson, Martin, *Patton. The Man behind the Legend, 1885–1945*, (Jonathan Cape, London, 1985).

Bonaparte, Napoleon, *Maxims of War*, 1831.
Boyle, J. A. (ed.), *The Mongol World Empire*, collected essays, Variorum, London, 1977.
Brown, D. Alexander, *Grierson's Raid*, University of Illionois Press, Urbana, 1954.
Butler, J. R. M., *History of the Second World War. UK Military Series. Grand Strategy*, vol. II, September 1939 to June, 1941, HMSO, London, 1957.
with J. M. A. Gwyer, vol. III, June 1941 to August 1942, HMSO, London, 1964.
Caesar, Julius, *De Bello Gallico* (*The Gallic War*), various editions.
Chernenko, Dr E. V., *The Scythians, 700–300 BC*, Osprey Men-at-Arms Series, London, 1983.
Churchill, Winston S., *The World Crisis: The Eastern Front*, London, Thornton Butterworth, 1931.
Cipolla, Carlo M., *Guns and Sails in the Early Phase of European Expansion, 1400–1700*, Collins, London, 1965.
Clausewitz, General Karl von, *Vom Kriege* (*On War*), Berlin and Leipzig, 1918, translated as *On War*, by Colonels J. J. Graham and F. N. Maude, 3 vols, Routledge & Kegan Paul, London, 1962. Translated as *On War*, by Michael Howard and Peter Paret, Princeton NJ, Princeton University Press, 1975, with introductory essays and commentary.
Coates, Colonel James Boyd, Jr, *Wound Ballistics*, Office of the Surgeon General, US Government Printing Office, Washington DC, 1962.
Cross, Robin, *The Bombers. The Illustrated Story of Offensive Strategy and Tactics in the Twentieth Century*, Bantam, London, 1987.
De Gaulle, General Charles, *Vers l'Armée de Métier* (*Towards the Professional Army*), Plon, Paris, 1973.
Dewar, G. A. B. and Boraston, Lt Col J. H., *Sir Douglas Haig's Command, December 19, 1915 to November 11, 1918*, 2 vols, Constable, London, 1922.
Doroshenko, S. S., *Lev Tolstoy, voyn i patriot: voyennaya sluzhba i voyennaya deyatel 'nost'* (*Leo Tolstoy, Warrior and Patriot: Military Service and Attainments*), Sovetskiy pisatel', Moscow, 1966.
Drea, Edward J., *Nomonhan: Japanese-Soviet Tactical Combat, 1939*, Leavenworth paper no. 2, Combat Studies Institute, Fort Leavenworth, 1981.
Duffy, Christopher, *The Army of Frederick the Great*, David & Charles, London, 1974.
Russia's Military Way to the West. Origins and Nature of Russian Military Power, 1700–1800. Routledge & Kegan Paul, London, 1981.
Frederick the Great. A Military Life, Routledge, London, 1988.
Dupuy, Ernest R. and Dupuy, Trevor N., *The Encyclopedia of Military History from 3,500 BC to the Present*, Macdonald, London, 1970.
Dupuy, Trevor N., *The Evolution of Weapons and Warfare*, Jane's, London, 1982.
Earle, Edward M., *Makers of Modern Strategy. Military Thought from Machiavelli to Hitler*, Princeton NJ, Princeton University Press, 1944.
Edmonds, Brigadier General J. E., and Wynne, Captain, G. C., *History of the Great War based on Official Documents*, by direction of the Historical Section of the Committee of Imperial Defence, Macmillan, London, various volumes during the 1920s and 30s.
Engels, Friedrich, see Marx.
Erickson, John, (ed.), *The Military Technical Revolution: Its Impact on Strategy and Foreign Policy*, Institute for the Study of the USSR, Pall Mall Press, London, 1966.
Stalin's War with Germany, vol. 1, *The Road to Stalingrad*, Weidenfeld & Nicholson, London, 1975. vol. 2, *The Road to Berlin*, Weidenfeld and Nicholson,

London, 1983.
Estienne, Jean Baptiste, *Étude sur les Missions des chars blindées en campagne*, (*Study of the Roles of Armoured Vehicles in the Field*) 25 May 1919.
Fall, Bernard B., *Hell in a Very Small Place. The story of Dien Bien Phu*, Pall Mall Press, London, 1967.
Fëdorov, Lt Gen Arty V. G., *K voprosu o date povavlenii artillerii na Rusi* (*On the Question of the Date of Artillery's Appearance in Russia*), Academy of Artillery Sciences, Moscow, 1949.
Ferré, Georges, *Le Defaut de l'Armure* (*The Lack of Armour*), Paris, 1948.
Ferrill, Arther, *The Origins of War. From the Stone Age to Alexander the Great*, Thames and Hudson, London, 1986.
Firth, C. H., *Cromwell's Army*, with a new Introduction by P. H. Hardacre, Methuen, London, 1962.
FM 100-5 Operations, Headquarters, Department of the Army, Washington DC, 20 August 1982.
Frederick the Great, *Instructions for his Generals*, Military Service Publishing Company, Harrisburg, Penn., 1944.
Freytag-Loringhoven, General Freiherr Dr von, *Deductions from the World War*, (translated from the German of 1917, with an introduction by 'J E M', Constable, London, 1918).
Generalship in the World War, translated from the German of 1920, US Army War College, Washington, DC, 1934.
Fromm, Erich, *The Anatomy of Human Destructiveness*, Holt, Reinhart, Winston, New York, 1973.
Fuller, John F. C., *The Reformation of War*, Hutchinson, London, 1923.
Foundations of the Science of War, Hutchinson, London, 1925.
On Future Warfare, Sifton Praed, London, 1928.
Memoirs of an Unconventional Soldier, Ivor, Nicholson & Watson, London, 1936.
Machine Warfare. An Enquiry into the Influences of Mechanics on the Art of War, Hutchinson, London, 1942.
Armament and History. A study of the Influence of Armament on History from the Dawn of Classical Warfare to the Second World War, Eyre & Spottiswoode, London, 1946.
Decisive Battles of the Western World, and their Influence upon History, 3 vols, Eyre & Spottiswoode, London, 1954–6.
Galaktionov, Divisional Commissar M., *Temp Operatsii* (*The Tempo of Operations*), 2 vols, Voyenizdat, Moscow, 1937.
Gale, General Sir Richard, *Kings at Arms. The Use and Abuse of Power in the Great Kingdoms of the East*, Hutchinson, London, 1971.
Gareyev, Col. Gen. Makhmut A., *M V Frunze – Voyenny teoretik* (*M V Frunze – Military Theorist*), Voyenizdat, Moscow, 1985.
Gayvronskiy, Col. Gen. F. F., *Evolyutsiya voyennogo iskusstva: etapy, tendentsii, printsipy (The Evolution of Military Art: Stages, Trends, Principles*, Voyenizdat, Moscow, 1987.
Giap, General Vo Nguyen, *People's War, People's Army*, Foreword by Roger Hilsman, profile by Bernard B. Fall, Praeger, New York, 1962.
Glantz, Colonel David, *August Storm: the Soviet 1945 Strategic Offensive in Manchuria*, Leavenworth Paper No. 7, US Government Printing Office, Washington DC, 1984.
August Storm: Soviet Tactical and Operational Combat in Manchuria, Leavenworth Paper No. 8, US Government Printing Office, Washington DC, 1984.

Golubev, A., *M V Frunze o kharaktere budushchey voyny* (*M V Frunze on the Character of Future War*), Voyenizdat, Moscow, 1931.
Gorshkov, Admiral of the Fleet Sergey M, *The Sea Power of the State*, Pergamon, Oxford, 1979, translated from the Russin *Morskaya moshch' gosudarstva*.
Greenfield, Kent R., *Army Ground Forces and the Air-Ground Battle Team, including Organic Light Aviation*, AGF Study No. 35, Washington DC, 1948.
Grendal', Col. Gen. Arty V. D., *Ogon' artillerii* (*Artillery Fire*), Voyenizdat, Moscow, 1926.
Guderian, General Heinz, *Panzer Leader*, with a foreword by B. H. Liddell Hart, trans. Constantine Fitzgibbon, Michael Joseph, London, 1952.
Halliday, Jon, and Cummings, Bruce, *Korea: the Unknown War*, Viking/Penguin, London, 1988.
Harris, Robert and Paxman, Jeremy, *A Higher Form of Killing: The Secret History of Gas and Germ Warfare*, Paladin, London, 1983.
Hartcup, Guy, *The War of Invention: Scientific Development, 1914–18*, Brassey's London, 1988.
Hastings, Max, *The Korean War*, Michael Joseph, London, 1987.
Hattaway, Herman and Jones, Archer, *How the North Won: a Military History of the Civil War*, University of Illinois Press, 1983.
Heinl, Robert, *Dictionary of Military and Naval Quotations*, US Naval Institute Press, Annapolis, Maryland, 1966.
Henderson, Lt Col G. F. R., *The Science of War. A Collection of Essays and Lectures, 1892–1903*, Longman Green and Co. Ltd., London, 1905.
Herzog, Chaim, *The War of Atonement*, Weidenfeld & Nicholson, London, 1975.
History of the Second World War. UK Military Series. Grand Strategy. See Butler, Howard.
Hogg, Ian V., *Barrage. The Guns in Action*, Macdonald, London, 1970.
German Artillery of World War II, Arms and Armour Press, London, 1975.
Horne, Alistair, *The Price of Glory. Verdun, 1916*, Macmillan, London, 1975, first published 1962.
House, Captain Jonathan M., *Towards Combined Arms Warfare: A Survey of Tactics, Doctrine and Organization in the 20th Century*, Combat Studies Institute, US Army Command and General Staff College, Fort Leavenworth, 1984.
Howard, Michael E., *History of the Second World War. UK Military Series. Grand Strategy*, vol IV, August, 1942 to September, 1943, HMSO, London, 1972.
Hughes, Maj. Gen. P. B., *Open Fire. Artillery Tactics from Marlborough to Wellington*, Antony Bird, Chichester, 1983.
Imperial War Museum, Department of Documents, Papers of Lt Gen. Sir Gifford March, file Martel 4/4, Letters of the Military Mission to Russia '43–'44.
Issawi, Charles, *An Arab Philosophy of History. Selections from the Prolegomena of Ibn Khaldun of Tunis, 1332–1406* John Murray, London, 1950.
Ivanin, Lt Gen. M. I., *Opisaniye zimnyago pokhoda v Khivu 1839–40 gg.* (*Description of the Winter Expedition to Khiva in 1839–40*), St Petersburg, 1874.
Jankovich, Miklos, *They Rode into Europe*, Harrap, London, 1971.
Jomini, Baron Antoine Henri de, *Précis de l'Art de la Guerre*, second edn, Librairie pour l'Art Militaire, les Arts et les Sciences, Paris, 1855, translated as *The Art of War*, by Captain G. H. Mendell and Captain W. P. Craighill, Lippincott, Philadelphia and Trubner, London, 1879.
Jones, Archer, *The Art of War in the Western World*, Harrap, London, 1987.
Jones, R. V., *Most Secret War*, Coronet, London, 1981.
Jones, Ralph, *Readings from Futures*, 1974–80, Westbury, Guildford, 1981.
Kahn, Herman, *On Thermonuclear War*, Princeton University Press, New Jersey

and Oxford University Press, London, 1960.
Kavtaradze, A. G., *Voyennye spetsialisty na sluzhbe Respubliki Sovetov, 1917–1920 gg.* (*'Military Specialists' in the Service of the Soviet Republic, 1917–20*), Nauka, Moscow, 1988.
Kaye, G. D., Grant, D. A., Edmond, E. J., *Major Armed Conflict. A Compendium of Major Interstate and Intrastate Conflict, 1720 to 1985*, Operational Research and Analysis Establishment, Department of National Defence, Ottawa, 1985, ORAE Report R 95.
Kazakov, Marshal Arty K. P., *Artilleriya i rakety* (*Artillery and Rockets*), Voyenizdat, Moscow, 1968.
Keegan, John, *Opening Moves. August, 1914*, Pan/Ballantine, London, 1973.
The Face of Battle, Cape, London, 1976.
(ed.), *The Times Atlas of the Second World War*, Times Books, London, 1989.
Kipp, Jacob W., *Mass, Mobility and the Red Army's Road to Operational Art, 1918–1936*, Soviet Army Studies Office, Fort Leavenworth, Texas, 1986.
Kirey, Lt Col. V. F., *Artilleriya ataki i oborony (Artillery of the Attack and Defence*, Voyenizdat, Moscow, 1936.
Kishmishev, Lt Gen O., *Pokhody Nadir Shakha v Gerat, Kandagar, Indiyu i sobytiya v Persii posle yego smerti* (*Nadir Shah's Expeditions to Herat, Kandahar and India, and Events in Persia after his death*), Military Historical Department attached to the Staff of the Caucasian Military District, Tiflis, 1889.
Klein, Burton, *Germany's Economic Preparations for War*, Harvard Economic Studies, vol. 109, Harvard University Press, Cambridge, Mass., 1959.
Kuhn, Thomas S., *The Structure of Scientific Revolutions*, (second edn, enlarged, *Foundations of the Unity of Science*, vol. II, no. 2, *International Encyclopedia of Applied Science*, University of Chicago, 1970), first edn 1962.
Lamb, Harold, *Genghis Khan, the Emperor of all Men*, Thornton Butterworth, London, 1928.
Tamerlane, the Earth Shaker, Thornton Butterworth, London, 1928.
Langlois, General H., *Lessons from two Recent Wars (The Russo-Turkish and South African Wars*), (General Staff, War Office, London, 1909), translated from the French of 1904.
Lawrence, Terence E., *Seven Pillars of Wisdom*, Jonathan Cape, London, 1935.
Crusader Castles, 1909, University of Oxford BA thesis; first general edition, Michael Haag, London, 1986.
Leakey, Richard E., The Making of Mankind, Abacus, Sphere, London, 1982.
Lenin, Vladimir I., *The Revolutionary Army and the Revolutionary Government* 27 June (10 July) 1905, Progress, Moscow, 1980.
Liddell Hart, Captain Sir Basil, *Great Captains Unveiled*, William Blackwood & Sons, Edinburgh and London, 1927.
Lider, Julian, *Military Theory. Concept, Structure, Problems*, Swedish Studies in International Relations, Gower, Aldershot, 1983.
Liskenne and Sauvan, *Mémoires sur l'Art Militaire*, Paris, *Bibliothèque Historique et Militaire*, 1851.
Lloyd, General Henry H., *History of the Late War in Germany* (1766).
Lockhart, Laurence, *Nadir Shah: A Critical Study based Mainly Upon Contemporary Sources*, Luzac, London, 1938.
Longacre, Edward G., *Mounted Raids of the Civil War*, A. S. Barnes, Cranbury, NJ, 1975.
Longworth, Philip, *The Art of Victory: The Life and Achievements of Generalissimo Suvorov*, Constable, London, 1965.
Lorenz, Konrad, *On Aggression*, Bantam, London, 1971.
Lucas, Lt Col. Pascal, *L'Evolution des Idées Tactiques en France et en Allemagne*

pendant la Guerre de 1914–18, Berger-Levrault, Paris, 1923.
Lumsden, Dr M., *Anti Personnel Weapons*, Stockholm International Peace Research Institute, Taylor & Francis, London, 1978.
Lupfer, Timothy T., *The Dynamics of Doctrine, The Changes in German Tactical Doctrine during the First World War*, Leavenworth Paper No. 4, US Government Printing Office, Washington DC, 1981.
Luvaas, Jay, (ed.), *Frederick the Great and the Art of War*, Free Press, New York, 1974.
MacArthur, General Douglas, *MacArthur on War*, (ed.) Frank C Waldorp, Duell, Sloan, and Pearce, New York, 1942.
Machiavelli, Niccolò, *The Art of War* (trans. of *Arte della guerra*, revised edition of the Ellis Farnworth translation, with introduction by Neal Wood) Bobbs-Merrill, Indianapolis, New York and Kansas, 1965.
Macksey, Kenneth, *Guderian. Panzer General*, Macdonald & Jane's, London, 1975.
McKenney, Janice E., *Field Artillery Lineage Series*, Washington DC, 1980.
Mahdi, Muhsin, *Ibn Khaldun's Philosophy of History. A Study in the Philosophical Foundation of Science of Culture*, George Allen & Unwin, London, 1957.
Maisky, Ivan, *'V Londone'* (*'In London'*) in N. I. Koritskiy, (ed.), *Marshal Tukhachevskiy. Vospominaniya druzyey i soratnikov* (*. . . Memoirs of Friends and Comrades-in-Arms*), Voyenizdat, Moscow, 1965.
Mao Tse-tung on Guerrilla Warfare, (ed.) Samuel B. Griffith, Praeger, New York and Washington, 1961.
Margelov, V. F., Lisov, I. I., Samoylenko, Ya, P., and Ivonin, V. I., *Sovetskye vozdushno-desantnye: voyenno-istoricheskiy ocherk* (*Soviet Airborne Forces: a Military Historical Sketch*), Voyenizdat, Moscow, 1980.
Marshall-Cornwall, General Sir James, *Grant as Military Commander* Batsford, London, 1970.
Martynov, E. A., *Strategiya v epokhu Napoleona I i v nashe vremya* (*Strategy in the era of Napoleon I and in our time*), General Staff Press, St Petersburg, 1894.
Marx, Karl and Engels, Friedrich, *Karl Marx and Friedrich Engels Correspondence, 1846–95*, (ed. and trans. Dona Torr), Martin Lawrence, London, 1934.
The Civil War in the United States, by Karl Marx and Friedrich Engels, Lawrence & Wishart, London, 1938.
Marx on China, 1853–56, articles from the New York *Daily Tribune*; introduction and notes by Dona Torr, Lawrence & Wishart, London, 1951.
Medical Statistics: Casualties and Medical Statistics of the Great War (History of the Great War), HMSO, London, 1931.
Meyerovich, G. I., Budanov, F. V., *Suvorov v Peterburge* (*Suvorov in St Petersburg*), Lenizdat, Leningrad, 1978.
Miksche, Ferdinand Otto, *Attack! A Study of Blitzkrieg Tactics*, Random House, New York, 1942.
Paratroops: the History, Organization and Tactical Use of Airborne Formations, Faber & Faber, London, 1943.
(with E. Combaux), *War Between Continents*, Faber & Faber, London, 1948.
Atomic Weapons and Armies, Faber & Faber, London, 1955.
The Military Balance, 1988–89, International Institute for Strategic Studies, London, 1989.
Millemet, Walter de, *De Regis Misericordia*, manuscript prepared for King Edward III of England, late 1326 or early 1327, Christ Church, Oxford.
Milward, Alan, *The German Economy at War*, University of London, Athlone Press, London, 1965.
Ministère de la Guerre, France, *Instruction Provisoire sur l'Emploi Tactique des Grandes Unités*, Paris, 1922.

Moltke, Field Marshal Helmuth von, the elder, *Moltke's Tactical Problems from 1858 to 1882*, ed. the Prussian General Staff Department for Military History, trans. Karl von Donat, Hugh Rees, London, 1903.

The Franco-German War of 1870–71, trans. Archibald Forbes, Harper & Bros, London, 1907.

Montecuccoli, Field Marshal Raimondo, Comte de, *Mémoires, ou Principes de l'Art Militaire*, (1712), *Commentarii Bellici . . . juncto Artis Bellicae Systemate . . .* Vienna, 1718.

Opere di Raimondo Montecuccoli, compiled and ed. Giuseppe Grassi, Stampa di Giuseppe Favali, Torino, 1821.

Moritz of Saxony, see Saxe, Field Marshal Maurice de.

Morris, Donald R., *The Washing of the Spears*, Cardinal, London, 1973.

Neznamov, General Aleksandr, *Sovremennaya voyna. Deystviya polevoy armii*, (*Modern War. Action of the Field Army*), Second edn, Moscow, 1912.

Nemecek, Vadlav, *The History of Soviet Aircraft from 1918*, translated from the Czech *Sovetska Letadla, Collins, London, 1986.*

Oman, Sir Charles, *History of the Art of War in the Sixteenth Century*, Methuen, London, 1937.

A History of the Art of War in the Middle Ages, Methuen Reprints, London, 1978.

O'Neill, Robert J., *General Giap, Politician and Strategist*, Cassell, Victoria, 1969.

The Strategy of General Giap Since 1964, Canberra Papers on Strategy and Defence, Australian National University Press, Canberra, 1969.

Palmer, Maj. Gen. Dave R., *Summons of the Trumpet. A History of the Vietnam War from a Military Man's Viewpoint*, Ballantine, New York, 1978.

Parker, Geoffrey. *The Military Revolution. Military Innovation and the Rise of the West, 1500–1800*, Cambridge University Press, 1988.

Pilsudski, Marshal of Poland Jozef, *Year 1920, and its Climax, Battle of Warsaw during the Polish Soviet War 1919–20, with the Addition of M. Tukhachevskiy's March Beyond the Vistula*, (Pilsudski Institute of London and America, London, and New York, 1972).

Phan Huy Le, *et al.* (eds.), *Our Military Traditions*, Vietnamese Studies Series, no. 55, Foreign Languages Publishing House, Hanoi, c. 1979.

Picht, Werner, *Vom Wesen des Krieges und vom Kriegswesen der Deutschen* (*On the Essence of War and the German Way of War*), Stuttgart, 1952.

Pike, Douglas, *PAVN: People's Army of Vietnam*, Brassey's, London, 1986.

Pokrovskiy, Maj. Gen. Technical Services G. I., *Science and Technology in Contemporary War*, trans. of a book, 1956, a pamphlet, 1957, and an article, 1957, introduction by Raymond L. Garthoff, Atlantic Books, London, 1959.

Popham, A. E., *The Drawings of Leonardo da Vinci*, Jonathan Cape, London, 1973.

Pratt, Edwin A., *The Rise of Rail Power in War and Conquest, 1833–1914*, P. S. King & Son, London, 1915.

Public Record Office (PRO) (UK), *WO 32/13087 (Kirke Report)*, 1932.

Defe II 1251 TWC (45), 1945.

Defe II 1252 TWC (46) (Revise) 1946.

PREM 3/89, Prime Minister to Ismay, D 217/4, July 1944.

Purnell's History of the Second World War, (Editor-in-Chief Sir Basil Liddell Hart, Editor Barrie Pitt), Phoebus Publishing/BPC Publishing, London, 1966/1972/1974/1979/1980.

Ritter, Gerhard, *The Schlieffen Plan: Critique of a Myth*, trans. Andrew and Eve Wilson, Oswald Wolff, London, 1958, (original German edn, Verlag R. Oldenbourg, Munich, 1956).

Rogers, J. M., and Ward, R. M., *Süleyman the Magnificent*, British Museum

Publications, London, 1988.
Ropp, Theodore, *War in the Modern World*, Collier, New York, 1962.
Rutherford, Ward, *The Ally. The Russian Army in World War I*, Gordon & Cremonesi, London, 1975.
Ste. Croix, Philippe de, *Airborne Operations. An Illustrated Encyclopedia of the Great Battles of Airborne Forces*, Salamander, London, 1978.
Salewski, Michael, *Zeitgeist und Zeitmaschine. Science Fiction und Geschichte* (*The Spirit of the Time and the Time Machine. Science Fiction and History*), DTV, Munich, 1986.
Savkin, V. Ye, *Basic Principles of Operational Art and Tactics*, translated from the Russian of 1972 under the auspices of the US Air Force, 1974, US Government Printing Office, Washington, 1982 edn.
Saxe, Field Marshal Maurice de, *Les Réveries ou Mémoires sur l'Art de la Guerre de Maurice, Comte de Saxe*, new edn, Paris, 1757.
Reveries, or Memoirs concerning the Art of War, anonymous translation [by Sir William Fawcett], Sands, Donaldson, Murray, and Cochran, Edinburgh, 1759.
The Art of War, Reveries and Memoirs, London, 1811.
Schlichting, General Sigismund von, *Taktische und Strategische Grundsätze der Gegenwart (Tactical and Strategic Principles of the Present)*, 1897–9.
Scott, Harriet Fast, and Scott, William F., *The Armed Forces of the USSR*, Westview, Boulder, Colorado, 1979.
Seagrave, Sterling, *Yellow Rain. (Chemical Warfare: the Deadliest Arms Race)*, Abacus, London, 1981.
Shaposhnikov, Boris M., *Vospominaniya. Voyenno-mauchnye trudy (Memoirs. Military Scientific Works)*, Voyenizdat, Moscow, 1974.
Sherman, Gen. William T., *From Atlanta to the Sea*, introduction by Basil Liddell Hart [the second half of Sherman's memoirs, 1863 to the end of the Civil War], Folio Society, London, 1961.
Simpkin, Richard, *Race to the Swift. Thoughts on Twenty-First Century Warfare*, Brassey's, London, 1985.
Simpkins, Michael, and Embleton, Donald, *The Roman Army from Caesar to Trajan*, revised edn, Osprey Men-at-Arms Series, London, 1984.
Sinor, Denis, (ed.), *Inner Asia and its Contacts with Medieval Europe*, Variorum reprints, London, 1977.
Sokolovskiy, Marshal of the Soviet Union V. D., *Soviet Military Strategy*, trans. from the Russian *Voyennaya strategiya* and ed. Harriet Fast Scott and William F. Scott, Macdonald & Jane's, London, 1975.
Sovetskaya voyennaya entsiklopediya (SVE) (Soviet Military Encyclopedia) 8 vols., Voyenizdat, Moscow, 1976–80.
Steth, Russell, *The Military Art of People's War. Selected Writings of Vo Nguyen Giap*, Monthly Review Press, New York and London, 1970.
Stone, Norman, *The Eastern Front, 1914–17*, Hodder & Stoughton, London, 1975.
Sun Tzu (possibly Sun Wu or Sun Pin), pronounced *Syun-dze* or *dzhe, The Art of War*, translated and with an introduction by Samuel B. Griffith, Foreword by B. H. Liddell Hart, Oxford University Press, 1963.
The Art of War, edited and translated by James Clavell, Hodder & Stoughton, London, 1981.
Suvorov, Field Marshal Aleksandr, *Nauka pobezhdat'* (*The Science of Winning*), first promulgated 1795, Voyenizdat, Moscow edn, 1980.
Svechin, Maj. Gen. A. A., *Strategiya (Strategy)*, second edn, Voyenizdat, Moscow, 1927.
The Taiping Revolution, compilation group for the 'History of Modern China'

Series, Foreign Languages Press, Peking, 1976.
Teng, S. Y., *The Taiping Rebellion and the Western Powers. A Comprehensive Survey*, Oxford University Press, 1971.
Terraine, John, *White Heat. The New Warfare, 1914–18*, (Sidgwick & Jackson, London, 1982.
Tolstoy, Count Leo, *War and Peace*, translated by Rosemary Edmonds, 2 vols, Penguin, Harmondsworth, 1957.
Towle, Philip N., 'The Influence of the Russo-Japanese War on British Naval and Military Thought, 1905–14', (PhD, King's College, University of London, 1973).
Toy, Sidney, *A History of Fortification from 3,500 BC to AD 1700*, Heinemann, London, 1955.
Triandafillov, Vladimir A., *Kharakter operatsii sovremennykh armii* (*The Character of the Operations of Modern Armies*), third edn, Voyenizdat, Moscow, 1936, first edn, 1929.
Tugwell, Maurice, *Airborne to Battle. A History of Airborne Warfare, 1918–1971*, Kimber, London, 1971.
Tukhachevskiy, Marshal of the Soviet Union Mikhail, *Izbrannye proizvedeniya (Collected Works)*, 2 vols, Voyenizdat, Moscow, 1964.
Turnbull, S. R., and McBride, Angus, *The Mongols*, Osprey Men-at-Arms Series, London, 1980.
Turnbull, Patrick, *Solferino: The Birth of a Nation*, Robert Hale, London, 1985.
Van Creveld, Martin, *Military Lessons of the Yom Kippur War: Historical Perspectives*, The Washington Papers, 3, Sage Policy Papers, Beverly Hills, Ca., 1975.
Command in War, Harvard University Press, Cambridge, Mass., 1985.
Vernadskiy, George, *A History of Russia*, vol. 3, The Mongols and Russia, Yale University Press, New Haven and Oxford University Press, London, 1953.
War Department (US), Office of the Chief of Staff, *US Official Reports on the Russo-Japanese War*, Washington DC, 1906.
War Office (UK), General Staff, *Reports of British Observers Attached to the Russian and Japanese Armies in the Field*, 3 vols, HMSO, London, 1907.
Weeks, John, with artwork by John Batchelor, *Airborne Soldier*, Blandford, Poole, 1982.
Whitton, Lt Col F. E., *Moltke*, (Makers of the Twentieth Century) Constable, London, 1921.
Windrow, Martin and Hook, Richard, *The Horse Soldier*, Oxford University Press, 1986.
Young, Brigadier Peter and Natkiel, Richard, *Atlas of the Second World War*, Weidenfeld & Nicholson, London, 1973.
Zolotarëv, V. A., *Rossiya i Turtsiya: voyna 1877–78 gg.* (*Russia and Turkey: the War of 1877–78*), Nauka, Moscow, 1983.

ARTICLES

Adams, Dwight L. and Newell, Clayton R., 'Operational art in the joint and combined arms arenas', *Parameters. Journal of the US Army War College*, June 1988, 33–9.
Beaumont, Roger, 'Nuclear warfare: the illusion of accurate forecasting', *Futures. The Journal of Forecasting and Planning*, February 1977, 53–8.
Bell, Sergeant Harry, 'Cavalry raids and the lessons they teach us' *Journal of the US Cavalry Association*, July 1908, 142–52. Translated from 'Employment of cavalry in independent action against the flank and rear of an enemy army', by Lt Col. Wenniger, Bavarian Cavalry, in *Austrian Cavalry Monthly*, April 1908.

Bellamy, Christopher D., 'The Russian artillery and the origins of indirect fire', *Army Quarterly and Defence Journal,* 2 parts, April 1982, 211–22, July 1982, 330–7.
'Heirs of Genghis Khan: the influence of the Tartar-Mongols on the Imperial Russian and Soviet Armies', *RUSI*, March, 1983, 52–60.
'The firebird and the bear. 600 years of the Russian artillery', *History Today*, September 1982, 16–20.
'Red star in the west: Marshal Tukhachevskiy and east–west exchanges on the art of war', *RUSI (Journal of the Royal United Services Institute for Defence Studies)*, December 1987, 63–73.
Clarke, I. F., 'Methods of prediction 1918–39', *Futures*, December 1970, 375–9.
Davies, Norman, 'The Soviet command and the Battle of Warsaw', *Soviet Studies* (Glasgow), April 1972, 573–8.
Dick, Charles J., 'Soviet operational art. Part 1: the fruits of experience, *International Defense Review*, 7, 1988, 755–61.
Dror, Yezhekel, 'War, violence and futures studies', *Futures*, February, 1974, 2–3.
Dupuy, Trevor N., 'Understanding war from a historical perspective', *Marine Corps Gazette*, June 1985, pp. 53–8.
Effendi, Col. Mohammed Y., 'The north-western routes and the invasions of the Indian Sub-Continent: a historical study in modern perspective', *Pakistan Army Journal*, June 1987, 2–14.
Essamé, Maj. Gen. H., 'The Suvorov legend', *Military Review*, January 1961.
Freytag-Loringhoven, General Freiherr Dr von, 'Wandlungen im operativen und taktischen Verfahren Napoleons verglichen mit denjenigen des Weltkriegs' ('Changes in Napoleon's operational and tactical methods compared with those of the [First] World War'), *Wiessen und Wehr (Military Thought)*, Berlin, 1921, 336–8.
Geysman, P. 'Uchrezhdeniye Imperatorskoy voyennoy akademii', ('Establishment of the Imperial Military Academy'), *Voyenny Sbornik (Military Collection)*, 11, 1908, 81–98 and 12, 1908, 133–56.
Hollingworth, Claire, 'The tasks of China's Minister Qin', *Jane's Defence Weekly*, 23 April 1988, pp. 788–9.
Howard, Michael, 'War and technology' (based on the Roskill Lecture given at Churchill College, Cambridge on 18 February 1986), *RUSI* December 1987, 17–22.
Jukes, Geoffrey, 'The strategic nuclear balance to 1983', *Futures*, December 1970.
'Kar', 'A personal view of the Yom Kippur War', *British Army Review*, August 1975, 12–18.
Kaye, G. D. and Solem, K. E., 'Futures studies and conflict, *Futures*, June 1979, 235–8.
Keenan, Bruce, 'The principles of war: a thesis for change', *US Naval Institution Proceedings*, December 1967.
Lockhart, Laurence, 'The navy of Nadir Shah', *Proceedings of the Iran Society*, London, 1936, 3–18.
Lomperis, Timothy J., 'Giap's Dream, Westmoreland's Nightmare', *Parameters*, June 1988, 18–32.
Martynov, Captain E. A. 'Neskol'ko slov v ob'yasneniye i razvitiye sochineniya "Strategiya..."' ('A few words in explanation and development of the work "Strategy"' . . .), *Voyenny Sbornik*, 7, 1894, 21–41; 8, 1894, 232–48.
Mirghani, Lt Col. I. E. 'Lessons from the Ogaden war', *British Army Review*, August 1981, 28–33.
Modiano, Mario, 'Muscle powered missile hits the bull's eye', *Sunday Times*, 2 August 1987, 29.

Murray, Professor Williamson, 'Force strategy, Blitzkrieg strategy and the economic difficulties: Nazi strategy in the 1930s', *RUSI*, March 1983, 39–43.

Nagaty, Brigadier Mohammed Ibrahim, 'Some lessons of the Ramadan war', *Pakistan Army Journal*, April 1975

Ogarkov, Marshal of the Soviet Union Nikolay, 'Zashchita sotsializma: opyt istorii i sovremennost' ('The defence of socialism: the experience of history and of the present'), *Krasnaya zvezda* (*Red Star*), 9 May 1984, 2–3.

Overy, R. J., 'From "Uralbomber" to "Amerikabomber": the Luftwaffe and strategic bombing', The Journal of Strategic Studies, September 1978, 154–78.

Petrov, A. 'Zadachi sovremennoy strategii...' ('Tasks of modern strategy'), *Voyenny Sbornik*, 5, 1894, 35–64.

Raach, Captain George T., 'Raid: an historical example, (Grierson's Raid of April 1863), *Armor*, May-June 1973, 8–13.

Ramanichev, Colonel N., 'Razvitiye teorii i praktiki boyevogo primeneniya vozdushno-desantnykh voysk v mezhvoyenny period' ('Development of the theory and practice of the military employment of airborne forces during the interwar period), *Voyenno-istoricheskiy zhurnal* (*Military Historical Journal*) (*VIZh*), 10, 1982, 72–7.

Rosser-Owen, David, 'NBC warfare and anti-NBC protection', *Armada International*, 1, 1984, 78–90.

Rubin, F. letter to *RUSI*, June 1983, p. 80, in response to Bellamy, Christopher D, 'Heirs of Genghis Khan . . .'

Sabin, Philip, A. G., 'World War Three: a historical mirage?', *Futures*, August 1983, 272–80.

Sandrart, Lt Gen. H. von, 'Operative Führung über die Gefechtstaktik hinaus', ('Operational command goes beyond combat tactics'), *Europäische Wehrkunde*, 9, 1987, 503–5.

Shaposhnikov, Boris M., '"Konnye massy" na flange armii. Ocherk deystvii russkoy konnitsy na levom beregu Visly v avguste 1914 g,' ('"Cavalry masses" on the army's flank: a sketch of the actions of the Russian Cavalry on the left bank of the Vistula in August 1914') *Voyennaya nauka i revolyutsiya*, (*Military Science and the Revolution*), Moscow, 1921, bk 1, July-August, 95–126.

Snow, Edgar, 'The Red Army's flying tank', *Saturday Evening Post*, 10 March, 1945, 18–19.

Starry, Don H. 'Principles of war', *Military Review*, 9, 1981.

Stolfi, R. J. S., 'Equipment for victory in France in 1940', *History*, vol. 55, no. 183, February 1970, 1–20.

Sunderland, Riley, 'Massed fires and the F.D.C. [Fire Direction Centre]', *Army*, 8, 1958.

Tret'yak, Army General I. M., 'Razgrom kvantunskoy armii n dal'nem vostoke' ('Defeat of the Kwantung army in the Far East'), *VIZh* 8, 1985, 9–19

Tukhachevskiy, Marshal Mikhail N., '*Novye voprosy voyny*' ('New questions of war'), *VIZh*, 2, 1962, 62–77.

Whaley, Barton, 'Towards a general theory of deception', *Journal of Strategic Studies*, March 1982, pp. 178–92.

Index

The prefix 'f' indicates a figure and/or relevant caption, 't' a table, and 'n' a footnote containing a point of substance not in the text, e.g. 255n49: page 255, note 49. Military ranks given are normally the last attained by individual.
Adm. = Admiral, FM = Field Marshal, MSU = Marshal of the Soviet Union, Gen. refers to all grades of general officer: Brigadier-, Major-, Lieutenant-, Colonel- and full General. Social distinctions, e.g. Baron, Prince, are omitted.